Valence theory

Valence theory

SECOND EDITION

J. N. Murrell

Professor of Chemistry
University of Sussex, England

S. F. A. Kettle

Reader in Chemistry
University of Sheffield, England

J. M. Tedder

Professor of Chemistry
University of St. Andrews, Scotland

JOHN WILEY & SONS LTD
London New York Sydney Toronto

Library of Congress Catalog Card No. 70–129161

ISBN 0 471 62688 0

Second printing July 1966

Third printing April 1967

Fourth printing January 1969

Fifth printing April 1976

Reproduced and printed by photolithography and bound in Great Britain at The Pitman Press, Bath

Preface to the first edition

This book has been written for honours chemistry undergraduates and for graduate students not specializing in theoretical chemistry. It is meant to bridge the gap between the semiqualitative picture given in Coulson's *Valence* and the formal mathematical account given in Eyring, Walter and Kimball's *Quantum Chemistry*—two books which we greatly admire.

It has been assumed that the student has acquired elsewhere an elementary knowledge of differential calculus and vector algebra. There are some points in the chapter on symmetry which need a slight knowledge of matrix algebra.

The book has been based on lectures given for the past four years at the University of Sheffield. Our experience has been that the student only obtains a full grasp of the subject when he has successfully tackled some problems, and we consider that the problems in this book are an essential part of the text; some of the problems provide proofs of statements made in the text. A student benefits most if he is given the minimum of help necessary to enable him to solve the problem. To take a step in this direction we have provided hints to the problems at the end of the book. The student should first try the problems without the hints, but with their help most students should be able to obtain the solutions.

The first five chapters are intended for first year students, since we feel that an early grasp of such concepts as the orbital and the electron-pair bond are necessary for any course in experimental chemistry. The more formal mathematical aspects of the book start in chapter 6 when it has been assumed that the student has received some mathematical training during his first year. These later chapters also assume a knowledge of other branches of chemistry.

Inevitably we have had to omit some topics on the fringe of valence theory. Spectroscopy and the theory of the solid state are the two major omissions but we feel that both these require books of their own for an adequate presentation. For the same reason we have made only scanty reference to the theory of intermolecular forces.

We have only given references to work which we think the student might wish to examine in more detail, and to acknowledge numerical data that we have quoted. There are plenty of advanced texts which provide a full list of references to the subject.

We would like to thank Drs. M. Child and M. Randič who gave a great deal of help with the whole manuscript. In addition we would like to express our appreciation of many friends and colleagues who gave valuable comments on individual chapters.

December 1964

J.N.M.
S.F.A.K.
J.M.T.

Preface to the second edition

This new edition has a few changes in the introductory chapters (1–5) which we hope will make them easier to follow, a few deletions and a few additions.

The main development in valence theory over the last four years has been that molecular orbital calculations are moving out of the era of the empirical Hamiltonian and approximation of integrals. In preparing the first edition we anticipated this development by giving the theory of the self-consistent-field MO method, and its zero-overlap modification for π-electron systems. In this edition we have given a very brief account of the numerical problems encountered in a non-empirical SCFMO calculation and have described the method of calculating overlap integrals—the simplest integral encountered in these calculations.

Non-empirical calculations are just beginning to be made on transition metal complexes. Their main value in the near future will be to test the results obtained from the empirical ligand field approach, but as yet they have not introduced any new concepts into the subject which we feel need emphasizing.

On the other hand the last five years have seen rapid developments in the treatment of organic molecules. The main advance has been in the treatment of σ-electrons both by the so-called extended Hückel methods and by the application of symmetry rules. We have made an extensive revision of the organic chemistry chapters to take account of these developments. The combined treatment of σ- and π-electrons in SCFMO calculations are now throwing some doubt on the deductions previously made from π calculations alone, but the results at present available must be considered somewhat tentative and have not been commented on in this edition.

In preparing the first edition we set a level which we thought suitable for students taking a specialist undergraduate course in chemistry and for the graduate student working in an experimental field. This second edition is aimed at the same two groups of readers and we hope the changes we have made will increase its usefulness.

January, 1970

J.N.M.
S.F.A.K.
J.M.T.

Contents

Symbols and abbreviations

Fundamental constants

e electronic charge, $4{\cdot}803 \times 10^{-10}$ esu
m electronic mass, $9{\cdot}110 \times 10^{-28}$ g
c velocity of light, $2{\cdot}998 \times 10^{10}$ cm sec^{-1}
h Planck's constant, $6{\cdot}626 \times 10^{-27}$ erg sec
k Boltzmann's constant, $1{\cdot}381 \times 10^{-16}$ erg degree^{-1}
a_0 Bohr radius, $0{\cdot}529$ Å
R Rydberg constant, 109737 cm^{-1} (for infinite nuclear mass)
N Avogadro's Number $6{\cdot}022 \times 10^{23}$ mole^{-1}

Operators

$\mathscr{H}$ the complete Hamiltonian or the complete electronic Hamiltonian
$\mathbf{H}$ a one-electron Hamiltonian not rigorously defined
$\mathbf{H}_i^c$ the core Hamiltonian (the terms in $\mathscr{H}$ which are functions of just the coordinates of electron i)
$\mathbf{F}$ the self-consistent field operator
$\mathbf{L}^2, \mathbf{L}_z, \mathbf{S}^2, \mathbf{S}_z, \mathbf{J}^2, \mathbf{J}_z, \mathbf{L}_+, \mathbf{L}_-$, etc. angular momentum operators
$\mathscr{A}$ the antisymmetrizing operator

$$\nabla^2 = \left(\frac{\partial^2}{\partial x^2} + \frac{\partial^2}{\partial y^2} + \frac{\partial^2}{\partial z^2}\right),$$ which occurs in the kinetic energy operator

Wave functions

ϕ an atomic orbital
ψ a molecular orbital
Ψ the complete wave function or a many-electron wave function
α, β spin wave functions
$\psi\alpha \equiv \psi,\ \psi\beta \equiv \bar{\psi}$ spin-orbitals

Integrals

$\int \ldots dv$ integration over the space coordinates of all the electrons
$\int \ldots ds$ integration over the spin coordinates
$\int \ldots d\tau$ integration over both space and spin coordinates

$\mathscr{H}_{rs} \equiv \int \Psi_r \mathscr{H} \Psi_s d\tau$
$\mathbf{H}_{rs} \equiv \int \psi_r \mathbf{H} \psi_s d\tau$ (or dv)
$S_{rs} \equiv \int \Psi_r \Psi_s d\tau$ or $\int \psi_r \psi_s d\tau$ (or dv)
$\alpha \equiv \int \phi_\mu \mathbf{H} \phi_\mu dv$, the Hückel coulomb integral
$\beta \equiv \int \phi_\mu \mathbf{H} \phi_\nu dv$, the Hückel resonance integral
Q the coulomb energy
A the exchange energy as defined in valence-bond theory
$J_{rs} \equiv \iint \psi_r(i) \psi_s(j) (1/r_{ij}) \psi_r(i) \psi_s(j) dv_i dv_j$
$K_{rs} \equiv \iint \psi_r(i) \psi_s(j) (1/r_{ij}) \psi_s(i) \psi_r(j) dv_i dv_j$
the two-electron coulomb and exchange integrals respectively
$\gamma_{\mu\nu} \equiv \iint \phi_\mu(i)\phi_\nu(j)(1/r_{ij})\phi_\mu(i)\phi_\nu(j) dv_i dv_j$, the two-electron integral between atomic orbitals used in the P-method

Other symbols

A electron affinity
c an expansion coefficient
F_μ free valence
$\mathscr{F}_\mu$ frontier-electron density
$i = \sqrt{-1}$
I ionization potential
I_σ and I_π inductive effect
I.R. irreducible representation
k rate constant
K equilibrium constant
L_μ localization energy
LCAO linear combination of atomic orbitals
$x = (\alpha - E)/\beta$, the dimensionless energy parameter of Hückel theory
MO molecular orbital
N a normalization coefficient
N_μ Dewar number
$p_{\mu\nu}$ π-bond order
P-method the method of Pariser, Parr and Pople, used for treating electron interaction in π-electron theory
q_μ π-electron charge at atom μ
R resonance effect
S superdelocalizability
SCF self-consistent field
V potential energy
Z nuclear charge
Z_μ Brown's reactivity index

δ_{rs} Kronecker delta ($=0$, $r \neq s$; $=1$, $r = s$)
Δ crystal-field splitting constant
ζ Slater orbital exponent
λ wavelength
ν frequency
π polarizability factor defined in Hückel theory
ρ electron density
χ character as used in group theory
Other symbols are defined as they occur in the text
Group theory symbols are defined in Chapter 8

Energy conversion table

	erg molecule^{-1}	kcal mole^{-1}	ev	cm^{-1}
1 erg molecule^{-1}	1	$1{\cdot}44 \times 10^{13}$	$6{\cdot}24 \times 10^{11}$	$5{\cdot}04 \times 10^{15}$
1 kcal mole^{-1}	$6{\cdot}95 \times 10^{-14}$	1	$4{\cdot}34 \times 10^{-2}$	350
1 ev	$1{\cdot}60 \times 10^{-12}$	23·1	1	8065
1 cm^{-1}	$1{\cdot}99 \times 10^{-16}$	$2{\cdot}86 \times 10^{-3}$	$1{\cdot}24 \times 10^{-4}$	1

This is a table of equivalence not identity. For example, the unit cm^{-1} is not an energy unit, but it is related to energy through the Planck–Einstein relationship.

Chapter 1

The foundations of atomic theory

The behaviour of macroscopic bodies is described by Newton's laws of classical mechanics. The behaviour of atomic particles is described by the laws of quantum mechanics. Both laws are founded on experiment. We cannot derive the classical equation

$$\text{force} = \text{mass} \times \text{acceleration}$$

but are satisfied that it gives the right answer to problems in classical mechanics. We shall treat the quantum-mechanical equations in the same light; the final test is that the solutions of these equations agree with experiment.

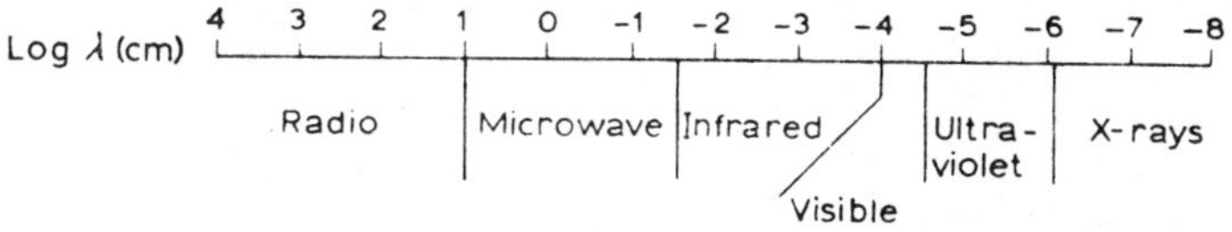

FIGURE 1.1. The electromagnetic spectrum

The properties of interference and diffraction of visible light show that it has a wave character. Maxwell showed that visible light was part of a general electromagnetic radiation which moves through space with a constant velocity (c) ($3{\cdot}0 \times 10^{10}$ cm per second in a vacuum) and whose amplitudes are harmonically oscillating electric and magnetic fields. This radiation is characterized by its frequency (ν) or wavelength (λ) which are connected by $c = \nu\lambda$. The electromagnetic spectrum is split up roughly as shown in figure 1.1.

A body emits or absorbs electromagnetic radiation to establish an equilibrium with its surroundings. At equilibrium the intensity of the radiation (known as black-body radiation) depends on temperature and follows the distribution shown in figure 1.2. Planck showed that this distribution could be understood by assuming that the body contained oscillators of all frequencies which could take up or give out radiation. *But*, agreement with experiment was only found if it was further assumed that an oscillator which emitted radiation of frequency ν could only take up energy in discrete amounts $h\nu$. h is called Planck's constant and is equal to $6{\cdot}6 \times 10^{-27}$ erg sec. $h\nu$ is called a quantum of energy. This was the first suggestion that atomic systems (the oscillators) could only exist in discrete energy states.

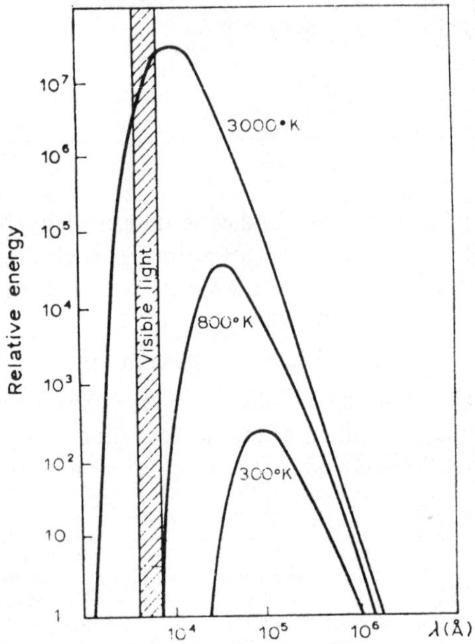

FIGURE 1.2. Energy distribution of black-body radiation as a function of temperature

When light falls on a metal electrons may be ejected. This is not surprising since light is an electromagnetic radiation. But the details of this photoelectric effect are not comprehensible in terms of the wave nature of light. It is found that no electrons are emitted unless the frequency of the light is above a certain minimum value, and above this value the number of ejected electrons is proportional to the intensity of the light but their energy is only a function of the frequency. Einstein showed that this result could be understood if light behaved like a stream of particles each with energy $h\nu$. He predicted that the kinetic energy of the ejected electrons would be given by

$$\tfrac{1}{2}mv^2 = h\nu - W, \tag{1.1}$$

where W is the minimum energy required to remove an electron from the metal (W is called the work function). This result was later confirmed by experiment. The results of the Planck black-body observation, and the Einstein photoelectric effect can be combined to link energy and frequency in the expression

$$E = h\nu = \frac{hc}{\lambda}. \tag{1.2}$$

This is called the Planck–Einstein relationship, and it applies to atomic oscillators and to light.

In 1911 Rutherford proposed his model of the atom, the basic ideas of which have remained unchanged. Atoms are approximately 10^{-8} cm in diameter (the unit 10^{-8} cm is called an Angstrom Å) but experiments on the scattering of α particles (He^{2+}) showed that the scattering centre in the atom is a small positively charged nucleus (diameter 10^{-13}–10^{-12} cm). This is surrounded by electrons which occupy the rest of the atomic volume. The mass of an electron is $5{\cdot}4 \times 10^{-4}$ times the mass of a proton.

The details of the structure of the nucleus can be left to the physicist. For chemical purposes we can say that the nucleus contains protons and neutrons, which have equal masses; the former has a charge equal but opposite to that of the electron and the latter is neutral. The neutron was discovered by Chadwick in 1932 although Rutherford had suspected its existence at an earlier date. In a neutral atom the number of electrons is equal to the number of protons and this number is the atomic number.

The mass number is the sum total of the protons and neutrons. It is possible to have different mass numbers for a given atomic number. These are called isotopes. For example, five isotopes of lead are known with mass numbers 206, 207, 208, 210 and 214. These all have 82 protons, with neutrons to make up the balance of the mass. The atomic weight of naturally occurring lead is 207·2†, ^{206}Pb and ^{208}Pb being the most abundant and being present in almost equal amounts.

According to classical mechanics a Rutherford atom cannot be a static system since, if this were so, the electrons would fall into the nucleus. But even a dynamic model with electrons moving around a positive charge is classically unstable because the electrons should interact with the electric field of the nucleus, radiate energy, and spiral into the nucleus. However, in the absence of external stimulation, atoms do not emit radiation, and when they are stimulated they emit only special discrete frequencies which are characteristic of the atom.

Hydrogen atoms can emit light at four wavelengths in the visible region of the spectrum; $\lambda = 6563, 4861, 4341, 4102$ Å. In 1895 Balmer made the

† The most recent scale is based on the most common isotope of carbon with an atomic number of 12.

inspired observation that these four numbers fitted the formula

$$\frac{1}{\lambda} = R\left[\frac{1}{2^2} - \frac{1}{n^2}\right](\text{cm}^{-1}), \tag{1.3}$$

where $n = 3, 4, 5$ and 6, and R, the Rydberg constant, equals 109678 cm^{-1}. Later, other lines were found in the infrared and ultraviolet, and Ritz showed their wavelengths could be fitted by similar formulae

$$\frac{1}{\lambda} = R\left[\frac{1}{n_1^2} - \frac{1}{n_2^2}\right]. \tag{1.4}$$

Five series in the hydrogen emission spectrum have been found which are named after their discoverers.

$n_1 = 1$	$n_2 = 2, 3, 4 \ldots$	Lyman
$n_1 = 2$	$n_2 = 3, 4, 5 \ldots$	Balmer
$n_1 = 3$	$n_2 = 4, 5, 6 \ldots$	Paschen
$n_1 = 4$	$n_2 = 5, 6, 7 \ldots$	Brackett
$n_1 = 5$	$n_2 = 6, 7, 8 \ldots$	Pfund

Other atoms do not have such a simple spectral pattern, but in all cases the lines can be fitted to a formula similar to (1.4), although n_1 and n_2 are not necessarily integers.

In 1913 Bohr proposed a theory of atomic structure which accounted for these spectral observations. His theory was based on classical mechanics with electrons orbiting the nucleus. But he introduced a quantum condition that only special orbits were allowed, in order to account for the fact that the electrons do not spiral into the nucleus. He further postulated that electrons could only jump from one orbit to another if the energy difference was made up by absorption or emission of radiation, the frequency of the radiation being related to the energy difference by the Planck–Einstein relationship.

By imposing conditions on the type of orbit and the type of electron jump permitted, Bohr was able to fit his theory to the spectrum of hydrogen, the integers which occur in the Ritz formula being associated with so-called quantum numbers in the Bohr theory. The real triumph, however, was not the explanation of the spectral pattern, but that the value of the Rydberg constant was given in terms of fundamental constants by†

$$R = \frac{2\pi^2 e^4 m}{h^3 c}, \tag{1.5}$$

which agreed with experiment to 0·02%.

† The standard symbols used in this book are defined in the front (page xi).

For ten years Bohr's theory was adapted to explain spectral phenomena—principally by Bohr himself and also by Sommerfeld. But it was essentially a patchwork theory in which quantum conditions did not arise naturally but were imposed to fit experiment. A more serious objection to this theory was that it could not explain the fine structure of spectral lines for atoms other than hydrogen.

From Maxwell's electromagnetic equations it can be shown that if electromagnetic radiation consists of packets of energy $h\nu$, then these packets have momentum $h\nu/c$. In 1923 Compton confirmed this by an experiment in which x-rays were scattered by the electrons in graphite to produce free electrons. By analysing the change in frequency of the x-rays as a function of the scattering angle he showed that the collision of the photon and the electron obeyed the conservation laws of classical mechanics for energy and momentum, and that the momentum of the x-rays was indeed $h\nu/c$ or h/λ. This again emphasized the corpuscular nature of light.

So far we have met nothing too extreme for the classical physicist to accept. The most puzzling problem is still the dual wave–particle nature of light. In diffraction and interference it behaves like a wave; in the photoelectric and Compton effects it behaves like a stream of particles.

In 1923 de Broglie linked up the ideas of relativity with those of Planck and Einstein. Relativity relates energy to mass, and $E = h\nu$ relates energy to frequency. Thus a combination of the two gives a relationship between mass and frequency; or in other words, matter, like light, has a wave nature. De Broglie predicted that a particle with momentum p has an associated wavelength

$$\lambda = \frac{h}{p}. \tag{1.6}$$

This is the same relationship that holds for the momentum of photons†.

† In classical non-relativistic mechanics, mass and momentum are linked by the expression

$$p = mv.$$

In relativistic mechanics this is replaced by

$$p = \frac{mv}{1 - v^2/c^2},$$

where m is now called the rest mass. A particle having mass and moving with the velocity of light therefore has infinite momentum. A photon, however, has a zero rest mass and a finite momentum given by equation (1.6).

For example, an electron accelerated through a potential difference of 100 volts has a kinetic energy of 100 electron volts (1 ev = $1{\cdot}6 \times 10^{-12}$ erg). For such an electron (mass $9{\cdot}1 \times 10^{-28}$ g) $p^2/2m = 1{\cdot}6 \times 10^{-10}$; $p = 5{\cdot}4 \times 10^{-19}$ g cm sec^{-1}; $\lambda = 1{\cdot}2 \times 10^{-8}$ cm.

To produce a diffraction pattern from a wave motion a system of diffracting centres with a regular spacing of the order of λ is needed. Gratings with spacings approximately 10^{-8} cm occur in ionic or metallic crystals. The de Broglie relationship was confirmed by Davisson and Germer (1927) and Thompson (1928) by the observation of diffraction patterns, analogous to x-ray patterns, from a stream of electrons striking a metallic crystal.

Experiment has therefore shown that both matter and radiation have dual wave–particle properties. They behave in the manner expected of a wave under some conditions and in the manner expected of a stream of particles under other conditions; it depends on the conditions imposed by the experiment. It must be emphasized, however, that we do not treat matter and radiation as different aspects of one all embracing entity. For example, an electron has charge and mass, but a photon has neither. The diffraction pattern produced when x-rays strike a crystal may have the same general appearance as the pattern produced by a stream of electrons striking the same crystal but they are not identical. The variations in intensity within the diffraction pattern are different in the two cases because the interaction between photons and matter is subject to different laws to the interaction between electrons and matter.

If an electron has wave properties its behaviour must be capable of description by a wave equation, just as the waves of light, sound, strings, etc. This equation was given by Schrödinger in 1926. This equation can either be presented outright as being the basic equation of quantum mechanics, and its solutions shown to agree with experiment; or the equation can be derived from a combination of basic postulates and experiment. We shall follow the second approach at this stage only because our experience is that this leads to a more ready acceptance of the equation by those meeting quantum mechanics for the first time. But it is important to realize that this basic equation of quantum mechanics cannot be derived from any of the equations of classical mechanics without introducing non-classical postulates, and that the standing of Schrödinger's equation rests ultimately on the fact that its solutions agree with experiment.

The equations of classical wave motion have certain special solutions, known as standing waves, which have points of zero amplitude (nodes) that do not change with time. These standing waves for a stretched string are the fundamental and overtone vibrations of the string. They may be labelled by numbers 0, 1, 2 ... representing the number of nodes in the string between the two ends. Schrödinger reasoned that the integers which occur in the Ritz formula, and which are the quantum numbers of the Bohr theory, could arise in a similar manner, that is, from the standing-wave solutions of the electron wave equation.

The general expression for the amplitude of a harmonic standing wave in one dimension (x) is

$$\Psi(xt) = A \sin 2\pi\left(\frac{x}{\lambda} + k\right) \sin 2\pi\nu t, \tag{1.7}$$

where k is a constant. Differentiating (1.7) twice we find that Ψ satisfies the differential equation

$$\frac{\partial^2 \Psi}{\partial x^2} = -\frac{4\pi^2}{\lambda^2}\Psi. \tag{1.8}$$

If we now introduce the momentum through the de Broglie relationship (1.6), this equation becomes

$$\frac{\partial^2 \Psi}{\partial x^2} = -\frac{4\pi^2 p^2}{h^2}\Psi. \tag{1.9}$$

The total energy is the sum of the potential and kinetic energies, hence

$$E = \frac{p^2}{2m} + V. \tag{1.10}$$

Replacing p^2 in (1.9) by equation (1.10) leads us to

$$\frac{\partial^2 \Psi}{\partial x^2} + \frac{8\pi^2 m}{h^2}(E - V)\Psi = 0. \tag{1.11}$$

This is Schrödinger's equation for the electron in one dimension. In three dimensions the equation becomes

$$\nabla^2\Psi + \frac{8\pi^2 m}{h^2}(E - V)\Psi = 0, \tag{1.12}$$

where ∇^2 (del squared) is an abbreviation for

$$\frac{\partial^2}{\partial x^2} + \frac{\partial^2}{\partial y^2} + \frac{\partial^2}{\partial z^2}. \tag{1.13}$$

We shall return to a more general derivation of Schrödinger's equation in chapter 6.

Ψ represents the amplitude of the electron wave, and is called the *wave function.* In 1926 Born suggested that Ψ was to be connected to some physically observable property of the electron through Ψ^2 dv (or more generally $\Psi^*\Psi$ dv if Ψ is a complex function)† being interpreted as the probability of finding the electron in the small volume dv. This is a more satisfactory interpretation than that given by Schrödinger himself, namely that the electron is a charge cloud distributed with a density Ψ^2. Born's interpretation is better because, as we shall see, quantum-mechanical equations are based on the electron being considered as a point particle and not as a charge cloud. However, the measurement of a time-average electron density in an atom or molecule cannot distinguish between these two interpretations of Ψ^2, so it is often convenient, and physically illuminating, to think of Ψ^2 as an electron cloud even though this is not strictly correct.

Some further conditions must be imposed on Ψ for it to be a physically acceptable function. For example, it must be a single-valued function (only one value at a point x, y, z, in space). Also, since there must be unit probability of finding the electron somewhere in space, if we integrate Ψ^2 over all space we must obtain 1. This is then another condition we impose on Ψ, the so-called *normalization* condition

$$\int \Psi^2 \,\mathrm{d}x\,\mathrm{d}y\,\mathrm{d}z = 1. \tag{1.14}$$

The statistical interpretation of Ψ^2 as a probability density is in accord with the wave nature of an electron, and the Heisenberg uncertainty principle (1927). This states that the position and momentum of an atomic particle cannot simultaneously be measured exactly; for in measuring its position we must disturb it and so change its momentum. Heisenberg showed that in one dimension the product of the uncertainties in position and momentum is approximately $h/2\pi$. In three dimensions

$$\Delta x.\Delta p_x \simeq \frac{h}{2\pi}, \qquad \Delta y.\Delta p_y \simeq \frac{h}{2\pi}, \qquad \Delta z.\Delta p_z \simeq \frac{h}{2\pi}. \tag{1.15}$$

If we specify exactly the momentum of an electron then its de Broglie wavelength is specified exactly but we have no knowledge of its position, and its wave function is an unbounded wave in space. If we specify the

† If Ψ is a function which contains the pure imaginary number $\mathrm{i} = \sqrt{-1}$, then Ψ^* the complex conjugate of Ψ is obtained by replacing i by $-$i in Ψ. For example, if $\Psi = a + ib$ then $\Psi^* = a - ib$; if $\Psi = \mathrm{e}^{\mathrm{i}x}$, $\Psi^* = \mathrm{e}^{-\mathrm{i}x}$. It follows that $\Psi^*\Psi$ is a real function, everywhere positive, so this is physically acceptable as the probability of finding an electron at a point in space.

position of the electron exactly then we have no knowledge of its momentum so that the wave function would be a superposition of waves with all values of $\lambda = h/p$, and these would constructively interfere at one point (the position of the electron), and would destructively interfere everywhere else.

In the next two chapters we shall investigate the solutions of equation (1.12) for two relatively simple cases: the electron in a constant potential, and the electron in a spherically symmetrical coulomb potential (the hydrogen atom). These will illustrate the basic properties of quantum-mechanical systems. Before discussing the finer points of the theory of valency in chapters 7–18 we shall look in chapter 6 at the fundamentals of quantum mechanics in more detail.

Suggestions for further reading

HOFFMANN, *The Strange Story of the Quantum*, Dover, 1947. A very readable account of the historical development of quantum theory in non-mathematical language.

BORN, *Atomic Physics*, Blackie and Son, 1945. Chapter 4 and parts of chapters 5 and 8 give a more detailed account of the mathematical background to this chapter.

Problems 1†

1.1 What energy jump must an electron make if it is to be accompanied by the emission of visible light (7500–4000 Å)?

1.2 It requires 2·48 ev to dissociate Cl_2 into atoms. What is the longest wavelength of light that could lead to photodissociation?

1.3 The work function of metallic sodium is 2·3 ev. What is the kinetic energy of the electrons emitted when sodium is irradiated by the strong mercury emission line at 2537 Å?

1.4 Roughly what fraction of a neutron beam would be scattered after passing through a crystal 1 cm thick?

1.5 What is the associated de Broglie wavelength of
   i The electrons referred to in problem (1.3)?
   ii A fast-bowled cricket ball (in America for 'cricket' read 'base' and for 'bowled' read 'pitched')?

1.6 Consider on the basis of the uncertainty principle the validity of the following model for the neutron: a proton and an electron treated as point particles separated by less than 10^{-13} cm and held together by a coulombic force.

† The values of fundamental constants are given in the front (page xi).

Chapter 2

The electron in a constant potential

As an example of a solution of Schrödinger's equation we can consider the case of an electron confined to a square-well potential of width a and infinite depth (figure 2.1). The mathematics is very simple for this

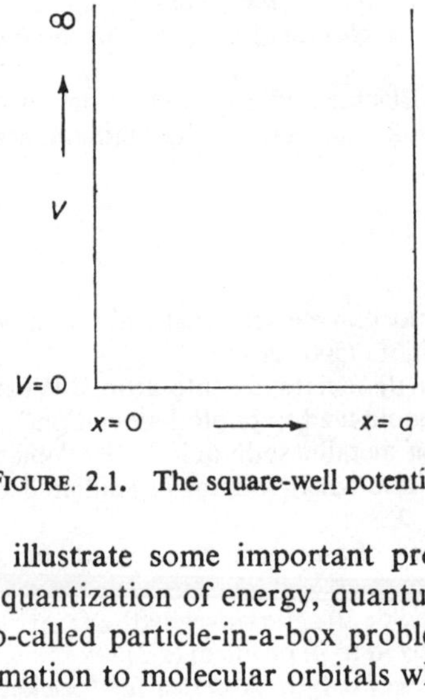

FIGURE. 2.1. The square-well potential

case, yet the results illustrate some important properties of quantum-mechanical systems: quantization of energy, quantum numbers and zero-point energy. This so-called particle-in-a-box problem is the basis of the free-electron approximation to molecular orbitals which we shall examine in chapter 15.

A solution of the one-dimensional Schrödinger equation (1.11)

$$\frac{d^2\Psi}{dx^2} + \frac{8\pi^2 m}{h^2}(E - V)\Psi = 0, \tag{2.1}$$

in a region of constant potential is

$$\Psi = N \sin\left\{\frac{2\pi}{h}(x + k)(2m(E - V))^{\frac{1}{2}}\right\}, \tag{2.2}$$

where k and N are constants of integration. This may be confirmed by substituting (2.2) into (2.1).

Inside the potential well $V = 0$, so that

$$\Psi = N \sin\left\{\frac{2\pi}{h}(x + k)(2mE)^{\frac{1}{2}}\right\}. \tag{2.3}$$

If $V = \infty$ then for finite energies Ψ must be zero (by inspection of (2.1)). Hence our solutions (2.3) are only acceptable (since they must be single-valued) if $\Psi = 0$ when $x = 0$ and $x = a$†. That is, Ψ must have the form

$$\Psi = N \sin\frac{\pi r x}{a}, \qquad \text{where } r = 1,2,3, \ldots \tag{2.4}$$

By comparing (2.3) and (2.4) we see that $k = 0$ and

$$E = \frac{h^2 r^2}{8ma^2}. \tag{2.5}$$

We see that the boundary conditions introduce quantization; only states are allowed whose energies satisfy equation (2.5). r is a quantum number which can be used to label the state of the electron and its energy: Ψ_r and E_r. We note that a state Ψ_0 does not exist (Ψ_0 is everywhere zero). The lowest state Ψ_1 has energy, which is called zero-point energy. Since $V = 0$ this must be entirely kinetic energy. In other words, even in the lowest energy state the electron is moving; as it must be, if we are not to violate the Heisenberg uncertainty principle. Ψ_r^2 represents the probability of finding the electron at a point in the range $x = 0$ to a. Since this must integrate to 1, we have the normalization condition (1.14)

$$\int_0^a N^2 \sin^2\frac{\pi r x}{a}\,\mathrm{d}x = 1, \tag{2.6}$$

which gives

$$N = \sqrt{\frac{2}{a}}. \tag{2.7}$$

Figure 2.2 illustrates our solutions. The functions Ψ_r look like the amplitudes of the fundamental (Ψ_1) and overtone (Ψ_2, Ψ_3, ...) vibrations of a stretched string.

† If V is finite and Ψ and E are to be finite, then $\mathrm{d}^2\Psi/\mathrm{d}x^2$ must be finite. $\mathrm{d}^2\Psi/\mathrm{d}x^2$ is only finite if Ψ and $\mathrm{d}\Psi/\mathrm{d}x$ are continuous functions of x. But if $V = \infty$, then (2.1) can have a solution with $\mathrm{d}^2\Psi/\mathrm{d}x^2$ infinite. That is, at a point of infinite potential the restriction that $\mathrm{d}\Psi/\mathrm{d}x$ be continuous may be relaxed.

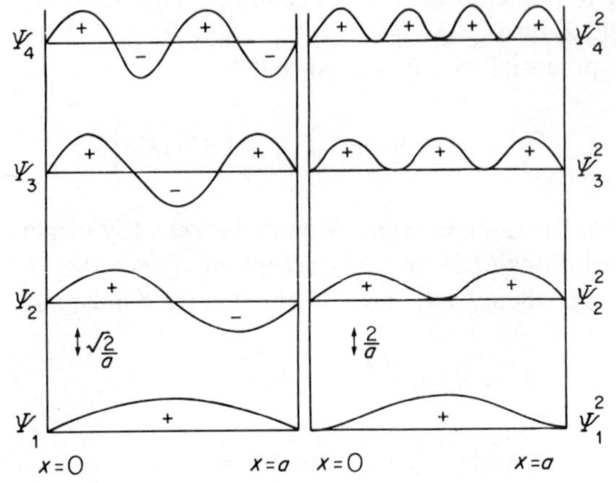

FIGURE 2.2. Wave functions and probability densities of the first four solutions of the 'electron-in-a-box' problem

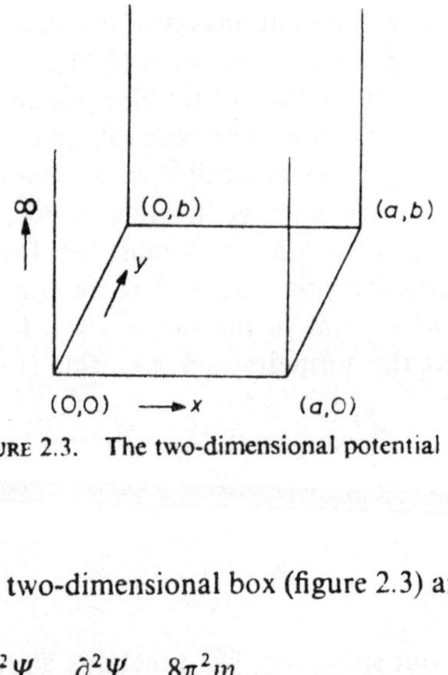

FIGURE 2.3. The two-dimensional potential well

We turn now to a two-dimensional box (figure 2.3) and solve

$$\frac{\partial^2 \Psi}{\partial x^2} + \frac{\partial^2 \Psi}{\partial y^2} + \frac{8\pi^2 m}{h^2}(E - V)\Psi = 0. \tag{2.8}$$

Whenever V can be written as a sum of one function of x and another function of y, the wave function Ψ can be written as a product of a function of x and one of y, and under these conditions the differential equation is said to be separable. In the simple case of $\dot{V} = 0$, the solution of (2.8) is the product of two one-dimensional solutions (2.3)

$$\Psi_{rs} = N \sin\left\{\frac{2\pi}{h}(x + k)(2mE_r)^{\frac{1}{2}}\right\} \cdot \sin\left\{\frac{2\pi}{h}(y + k')(2mE_s)^{\frac{1}{2}}\right\}, \qquad (2.9)$$

where $E = E_r + E_s$; and k, k' and N are again constants.

Applying the boundary conditions as in the one-dimensional problem we obtain two quantum numbers r and s, which are integers, and we find

$$E_r = \frac{h^2 r^2}{8ma^2}, \qquad E_s = \frac{h^2 s^2}{8mb^2}, \qquad N = \frac{2}{(ab)^{\frac{1}{2}}}.$$

The lowest state is

$$\Psi_{11} = N \sin\frac{\pi x}{a} \cdot \sin\frac{\pi y}{b}; \qquad E_{11} = \frac{h^2}{8m}\left(\frac{1^2}{a^2} + \frac{1^2}{b^2}\right)$$

and the next highest states are

$$\Psi_{12} = N \sin\frac{\pi x}{a} \cdot \sin\frac{2\pi y}{b}; \qquad E_{12} = \frac{h^2}{8m}\left(\frac{1^2}{a^2} + \frac{2^2}{b^2}\right)$$

$$\Psi_{21} = N \sin\frac{2\pi x}{a} \cdot \sin\frac{\pi y}{b}; \qquad E_{21} = \frac{h^2}{8m}\left(\frac{2^2}{a^2} + \frac{1^2}{b^2}\right).$$

For a square box ($a = b$), Ψ_{12} and Ψ_{21} have the same energies and are said to be *degenerate*. In general, the number of quantum numbers equals the number of dimensions of the box, and this is true even if the box is not rectangular. We shall see that the energy levels of an electron in an atom (a three-dimensional problem with spherical symmetry) are characterized by three space quantum numbers.

Suggestions for further reading

BOHM, *Quantum Theory*, Prentice-Hall, 1960. Chapter 11 gives a very complete account of the solutions of the wave equations for square-well potentials.

Problems 2

2.1 What quantization exists for a particle in a constant potential space with no boundaries?

2.2 What is the lowest kinetic energy of an electron confined to a cube of sides 10^{-13} cm?
Compare the answer with that of problem 1.6.

2.3 Show that the solutions of the Schrödinger equation for an electron moving on a circular ring of radius a in a constant zero potential are

$$\Psi_r = \sqrt{\frac{1}{\pi}}\sin r\vartheta, \quad \Psi_r' = \sqrt{\frac{1}{\pi}}\cos r\vartheta,$$

$$E_r = E_r' = \frac{h^2 r^2}{8\pi^2 m a^2}, \quad r = 0, 1, 2, \ldots$$

where ϑ measures the angle of the radius vector from any chosen origin. Does the lowest state violate the uncertainty principle?

Chapter 3

The hydrogen atom

In this chapter we shall examine the solutions of the Schrödinger equation for the simplest atom. These solutions will lead us to the concept of an *atomic orbital*, which is fundamental to all modern theory of valency.

The hydrogen atom consists of an electron and a proton which attract each other according to Coulomb's law. If r is the distance between the two particles, their potential energy is $-e^2/r$. This is the potential energy which must be inserted into the Schrödinger equation (1.12). Because the proton is much heavier than the electron, we can simplify the problem by saying that, as far as the electron is concerned, the proton is at the centre of mass and at rest. The Schrödinger equation for the electron is then†

$$\nabla^2\Psi + \frac{8\pi^2 m}{h^2}\left(E + \frac{e^2}{r}\right)\Psi = 0. \tag{3.1}$$

The solutions of this equation (Ψ) are functions in three-dimensional space and are called atomic orbitals. We give here only a brief outline of the mathematics by which the equation is solved and refer the reader who wants a fuller treatment to any of the standard textbooks on quantum mechanics.

If we were to try and solve equation (3.1) in cartesian coordinates then it would be necessary to write the potential energy term as $-e^2/(x^2+y^2+z^2)^{1/2}$ and this clearly makes the equation look a bit complicated. It is in fact much easier to work with a set of coordinates called polar coordinates which take full advantage of the spherical symmetry of the potential. These are defined in figure 3.1. r is the distance of the electron from the proton, ϑ is the angle which r makes with the z axis, and φ is

† The centre of mass is not quite at the proton, but to correct for this, m in (3.1) can be replaced by the so-called reduced mass of the proton and electron $mM/(m+M)$. Since the proton mass M is much greater than the electron mass, this reduced mass is very close to m.

the angle which the projection of r on the xy plane makes with the x axis. The cartesian and polar coordinates are related by the following expressions

$$z = r \cos \vartheta;$$
$$x = r \sin \vartheta \cos \varphi;$$
$$y = r \sin \vartheta \sin \varphi;$$
$$r^2 = x^2 + y^2 + z^2. \tag{3.2}$$

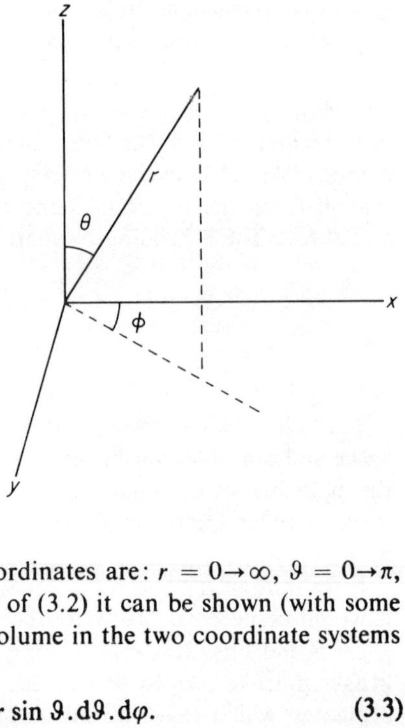

FIGURE 3.1. Polar coordinates

The limits of integration of these coordinates are: $r = 0 \rightarrow \infty$, $\vartheta = 0 \rightarrow \pi$, $\varphi = 0 \rightarrow 2\pi$. By partial differentiation of (3.2) it can be shown (with some difficulty) that the small element of volume in the two coordinate systems are related by

$$\mathrm{d}x\, \mathrm{d}y\, \mathrm{d}z = r^2\, \mathrm{d}r \sin \vartheta . \mathrm{d}\vartheta . \mathrm{d}\varphi. \tag{3.3}$$

The term $\nabla^2\Psi$ in polar coordinates looks rather more complicated than it does in cartesian coordinates (c.f. equation 1.13)

$$\nabla^2\Psi = \frac{1}{r^2}\left\{\frac{\partial}{\partial r}\left(r^2 \frac{\partial \Psi}{\partial r}\right) + \frac{1}{\sin^2 \vartheta}\frac{\partial^2 \Psi}{\partial \varphi^2} + \frac{1}{\sin \vartheta}\frac{\partial}{\partial \vartheta}\left(\sin \vartheta \frac{\partial \Psi}{\partial \Theta}\right)\right\} \tag{3.4}$$

However, the resulting Schrödinger equation

$$\frac{\partial}{\partial r}\left(r^2\frac{\partial \Psi}{\partial r}\right)+\frac{1}{\sin^2\vartheta}\frac{\partial^2\Psi}{\partial\varphi^2}+\frac{1}{\sin\vartheta}\frac{\partial}{\partial\vartheta}\left(\sin\vartheta\frac{\partial\Psi}{\partial\vartheta}\right)+\frac{8\pi^2 mr^2}{h^2}\left(E+\frac{e^2}{r}\right)\Psi=0 \tag{3.5}$$

can be solved fairly readily by writing Ψ as a product of two functions, one a function of the radial coordinate, and the other a function of the angular coordinates, as follows

$$\Psi = R(r)\,Y(\vartheta,\varphi) \tag{3.6}$$

If this is substituted into (3.5) and the equation divided throughout by Ψ it may be put in the form

$$\frac{1}{R}\frac{\partial}{\partial r}\left(r^2\frac{\partial R}{\partial r}\right)+\frac{8\pi^2 mr^2}{h^2}\left(E+\frac{e^2}{r}\right)=-\frac{1}{Y}\left\{\frac{1}{\sin^2\vartheta}\frac{\partial^2 Y}{\partial\varphi^2}+\frac{1}{\sin\vartheta}\frac{\partial}{\partial\vartheta}\left(\sin\vartheta\frac{\partial Y}{\partial\vartheta}\right)\right\} \tag{3.7}$$

On the left we have a function of r only and on the right a function of ϑ and φ only. Because r, ϑ and φ are independent variables we can argue that both sides must be equal to a constant, independent of r, ϑ and φ. If this were not so then changing r for constant ϑ and φ would change the left-hand side and therefore, from the equality, change the right-hand side also; this would be nonsense as the right-hand side is independent of r. If we define this constant to be λ then we have factorized the Schrödinger equation (3.7) into the following two equations

$$\frac{1}{\sin^2\vartheta}\frac{\partial^2 Y}{\partial\varphi^2}+\frac{1}{\sin\vartheta}\frac{\partial}{\partial\vartheta}\left(\sin\vartheta\frac{\partial Y}{\partial\vartheta}\right)+\lambda Y=0. \tag{3.8}$$

$$\frac{1}{r^2}\frac{\partial}{\partial r}\left(r^2\frac{\partial R}{\partial r}\right)+\left\{\frac{8\pi^2 m}{h^2}\left(E+\frac{e^2}{r}\right)-\frac{\lambda}{r^2}\right\}R=0. \tag{3.9}$$

The first of these equations must be solved subject to the boundary condition that Y is unchanged if either ϑ or φ is increased by 2π: this is clearly necessary if Ψ is to be a single-valued function. The solutions of (3.8) with this restriction are well known functions of mathematics called spherical harmonics. They arise in any problem of classical or quantum mechanics where the potential energy is a spherically symmetric function. Individual spherical harmonics are identified by two numbers l and m

which we call quantum numbers†. l can be zero or any positive integer, and for a given l, m can take the $2l+1$ different values

$$l,\ l-1,\ l-2,\ \ldots .\ 0,\ \ldots .\ -l+2,\ -l+1,\ -l$$

The constant λ in (3.8) and (3.9) has the value $l(l+1)$. It is important to note that m does not appear in (3.9) so that the radial part of the wave function is independent of m.

The atomic orbitals are labelled by the l values of the spherical harmonics according to the following scheme:

$$l = 0, 1, 2, 3, 4 \ldots$$

label s, p, d, f, g ...

The first four labels have historical significance being connected with the nature of atomic spectral lines (s = sharp, p = principal, d = diffuse, f = fundamental). The following are the most important spherical harmonics for our purposes

s orbitals:

$$Y_{00} = \frac{1}{2\sqrt{\pi}}$$

p orbitals:

$$Y_{10} = \frac{\sqrt{3}}{2\sqrt{\pi}} \cos \vartheta$$

$$Y_{1\pm1} = \frac{\sqrt{3}}{2\sqrt{2\pi}} \sin \vartheta\, e^{\pm i\varphi}$$

d orbitals:

$$Y_{20} = \frac{\sqrt{5}}{4\sqrt{\pi}} (3 \cos^2 \vartheta - 1)$$

$$Y_{2\pm1} = \frac{\sqrt{15}}{4\sqrt{2\pi}} \sin 2\vartheta\, e^{\pm i\varphi}$$

$$Y_{2\pm2} = \frac{\sqrt{15}}{4\sqrt{2\pi}} \sin^2 \vartheta\, e^{\pm 2i\varphi} \qquad (3.10)$$

Note that the m quantum number appears as an integer in the exponential function. These functions, being complex, have the disadvantage that we cannot draw pictures of them in real space. However, we can obtain functions which are equally good solutions of equation (3.4) by taking any linear combination of spherical harmonics having *the same l value* (this will be justified in chapter 6). In this way we can obtain real functions. For example, using the well known relationship

$$\cos \varphi = \tfrac{1}{2}(e^{i\varphi} + e^{-i\varphi}) \qquad (3.11)$$

† l and m have been called the azimuthal and magnetic quantum numbers from their analogy to similar numbers which appeared in the Bohr theory. In this book we shall refer to them as angular momentum quantum numbers.

We see that

$$\sqrt{\tfrac{1}{2}}(Y_{11} + Y_{1-1}) = \frac{\sqrt{3}}{2\sqrt{\pi}} \sin\vartheta \cos\varphi. \tag{3.12}$$

Now $\sin\vartheta\cos\varphi$ is the angular dependence of the x component of r (see 3.1) and for this reason the combination (3.12) is labelled as a p_x atomic orbital. The most convenient set of real functions which describes the angular dependence of s, p and d atomic orbitals is given in table 3.1. The complete atomic-orbital wave functions are obtained by multiplying these angular functions by the appropriate radial function $R(r)$.

TABLE 3.1. The angular dependence of atomic orbitals

Angular variation of			
$s =$	Y_{00}	$= 1/2\sqrt{\pi}$	s orbitals
$p_z =$	Y_{10}	$= (\sqrt{3}/2\sqrt{\pi}) \cos\vartheta$	p orbitals
$p_x =$	$\sqrt{\tfrac{1}{2}}(Y_{11} + Y_{1-1})$	$= (\sqrt{3}/2\sqrt{\pi}) \sin\vartheta\cos\varphi$	
$p_y =$	$-\mathrm{i}\sqrt{\tfrac{1}{2}}(Y_{11} - Y_{1-1})$	$= (\sqrt{3}/2\sqrt{\pi}) \sin\vartheta\sin\varphi$	
$d_{z^2} =$	Y_{20}	$= (\sqrt{5}/4\sqrt{\pi})(3\cos^2\vartheta - 1)$	d orbitals
$d_{xz} =$	$\sqrt{\tfrac{1}{2}}(Y_{21} + Y_{2-1})$	$= (\sqrt{15}/4\sqrt{\pi}) \sin 2\vartheta\cos\varphi$	
$d_{yz} =$	$-\mathrm{i}\sqrt{\tfrac{1}{2}}(Y_{21} - Y_{2-1})$	$= (\sqrt{15}/4\sqrt{\pi}) \sin 2\vartheta\sin\varphi$	
$d_{x^2-y^2} =$	$\sqrt{\tfrac{1}{2}}(Y_{22} + Y_{2-2})$	$= (\sqrt{15}/4\sqrt{\pi}) \sin^2\vartheta\cos 2\varphi$	
$d_{xy} =$	$-\mathrm{i}\sqrt{\tfrac{1}{2}}(Y_{22} - Y_{2-2})$	$= (\sqrt{15}/4\sqrt{\pi}) \sin^2\vartheta\sin 2\varphi$	

In calculations using atomic orbitals one may use either the real or complex set, whichever is the more convenient (i.e. leads to the simpler mathematics). In calculations involving magnetic fields, for example, the complex set would normally be used because the real functions do not have well-defined m quantum numbers (except for $m = 0$), and hence the z component of the angular momentum of the electron is not well defined.

We turn now to the radial part of the wave function, which is determined by (3.9). This has to be solved subject to the boundary condition that R tends to zero as $r \to \infty$. If this condition were not fulfilled then the electron would not be bound to the nucleus. The solutions of (3.9) have the form

$$R_{nl}(\rho) = \rho^l L_{n+l}^{2l+1}(\rho) \exp(-\rho/2) \tag{3.13}$$

where

$$\rho = \left(\frac{8\pi^2 me^2}{nh^2}\right) r. \tag{3.14}$$

These functions are identified by the quantum number l and a further quantum number n which is an integer whose values are restricted by the condition

$$n \geqslant l+1 \tag{3.15}$$

n is called the principal quantum number, $L_{n+l}^{2l+1}(\rho)$ is a member of a set of finite polynomials in ρ, called the associated Laguerre polynomials. For the record these are defined as solutions of the differential equation

$$\frac{d^2L}{d\rho^2} + (2l+2-\rho)\frac{dL}{d\rho} + (n-l-1)\,L = 0. \tag{3.16}$$

From the restriction on the l quantum number (3.15) one can see that if $n = 1$, l can only be 0; if $n = 2$, l can be 0 or 1; if $n = 3$, l can be 0, 1 or 2. The atomic orbitals may then be designated by their principal quantum number and their spectroscopic symbol in the following sequence

1*s*; 2*s*, 2*p*; 3*s*, 3*p*, 3*d*; etc.

The orbitals with $n = 1$ are said to make up the *K* shell; $n = 2$ the *L* shell; $n = 3$ the *M* shell, etc. The radial wave functions for these orbitals are given in table 3.2.

TABLE 3.2. The radial wave functions for one-electron atoms $R_{nl}(\rho)$

$n \backslash l$	0	1	2
1	$2e^{-\rho/2}$	—	—
2	$(1/2\sqrt{2})(2-\rho)\, e^{-\rho/2}$	$(1/2\sqrt{6})\, \rho e^{-\rho/2}$	—
3	$(1/9\sqrt{3})(6-6\rho+\rho^2)\, e^{-\rho/2}$	$(1/9\sqrt{6})(4\rho-\rho^2)\, e^{-\rho/2}$	$(1/9\sqrt{30})\rho^2\, e^{-\rho/2}$

Equation (3.9) which determines the radial part of the wave function also determines the orbital energy E. We emphasize again that it contains the quantum number l but not m. It follows that the energy of an orbital must be independent of m, so that s orbitals are non-degenerate, p are three-fold degenerate, d are five-fold degenerate, etc. On the other hand, one would expect that E would depend on l. It happens not to, but we shall see later that in this respect hydrogen and hydrogen-like ions are a special case. The orbital energy only depends on the principal quantum number according to the expression

$$E_n = -\frac{1}{2n^2}\left(\frac{4\pi^2 me^4}{h^2}\right). \tag{3.17}$$

There are also solutions of (3.9) for which E is positive. These correspond to the situation where the electron is not bound to the nucleus and the wave functions do not have to satisfy the boundary condition $\Psi \rightarrow 0$ as $r \rightarrow \infty$. As a result the energy levels form a continuum (cf. the particle in free space, problem 2.1). These unbound states of the hydrogen atom are not particularly relevant to valency theory so we shall not deal with them further.

Equations (3.1)–(3.9) can be generalized for other one electron atoms like He^+ and Li^{2+} if we replace e^2/r by Ze^2/r, Z being the nuclear charge. The dimensionless quantity ρ can then be defined by

$$\rho = \frac{2Zr}{na_0}, \tag{3.18}$$

where a_0 is a characteristic length, which is called the Bohr radius because it was the radius of the lowest energy orbit in the Bohr theory of the hydrogen atom. In terms of fundamental constants it is given by

$$a_0 = \frac{h^2}{4\pi^2 me^2}. \tag{3.19}$$

The wave functions of table 3.2 then carry over for all one-electron atoms if ρ is defined by (3.18), but if they are to be normalized according to the condition

$$\int_0^\infty R_{nl}^2 r^2 \mathrm{d}r = 1, \tag{3.20}$$

then they must all be multiplied by $(Z/a_0)^{3/2}$. The orbital energies written in terms of a_0 are

$$E_n = -\frac{1}{2n^2}\left(\frac{e^2}{a_0}\right). \tag{3.21}$$

It is often convenient in quantum mechanical calculations to work in a set of units whose values are scaled to the fundamental physical quantities. In this way a calculated energy will not be changed if say the electronic charge, mass, or the velocity of light were at some time in the future given new values. These are called atomic units and are defined by the following choice:

a_0 is the unit of length called the Bohr. The mass of the electron m is the unit of mass. The charge on the electron is the unit of charge. $h/2\pi$ is the unit of action (energy $\times$ time). Note that the atomic unit of time is not the second if $h/2\pi = 1$. The Schrödinger equation (1.13) for an electron moving in a potential V can then be put in the following simple form

$$(-\tfrac{1}{2}\nabla^2 + V)\,\Psi = E\Psi. \tag{3.22}$$

For one-electron atoms $V = -Z/r$, and the orbital energies are, in atomic units (or Hartrees)

$$E_n = -\frac{Z^2}{2n^2}\,. \tag{3.23}$$

If under the stimulus of electromagnetic radiation an electron jumps from an orbital with principal quantum number n_1 to one with n_2, then this must be accompanied by absorption or emission of light in order that energy shall be conserved.

$$E_{n_2} - E_{n_1} = \frac{Z^2e^2}{2a_0}\left[\frac{1}{n_2^2} - \frac{1}{n_1^2}\right] = h\nu = \frac{hc}{\lambda}. \tag{3.24}$$

This is in agreement with the Ritz formula (1.4), and the value of the Rydberg constant is the same as given by the Bohr theory (1.5).

The complete wave functions for one-electron atoms are given in table 3.3.

In figure 2.2 we showed the wave functions of the electron in a one-dimensional constant potential: this was a two-dimensional figure. If we had done the same for the corresponding two-dimensional potential then we would have sketched a series of surfaces in three-dimensional space. However, atomic orbitals are functions of three variables and one would therefore need a four-dimensional space to illustrate the complete behaviour of the wave function in one figure. As it is beyond our ability to show such a thing on our two dimensional page we restrict ourselves to showing figures which represent the variation of the wave function with either the radial coordinate, keeping ϑ and φ fixed, or with ϑ and φ keeping r fixed.

TABLE 3.3. Normalized atomic-orbital wave functions for one-electron atoms, ($\rho = 2Zr/na_0$) all to be multiplied by $(Z/a_0)^{3/2}(\pi^{-1/2})$

n	l	$\lvert m \rvert$	Wave function
1	0	0	$1s = e^{-\rho/2}$
2	0	0	$2s = (32)^{-1/2}(2-\rho)\, e^{-\rho/2}$
2	1	0	$2p_z = (32)^{-1/2}\rho\, e^{-\rho/2} \cos\vartheta$
2	1	1	$2p_x = (32)^{-1/2}\rho\, e^{-\rho/2} \sin\vartheta\cos\varphi$; $2p_y = (32)^{-1/2}\rho\, e^{-\rho/2} \sin\vartheta\sin\varphi$
3	0	0	$3s = (972)^{-1/2}(6 - 6\rho + \rho^2)\, e^{-\rho/2}$
3	1	0	$3p_z = (648)^{-1/2}(4\rho - \rho^2)\, e^{-\rho/2} \cos\vartheta$
3	1	1	$3p_x = (648)^{-1/2}(4\rho - \rho^2)\, e^{-\rho/2} \sin\vartheta\cos\varphi$; $3p_y = (648)^{-1/2}(4\rho - \rho^2)\, e^{-\rho/2} \sin\vartheta\sin\varphi$
3	2	0	$3d_{z^2} = (2592)^{-1/2}\rho^2\, e^{-\rho/2} \sqrt{\tfrac{1}{3}}(3\cos^2\vartheta - 1)$
3	2	1	$3d_{xz} = (2592)^{-1/2}\rho^2\, e^{-\rho/2} \sin 2\vartheta\cos\varphi$; $3d_{yz} = (2592)^{-1/2}\rho^2\, e^{-\rho/2} \sin 2\vartheta\sin\varphi$
3	2	2	$3d_{x^2-y^2} = (2592)^{-1/2}\rho^2\, e^{-\rho/2} \sin^2\vartheta\cos 2\varphi$; $3d_{xy} = (2592)^{-1/2}\rho^2\, e^{-\rho/2} \sin^2\vartheta\sin 2\varphi$

We start first with the $1s$ orbital. This is spherically symmetric and decreases exponentially with distance from the nucleus. A graph of Ψ_{1s} as a function of r is shown in figure 3.2. The probability density for an electron in this orbital is Ψ_{1s}^2. The probability of finding an electron at a distance r from the nucleus is obtained by multiplying the probability density by the area of the spherical shell of radius r; this function, $4\pi r^2\Psi_{1s}^2$, and Ψ_{1s}^2 are also shown in figure 3.2.

The $1s$ orbital has no variation with the polar angles. We could therefore represent it by contour surfaces, which would be concentric spheres, or by contours on a plane passing through the nucleus which will be concentric circles as in figure 3.3a. We can also draw a spherical boundary surface such that nearly all the electron density (say 90%) is found inside the surface: in a plane this is figure 3.3b.

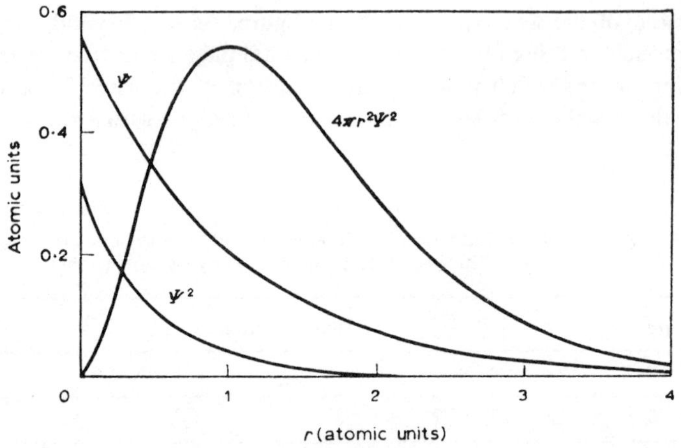

FIGURE 3.2. Wave functions and probability densities of the lowest energy state (1*s*) of the hydrogen atom

As the probability density for any orbital is everywhere positive there is one way in which its complete spatial variation can be represented qualitatively in one figure, and that is by the use of shading. Figure 3.3c shows a probability cloud for a 1*s* orbital if we imagine this being extended into three dimensions. In other words we are trying to represent a spherically symmetric cloud whose density gets less as we move away from the centre.

The radial variations of the most important atomic orbitals are illustrated in figure 3.4. There are several important points to note.

1. The orbitals increase in size as n increases. This is not immediately apparent from the figure since we have used ρ as abscissa instead of r.
2. Only *s* orbitals have a finite density at the nucleus. This is important in the coupling between electron spin and nuclear spin which shows up in magnetic resonance spectroscopy.
3. The number of times a function becomes zero (i.e. has a node) between 0 and ∞ is $n - l - 1$.
4. Amongst orbitals with the same n, those with the smaller l have the greater electron density close to the nucleus, but their principal maximum is further out.

To show the angular variation of the atomic orbitals we want graphs of Ψ against ϑ and φ for constant r. This is best done by the use of polar diagrams which are constructed according to the following recipe.

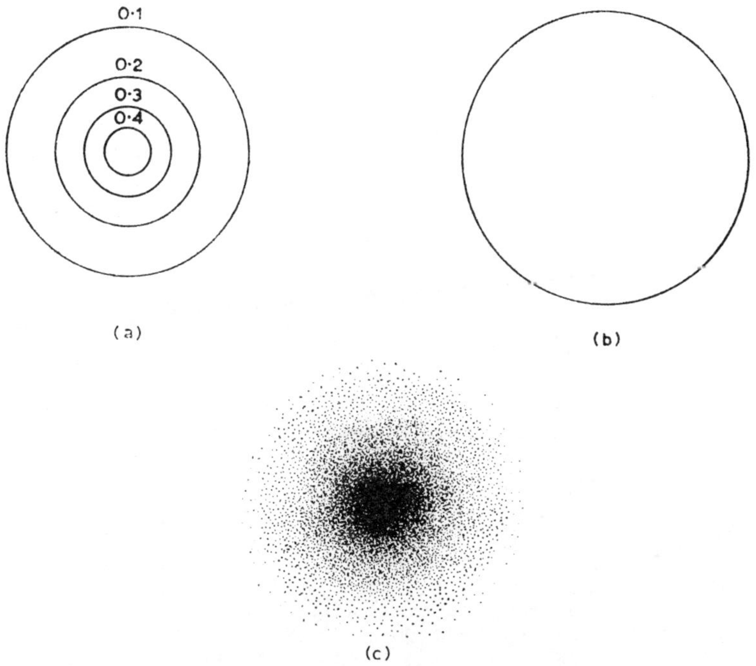

FIGURE 3.3. Different methods of representing the hydrogen 1s orbital (a) contours (b) boundary surface (c) probability cloud

A point is taken along a radius vector whose distance from the origin is equal to the modulus of the function for the ϑ and φ defined by that vector. These points define surfaces which are illustrated in figure 3.5. Sections through these surfaces in the xz plane are shown in figure 3.6. The sign of the angular function along a particular direction is indicated in each closed volume.

The s function is a sphere. The others all have special directions in space where the functions have a large value, and these special directions play an important role in the theory of valency. The p functions are pairs of spheres touching at the origin; the three p functions have identical shapes but they point along different coordinate axes. The d functions, except for d_{z^2}, have four pear-shaped lobes pointing towards the corners of a square. The d_{z^2} orbital has two larger pear-shaped lobes pointing along the z axis and a central collar with a figure of eight cross section.

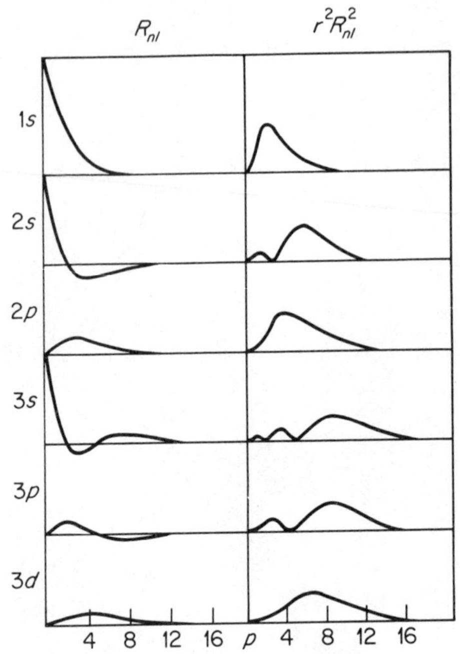

FIGURE 3.4. Radial wave functions and probability densities for the hydrogen atomic orbitals

Since the three p orbitals only differ by their directions in space we can see physically that they must have the same energy, a point we noted earlier from the fact that the quantum number m does not appear in the radial equation. Since d_{z^2} looks different from the other four d orbitals it is not immediately obvious that the five d orbitals are essentially equivalent. By equivalent we mean that if the three coordinate axes in space are equivalent then any measurement made on an electron in a d orbital does not depend on which d orbital it is occupying. The difference between the d orbitals is only apparent. d_{z^2} can be written as a linear combination of two orbitals which have the same shape as the other four d orbitals, namely $d_{z^2-x^2}$ and $d_{z^2-y^2}$, but which are not independent functions ($z^2 - x^2 + z^2 - y^2 = 3z^2 - r^2$ which is the angular dependence of the d_{z^2} orbital).

Probably the best pictorial representation of a wave function which combines some aspects of both the angular and radial variation is a

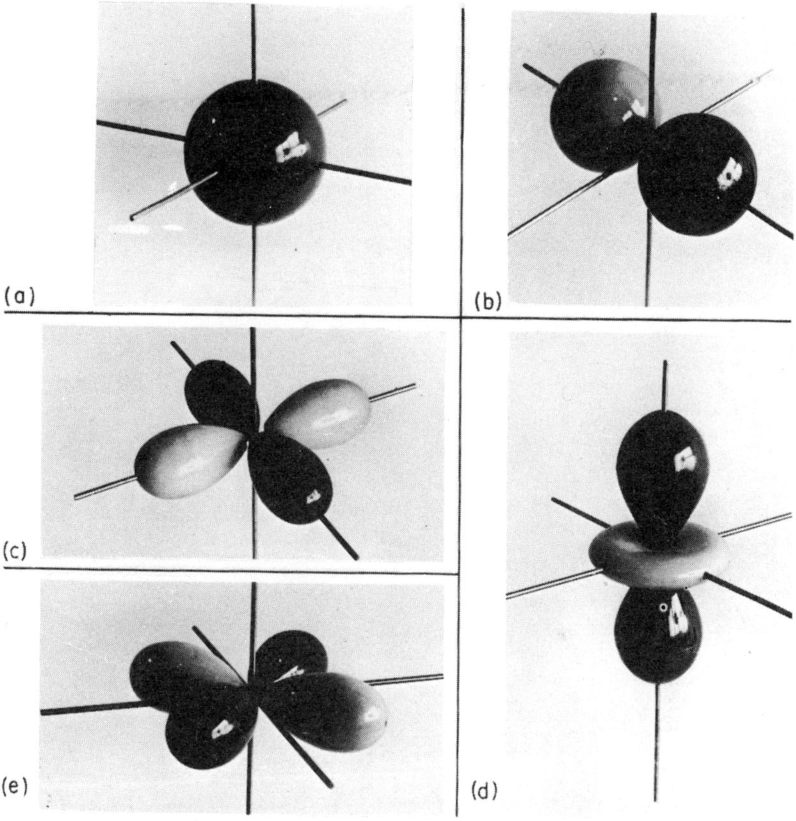

FIGURE 3.5. Polar models of orbitals: (a) s; (b) p; (c) $d_{x^2-y^2}$; (d) d_{z^2}; (e) $d_{xy}(d_{xz}, d_{yz})$

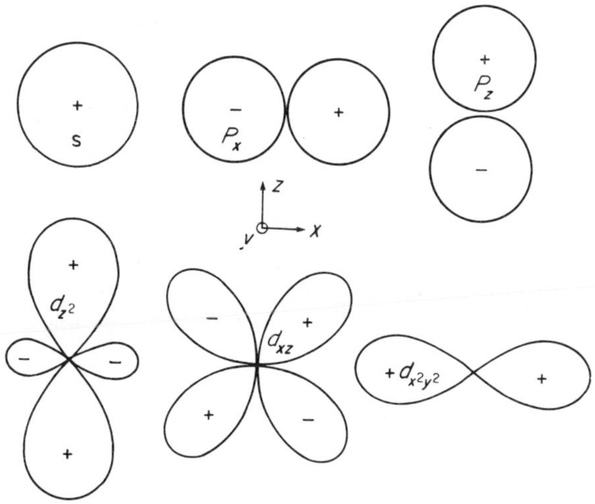

FIGURE 3.6. Polar diagrams for *s*, *p* and *d* orbitals

contour diagram as illustrated for a $2p_z$ orbital in figure 3.7. But a very convenient, although approximate, representation is given by the angular functions shown in figure 3.6. These will be found in many places throughout this book.

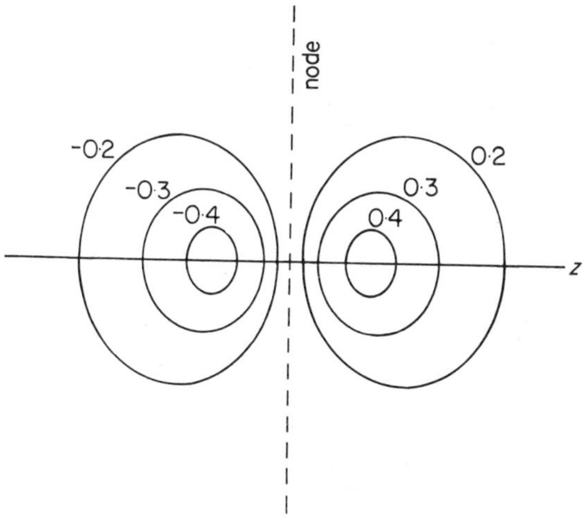

FIGURE 3.7. Contours on the *xz* plane for a p_z orbital

We have seen that a wave function can have positive and negative regions. However, the observable properties of an electron depend not on Ψ but on Ψ^2 (or more generally $\Psi^*\Psi$) and this is always positive. What then is the significance, if any, of the signs? We draw an analogy here with the amplitude of a light wave. This amplitude can be positive or negative but the sign is only important when two waves interfere, for then the relative signs of their amplitudes determine whether the interference is constructive or destructive. Similarly, we shall see in the theory of the chemical bond that it is the relative signs of two overlapping atomic orbitals which are important and not the absolute signs of any one orbital.

It must be remembered that the potential energy terms in the Schrödinger equation were obtained by treating the electron as a point particle. The electron is not a diffuse charge cloud. The physical significance of the wave function is that Ψ^2, evaluated at a point in space, gives the probability that the electron should be found at that point. This is the Born interpretation of the wave function which we described in chapter 1.

Finally, a comment on the relationship between the solutions of Schrödinger's equation for the hydrogen atom and the Bohr theory. Both theories lead to the same expression for the energy, but the wave functions from Schrödinger's equation are capable of explaining other properties of the electron, notably the probabilities of absorption and emission of radiation, which the Bohr theory is not. Quantum numbers appear naturally from the Schrödinger equation, through the boundary conditions, but have to be inserted in an arbitrary manner into the Bohr theory. The electrons in the Bohr theory occupy planet-like *orbits*—from the Schrödinger approach they occupy delocalized *orbitals*. A massive body of experimental evidence (e.g. electron densities from x-ray diffraction) shows that the latter is the correct picture.

Problems 3

3.1 Draw a polar diagram of the function

$$Y_{30} = (2\cos^3\vartheta - 3\cos\vartheta\sin^2\vartheta)$$

in the xz plane. This is the angular variation of one of the f orbitals (unnormalized).

3.2 If Ψ_x, Ψ_y and Ψ_z are wave functions for the three p orbitals directed along the x, y and z axes respectively, what is the wave function for a p orbital directed along the (1, 1, 1) vector (which makes an equal angle with the x, y and z axes)?

3.3 Show that the product of the hydrogen 1*s* and 2*s* atomic orbitals integrated over all space is zero. This is a general result for any product of different orbitals; it is called the *orthogonality* condition.

3.4 Calculate the mean value of $1/r$ for an electron in the 1*s* orbital of a one-electron atom, and hence deduce the mean value of the potential energy experienced by this electron. Show that the average kinetic energy is equal to the total energy with the sign changed. This result, known as the virial theorem, applies to systems of particles in equilibrium interacting with coulombic forces. It holds both in classical and quantum mechanics.

Chapter 4

Many-electron atoms and the periodic table

An electron in an atom with more than one electron is influenced by the potential field of the nucleus and the field due to the other electrons. For example, the potential for an electron labelled 1 will be

$$V_1(\mathbf{r}_1, \mathbf{r}_i) = -\frac{Ze^2}{r_1} + \sum_{i \neq 1} \frac{e^2}{r_{i1}}, \tag{4.1}$$

where r_1 is the distance of electron 1 from the nucleus, and r_{i1} is the distance between electron i and electron 1. To a good approximation we can replace V_1 by a potential in which the electron repulsion terms have been averaged over the positions of electrons i. This is then only a function of the vector $\mathbf{r}_1$.

$$V_1(\mathbf{r}_1) = -\frac{Ze^2}{r_1} + \overline{\sum_{i \neq 1} \frac{e^2}{r_{i1}}}. \tag{4.2}$$

In general, $V(\mathbf{r}_1)$ will not have spherical symmetry since the electrons i will be in orbitals (p, d, etc.) which do not have spherical symmetry. It is then necessary to make the further assumption that $V(\mathbf{r}_1)$ can be averaged over all orientations in space to obtain a central potential $V(r_1)$, and atomic orbitals can be obtained of the form (3.5) by substituting this average potential into expression (3.1) in place of e^2/r.

This method of reducing the many-electron Schrödinger equation to the form (3.1), and the techniques for solving the equation, were introduced by Hartree and Fock. It is known as the self-consistent field (SCF) method. An approximate set of atomic orbitals is chosen and from these the average potential acting on each electron is calculated. These potentials are then used to calculate new orbitals from which better approximations to the average potentials are obtained. The process is repeated until a set of orbitals reproduces the potentials which gave these orbitals. The mathematical basis of the method is described in chapter 10.

The atomic orbitals that are obtained by the SCF method are either given as a table of numbers (the radial part of the wave function tabulated as a function of r), or they may be expressed algebraically in the form

$$\Psi = \left(\sum_{n,\zeta} c_{n\zeta} r^n e^{-\zeta r}\right) Y_{lm}(\vartheta, \varphi). \tag{4.3}$$

where the cs are numbers. The orbitals obtained for many-electron atoms look similar to the hydrogen-like orbitals, and so may be conveniently labelled in the same way: $1s$; $2s$, $2p$; etc. However, the principal quantum number n has significance only in that $n - l - 1$ is the number of radial nodes between 0 and ∞. There is no simple algebraic formula for the energy in which n appears as an integer.

A considerable amount of work has been done using very approximate wave functions of the type (4.3) in which only one term in the summation is taken. These are of the type

$$\Psi_{nlm} = \left(\frac{(2\zeta)^{2n+1}}{(2n)!}\right)^{\frac{1}{2}} r^{n-1} e^{-\zeta r} \, Y_{lm} \, (\vartheta, \varphi) \tag{4.4}$$

where n is the principal quantum number. They are known as Slater-type orbitals (STO's) as the set which were mainly in use until the 1960s were determined according to a recipe proposed by Slater.

Note that STO's have no radial nodes so that, except for the orbitals $1s$, $2p$, $3d$, etc., they are not a good representation of an SCF atomic orbital. Also two of these orbitals which only differ in their n quantum are not orthogonal. It has been found however that if an orbital is orthogonalized to all the orbitals of smaller n according to the Schmidt procedure (page 106) then the resulting orthogonal set are quite a good approximation to the SCF orbitals. For example, the orthogonalized $2s$ orbital has the form

$$\Psi'_{2s} = c_1 e^{-\zeta_1 r} + c_2 r e^{-\zeta_2 r} \tag{4.5}$$

where the coefficients c_1 and c_2 are chosen such that the function is normalized and is orthogonal to the $1s$ orbital, $\exp(-\zeta_1 r)$.

The value of the exponent ζ is determined by the variation theorem which will be discussed in chapter 6. These exponents should ideally be optimized for each problem in which they are used, but the more usual practice is to use those optimized for the ground state wave function of the atom. Table 4.1 gives the exponents for some of the lighter atoms determined by this criterion. It is important to realize that these simple orbitals will not be sufficient for an accurate calculation, and if, in the molecule, an atom is very different from its neutral ground state, then they may not even give qualititatively useful results. This is particularly true of the transition metals which in a molecule can exist in several oxidation states. Two term expansions of the type (4.3) (double zeta orbitals) are more widely used for accurate calculations.

TABLE 4.1. Best values of ζ for the ground states of neutral atoms

Z	1s	2s	2p	3s	3p	4s	3d	4p
2.	1·6875							
3.	2·6906	0·6396						
4.	3·6848	0·9560						
5.	4·6795	1·2881	1·2107					
6.	5·6727	1·6083	1·5679					
7.	6·6651	1·9237	1·9170					
8.	7·6579	2·2458	2·2266					
9.	8·6501	2·5638	2·5500					
10.	9·6421	2·8792	2·8792					
11.	10·6259	3·2857	3·4009	0·8358				
12.	11·6089	3·6960	3·9129	1·1025				
13.	12·5910	4·1068	4·4817	1·3724	1·3552			
14.	13·5745	4·5100	4·9725	1·6344	1·4284			
15.	14·5578	4·9125	5·4806	1·8806	1·6288			
16.	15·5409	5·3144	5·9885	2·1223	1·8273			
17.	16·5239	5·7152	6·4966	2·3561	2·0387			
18.	17·5075	6·1152	7·0041	2·5856	2·2547			
19.	18·4895	6·5031	7·5136	2·8933	2·5752	0·8738		
20.	19·4730	6·8882	8·0207	3·2005	2·8861	1·0995		
21.	20·4566	7·2868	8·5273	3·4466	3·1354	1·1581	2·3733	
22.	21·4409	7·6883	9·0324	3·6777	3·3679	1·2042	2·7138	
23.	22·4256	8·0907	9·5364	3·9031	3·5950	1·2453	2·9943	
24.	23·4138	8·4919	10·0376	4·1226	3·8220	1·2833	3·2522	
25.	24·3957	8·8969	10·5420	4·3393	4·0364	1·3208	3·5094	
26.	25·3810	9·2995	11·0444	4·5587	4·2593	1·3585	3·7266	
27.	26·3668	9·7025	11·5462	4·7741	4·4782	1·3941	3·9518	
28.	27·3526	10·1063	12·0476	4·9870	4·6950	1·4277	4·1765	
29.	28·3386	10·5099	12·5485	5·1981	4·9102	1·4606	4·4002	
30.	29·3245	10·9140	13·0490	5·4064	5·1231	1·4913	4·6261	
31.	30·3094	11·2995	13·5454	5·6654	5·4012	1·7667	5·0311	1·5554
32.	31·2937	11·6824	14·0411	5·9299	5·6712	2·0109	5·4171	1·6951
33.	32·2783	12·0635	14·5368	6·1985	5·9499	2·2360	5·7928	1·8623
34.	33·2622	12·4442	15·0326	6·4678	6·2350	2·4394	6·1590	2·0718
35.	34·2471	12·8217	15·5282	6·7395	6·5236	2·6382	6·5197	2·2570
36.	35·2316	13·1990	16·0235	7·0109	6·8114	2·8289	6·8753	2·4423

CLEMENTI and RAIMONDI, *J. Chem. Phys.*, **38**, 2686 (1963) Double-zeta functions have been published by CLEMENTI in *I.B.M. J. Res. Develop.* **9**, 2 (1965). HUZINAGA and ARNAU, *J. Chem. Phys.*, **53**, 451 (1970).

For many-electron atoms, orbitals with the same values for n but different values for l are not degenerate. For the same value of n, the energy increases as l increases. The reason is, that as l increases the orbital penetrates less into the region close to the nucleus so that the average screened nuclear charge experienced by an electron decreases as l increases. This is illustrated in figure 4.1. The potential of K^+, calculated by a SCF method, is plotted as a function of r. The radial electron densities $4\pi^2r^2R^2$ for hydrogen-like $3s$, $3p$, and $3d$ orbitals are shown (cf. figure 3.4), and it is clear that an electron in the $3s$ orbital would experience the greatest potential, and so have the lowest energy, and one in the $3d$ orbital would experience the least potential and so have the greatest energy.

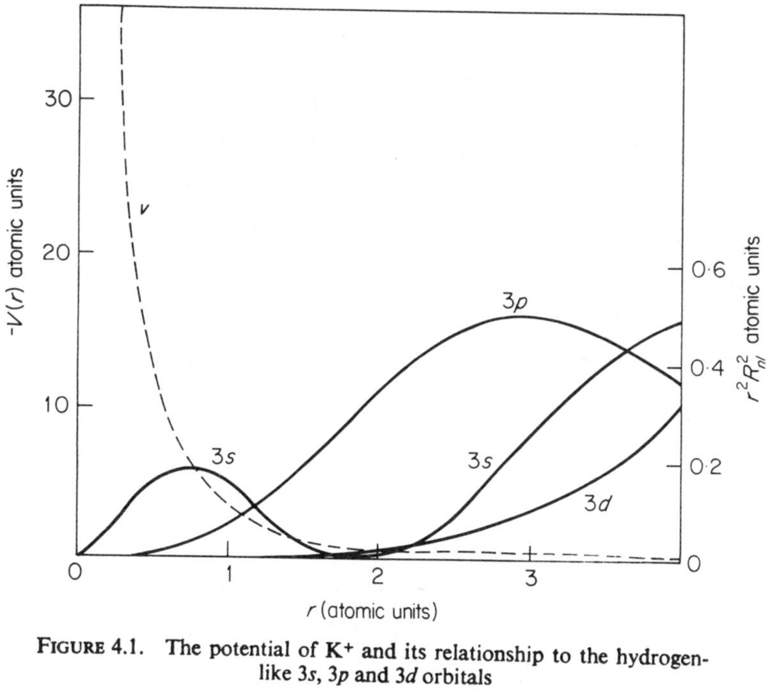

FIGURE 4.1. The potential of K^+ and its relationship to the hydrogen-like $3s$, $3p$ and $3d$ orbitals

We shall see later that the energy differences between orbitals of the same n but different l values play an important role in the theory of atomic structure, and in the theory of the chemical bond. For future reference, then, we quote here the separation of the 2*s* and 2*p* orbitals of atoms in the first row of the periodic table.

TABLE 4.2. Average energy differences between 2*s* and 2*p* orbitals (ev)

	Li	Be	B	C	N	O	F
$E(2p)$–$E(2s)$	1·85	3·36	5·75	8·77	12·39	16·53	21·54

The energy of an orbital is not a well-defined quantity as it depends on the distribution of the other electrons in the atom. The numbers quoted here are from ionization potentials averaged over all atomic states with appropriate weighting for the state degeneracy.

In 1925 Uhlenbeck and Goudsmit proposed that the electron was a spinning particle and as such had a spin angular momentum and an associated magnetic moment. This enabled them to explain some features of the fine structure of atomic spectral lines, but they found it necessary to postulate that the spin was quantized so that in an external magnetic field the magnetic moment of the spinning electron was either

$$+\tfrac{1}{2}\left(\frac{eh}{2\pi mc}\right) \quad \text{or} \quad -\tfrac{1}{2}\left(\frac{eh}{2\pi mc}\right)$$

relative to the direction of the external field. The results of an earlier experiment by Stern and Gerlach (1922) were seen to confirm this; they had passed a beam of silver atoms through an inhomogenous magnetic field and found that it split into two beams as though the atoms had one or the other of the magnetic moments.

The spin of the electron is described by another quantum number s which can have the values $+\frac{1}{2}$ or $-\frac{1}{2}$. It will be seen in chapter 9 that these quantum numbers represent the angular momentum of the spinning electron in units of $h/2\pi$.

We now have four electron quantum numbers: n, l, m and s. In 1925 Pauli expounded his *exclusion principle* which is that no two electrons in an atom can have the same four quantum numbers. This means that an orbital can be occupied by one electron with either $+\frac{1}{2}$ or $-\frac{1}{2}$ spin; or by two electrons, providing they have different spins. Here is a limiting factor, having no classical analogue, which completely excludes the possibility of two electrons having the same spin occupying the same orbital.

In 1869 Mendeleef pointed out that 'The physical and chemical properties of the elements and their compounds are a periodic function of the atomic weights'. He was able to predict the properties of then unknown elements (e.g. germanium) with remarkable precision. Apart from the replacement of atomic weight by atomic number as the basis of the ordering, and the insertion of new elements, the modern periodic classification of the elements is basically that of Mendeleef. We give a modern representation of this classification in table 4.3. The elements increase by one in atomic number if the table is read in normal fashion, i.e. left to right down the page.

We can understand the periodicity of the elements from the energy pattern of atomic orbitals and the Pauli exclusion principle.
The explanation is based on the following rules:

1. In the ground states of atoms the electrons occupy the orbitals of lowest energy, subject to the exclusion principle. Starting with hydrogen we can then deduce the electron configuration of the elements by adding electrons to the orbitals one at a time; this is called the building up (or aufbau) principle.
2. np orbitals have a higher energy than ns but much lower than $(n+1)s$.
3. nd orbitals have about the same energy as $(n+1)s$ but are lower than $(n+1)p$.
4. nf orbitals have about the same energy as $(n+1)d$.

The $1s$ orbital can take one electron (H) or two electrons with opposite spins, we then have a complete K shell (He). The next electron added must go into the $2s$ orbital (Li) and this can take another electron (Be). The next electron goes into a $2p$ orbital (B), and the structures of the rest of the first row elements C to Ne are built up by adding successive electrons to the $2p$ sub-shell until this is full (six electrons in all). We then have a complete L shell and the atom (Ne), one of the so-called inert or noble gases.

The eleventh electron goes into the $3s$ orbital (Na). This atom has one electron outside a complete shell, like Li and all the other group 1a elements of the periodic table. This extra electron determines the chemical properties of the atom so that all the 1a elements are chemically very similar.

When the $3s$ and $3p$ sub-shells have been filled, the next electron goes into the $4s$ orbital (K) rather than the $3d$. From spectroscopy it can be shown that the $3d$ orbital of K is about 2·7 ev higher in energy than the $4s$. After the $4s$ orbital has been filled the electrons start to go into the $3d$ orbitals; elements with an incomplete d sub-shell are known as transition metals. The separation of the $3d$ and $4s$ orbitals of the first transition metal series is very small and there are some cases where the ground state of the atom does not have a filled $4s$ orbital. For example, Cr has the ground

TABLE 4.3. A periodic classification of the elements

	The main group elements 1a 2a 3b 4b 5b 6b 7b 0	*The transition metals* 3a 4a 5a 6a 7a 8 1b 2b	*The rare earths* / *The actinides*
1s	H He		
2s	Li Be		
2p	B C N O F Ne		
3s	Na Mg		
3p	Al Si P S Cl Ar		
4s	K Ca		
3d		Sc Ti V Cr Mn Fe Co Ni Cu Zn	
4p	Ga Ge As Se Br Kr		
5s	Rb Sr		
4d		Y Zr Nb Mo Tc Ru Rh Pd Ag Cd	
5p	In Sn Sb Te I Xe		
6s	Cs Ba		
4f			La Ce Pr Nd Pm Sm Eu Gd Tb Dy Ho Er Tm Yb
5d		Lu Hf Ta W Re Os Ir Pt Au Hg	
6p	Tl Pb Bi Po At Rn		
7s	Fr Ra		
5f			Ac Th Pa U Np Pu Am Cm Bk Cf Es Fm Md No

state configuration $4s3d^5$ and not $4s^23d^4$; Cu is $4s3d^{10}$ and not $4s^23d^9$. Completely filled or half-filled d sub-shells give rise to an extra stability.

After the $3d$ sub-shell has been completed, the $4p$ is filled and we have the elements Ga–Kr which are like the corresponding elements of the first period.

The pattern of the $4s$, $3d$ and $4p$ electrons is repeated through the $5s$, $4d$ and $5p$ sub-shells. Then, after the $6s$ orbital has been filled, electrons start to fill the $4f$ orbitals. The elements containing an incomplete $4f$ sub-shell are called the rare earths and they behave chemically very like each other. As in the case of the transition metal series, the rare earths show a near-degeneracy of their orbitals. In their ground configurations they all have one $5d$ electron followed by an incomplete $4f$ sub-shell: the first $5d$ electron appears at La and then no more electrons are added to this $5d$ sub-shell until we reach Hf.

Lastly, after filling another transition metal series ($5d$) and then filling the $6p$ and $7s$ orbitals, the $5f$ orbitals are filled giving the actinide series. After uranium (element 92) the elements are only produced by neutron bombardment of the heavy elements, and are mostly short-lived†.

The chemistry of the elements is usually divided into two main sections: the chemistry of the transition elements and the chemistry of the others—the main group elements. The two sets have little in common although some similarities do exist in the higher valence states. Chromates and sulphates, for example, are often isomorphous although the chemistry of chromium is quite different from that of sulphur. The chemistry of the first row transition metals is largely that of their octahedral complexes and this will be dealt with in chapter 13. Corresponding elements of the second and third row transition series have much in common—Zr and Hf, for example, have almost identical chemical properties. This similarity is caused by the 'lanthanide contraction', a diminution in the radius of similarly charged ions from La to Lu as the f orbitals are filled. This has the effect of making the radius of Hf^{4+} almost identical to that of Zr^{4+} instead of being greater.

The main group elements, those with incomplete s and p sub-shells, divide themselves into two groups. The elements on the left-hand side of the table e.g. Li, Cs, Sn, readily lose electrons and occur as cations; those on the right, e.g. O, F, I, readily gain electrons and occur commonly as anions.

† For the historical background to the periodic classification of the elements the reader is referred to the following papers in the *Journal of Chemical Education:* **11,** 27, 217, 288 (1934); **16,** 394 (1939); **25,** 658, 662 (1948); **33,** 69 (1956); **34,** 30 (1957).

Elements in groups 2a and 3b have low lying empty *p* orbitals which are conveniently filled by the lone pairs of electrons present in many compounds of groups 5b and 6b. A large number of complexes exist between members of these pairs of groups.

Almost all elements form compounds with hydrogen. These range from the salt-like 'ionic' hydrides such as NaH to transition metal hydrides such as $HCo(CO)_4$. Of particular interest are the hydrides of boron which are termed 'electron deficient' since they appear to contain more bonds than the electronic structure will allow. These compounds will be discussed in chapter 14.

Compounds of xenon with fluorine and oxygen are now well known, but apart from some similar compounds of krypton and radon, the inert gases are devoid of chemical reactivity.

Chapter 5

Basic principles of the theory of valence

The criteria for a satisfactory theory of valence are as follows:

1. To explain why some atoms combine and others do not (e.g. $2H \rightarrow H_2$ but $2He \nrightarrow He_2$), and to calculate bond energies.
2. To explain why some atoms combine in definite proportions but not others (e.g. CH_4 is a stable molecule but CH_2, CH_5, etc. are not).
3. To explain, and if possible predict, bond lengths and bond angles.

In this chapter we shall make a preliminary investigation of these criteria.

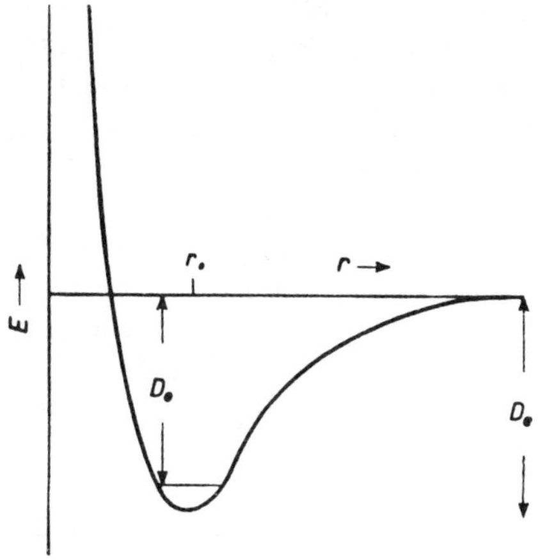

FIGURE 5.1. The potential energy curve for a diatomic molecule

Atoms combine because their total energy is lowered in the process, and the dimensions of a molecule are fixed by that arrangement of the atoms with the lowest energy. If we plot the sum of the electronic energy and nuclear repulsion energy (E) of a stable diatomic molecule as a function of internuclear distance (r), we get the curve shown in figure 5.1. This is the potential energy curve. It governs the vibrations of the molecule and will be examined in more detail in chapter 7. At small internuclear distances the energy is dominated by the nuclear repulsion energy, which approaches infinity as $1/r$. r_0 is the equilibrium bond length but we cannot say the nuclei are fixed at this position for this is contrary to the uncertainty principle; rather, even in their lowest energy state the atoms are vibrating, and r_0 is the mean separation of the nuclei. D_e is the electronic dissociation energy, but the measured dissociation energy is less than this by the zero-point vibrational energy $(D_e - D_0)$.

Experimentally it is clear that there are roughly two types of chemical bond: ionic and covalent. Molecules having ionic bonds may dissociate into ions in polar solvents; those having covalent bonds do not. In the Lewis theory of valence the atoms in an ionic bond are held together by coulombic forces; those in a covalent bond are held by a shared pair of electrons, some unspecified force being involved (this definition includes the so-called coordinate bond). We shall see that this force can be explained by quantum mechanics, and that the terms ionic and covalent can be taken as describing the extremes of bond type with real bonds being intermediate between the two.

We shall first examine the classical electrostatic energy between two atoms. The force between two neutral atoms which are so far apart that their electron clouds do not overlap significantly is zero (the potential outside a spherical charge distribution is the same as if the net charge were placed at the centre). When the electron clouds overlap slightly, there is a small attraction, but when the electrons of one atom overlap the nucleus of the other, repulsion occurs, and this increases rapidly as the two nuclei come together (all this is found from classical electrostatics). Although a binding energy can be obtained from this electrostatic interaction, it is rather small; for H_2 it is only about 5% of the total binding energy.

The attractive force between a positive and a negative ion comes into play before the electron clouds overlap and hence the binding energy in this case is large; the energy obtained by integrating the force from infinity is

$$E(r) = -\int_r^\infty F(r)\,\mathrm{d}r. \tag{5.1}$$

But the dissociation energy of an ionic bond does not depend simply on the attractive energy of the ions since there is no molecule, in the gas phase, for which the lowest energy dissociation products are ions. The atom having the lowest ionization potential is Cs ($I = 3{\cdot}86$ ev) and the one having the greatest electron affinity is Cl ($A = 3{\cdot}83$ ev), so that even CsCl would energetically prefer to dissociate to atoms rather than ions†. Figure 5.2 shows the potential energy curves for NaCl based on the classical electrostatic model. Ionic binding only occurs if the coulombic attraction of the ions can outweigh the energy $I - A$ required to form free ions.

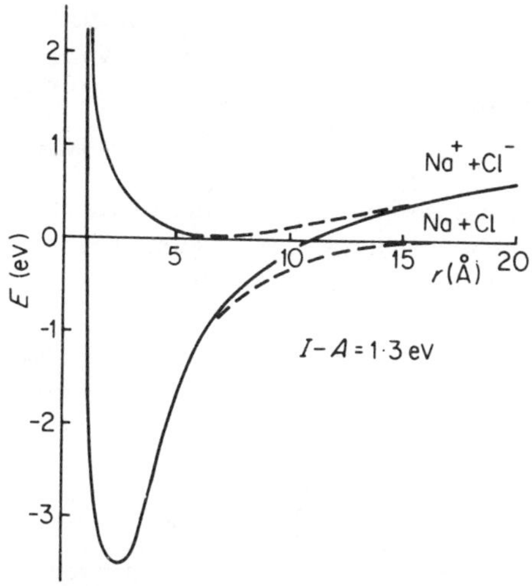

FIGURE 5.2. Potential energy curves for NaCl

The reason NaCl dissociates to ions in polar solvents is that the ions have a much larger binding energy to the solvent (solvation energy) than does the undissociated molecule. This energy is sufficient to counterbalance the 5 ev which is the gas-phase ionic dissociation energy. The solvation energy is mainly electrostatic in origin.

In figure 5.2 the potential energy curves for the ions and the neutral atoms are shown to cross (at the point where $I - A = e^2/r$). As we shall see in later chapters, the state of a molecule cannot really be labelled as being purely ionic, or as having no ionic character, but there is always some intermediate situation. At best we can say that a bond has some percentage of ionic character. One consequence of this mixture of ionic and covalent character is that if the potential energy curves for the states dissociating into $Na^+ + Cl^-$ and Na + Cl are calculated accurately, then it is found that they do not cross but follow the dotted lines shown in the figure. This

† The ionization potential is the energy required to remove an electron from an atom. The electron affinity is the energy gained by an atom when an electron is added to it. The electron affinity is the ionization potential of the negative ion.

is an important general result of quantum mechanics known as the non-crossing rule, which we shall prove in chapter 10. It means that even though the molecule NaCl in its lowest energy state is predominantly ionic, nevertheless, its lowest energy dissociation products in the gas phase will be neutral atoms.

If an atom has a high ionization potential it will have little tendency to lose electrons to other atoms; if it has a high electron affinity it will have a tendency to gain electrons. The overall tendency to increase its charge is determined by the sum of these two quantities. Mulliken suggested that

$$\chi = \tfrac{1}{2}(I + A) \qquad (5.2)$$

would best form the basis for an electronegativity scale of the elements. The ionic character of a bond AB is then determined by $|\chi_A - \chi_B|$. Table 5.1 shows part of the electronegativity scale.

TABLE 5.1. The Mulliken electronegativity scale[a]

H 7·17	Li 2·96	Be 2·86	B 3·83	C 5·61	N 7·34	O 9·99	F 12·32
	Na 2·94	Mg 2·47	Al 2·97	Si 4·35	P 5·72	S 7·60	Cl 9·45

[a] PRITCHARD and SKINNER, *Chem. Rev.*, **55**, 745 (1955).

Note that the electronegativity increases roughly from left to right and decreases down the groups. F is the most electronegative atom even though Cl has a greater electron affinity. Another electronegativity scale, due to Pauling, will be discussed in chapter 11.

Although electronegativity scales have been qualitatively very useful to chemists for the discussion of ionic character, they are quantitatively not very reliable. For example, the dipole moments of the hydrogen halides are well correlated with the electronegativity of the halide, but this correlation does not apply very well to other molecules. It is also clear that the electronegativity of an atom depends on the type of bond it is forming. For example, acetylene is a much stronger acid than methane, so we must deduce that a carbon atom in acetylene is more electronegative than that in methane.

We turn now to the covalent bond. There are two theories which are used to explain this: molecular-orbital (MO) theory and valence-bond (VB) theory. Both explain why the Lewis electron-pairing scheme leads to a stable bond, but at first sight the two theories look quite different. We shall see in later chapters, when the theories are discussed in detail, that in their simplest forms they are very approximate and only describe limiting cases of the covalent bond, and that a better description is obtained by going somewhere between the two.

The MO theory is a natural extension of the treatment of electrons in atoms to electrons in molecules. If atomic orbitals exist, then why not molecular orbitals? Since the electrons in atoms are only bounded at infinity we expect these molecular orbitals to extend over the whole molecule. Electrons can occupy these 'delocalized' orbitals subject to the same restriction, as for atoms, that not more than two electrons can occupy each orbital.

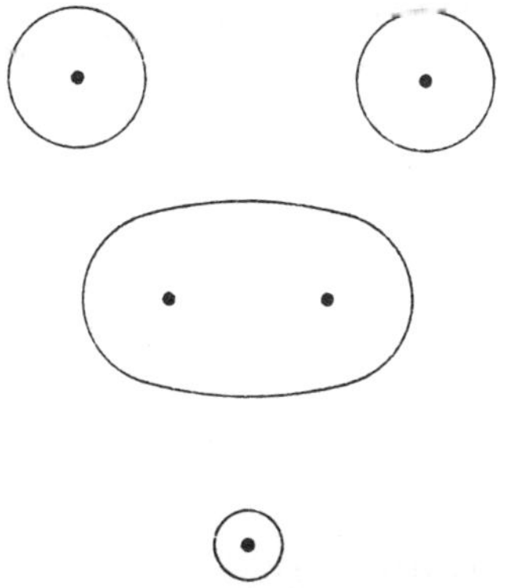

FIGURE 5.3. Correlation between the 1*s* orbitals of H and He and the molecular orbital of H_2

To explain why H_2 has a lower energy than two hydrogen atoms we only have to explain why the lowest molecular orbital of H_2 has a lower energy than the 1*s* orbital of H. The explanation is that the electrons in the molecular orbital of H_2 are in a potential from two protons instead of one. For comparison, the total electronic energy of He is 74 ev, that of two hydrogen atoms is 27 ev. The repulsion of the nuclei in H_2 (20 ev) is not sufficient to counterbalance this decrease in the potential energy of the electrons.

However, the change in energy of two hydrogen atoms when they form a bond is only 18% of their total energy, so it is fair to conclude that the orbitals of H_2 are fairly closely related to the atomic orbitals of the hydrogen atoms. Certainly, close to one nucleus an electron will be so dominated by the potential field of that nucleus that the molecular orbital must look like an atomic orbital. This is the basis of the linear combination of atomic orbital (LCAO) approximation to the molecular orbitals, which will be elaborated in chapter 10. Pictorially we can represent the formation of the lowest molecular orbital of H_2 from the two hydrogen 1*s* orbitals as in figure 5.3. If the two nuclei were to coalesce, we would have the 1*s* orbital of He.

The VB theory is a direct translation of the Lewis theory into the language of quantum mechanics. The electrons in a molecule are still supposed to occupy atomic orbitals (rather than molecular orbitals) but allowance is made for the fact that if two atomic orbitals overlap one another then we cannot be certain in which orbital an electron is to be found because the electrons are indistinguishable. We must write a wave function for the molecule which allows for the electrons to be found in either atomic orbital. In other words, our wave function describing this electron pair must allow for electron delocalization, which, as we have seen for the MO theory, is a stabilizing effect.

In both MO and VB theories a large overlap of atomic orbitals leads to electron delocalization and a strong bond. The overlap of atomic orbitals therefore plays the central role in both theories of the covalent bond. This overlap is measured quantitatively by the overlap integral; for two atomic orbitals ϕ_a and ϕ_b this is

$$S_{ab} = \int \phi_a \phi_b \, dv, \tag{5.3}$$

where the integration symbols stand for integration over all space.

Figure 5.4(a) illustrates some orientations of atomic orbitals which lead to good overlap and hence strong bonds. It should be noted that only the relative signs of the wave functions arise in calculating S_{ab}. We shall see later when the MO theory is dealt with in more detail that if S_{ab} is negative, as in figure 5.4(b), this leads to an increase in energy of the system (an *anti-bonding* situation). In some situations two orbitals may be overlapping but the overlap integral will be zero because the positive and negative overlap regions cancel out. This is illustrated in figure 5.4(c). If $S_{ab} = 0$ we have a *non-bonding* situation.

Electron spin plays a vital role in the theory of the covalent bond. In the MO description of H_2 the lowest molecular orbital is occupied by two electrons which must have opposite spins. If two hydrogen atoms come together with electrons of the same spin, then only one of these electrons can go into the lowest molecular orbital, and the other must go into a higher energy orbital. This second orbital is strongly antibonding and the overall effect is a repulsive energy state. The same thing is found in the VB theory. If the electrons in overlapping atomic orbitals have the same spin, then this leads to a state of repulsion. The repulsion of two electrons with the same spin is in accord with the Pauli exclusion principle that no two electrons can have the same spatial wave function if they have the same spins. When two atomic orbitals overlap, we are tending towards a situation in which both electrons have the same space function, and this is only possible if the electron spins are opposed. The reason why the inert gases do not form many compounds can now be seen. Their atomic

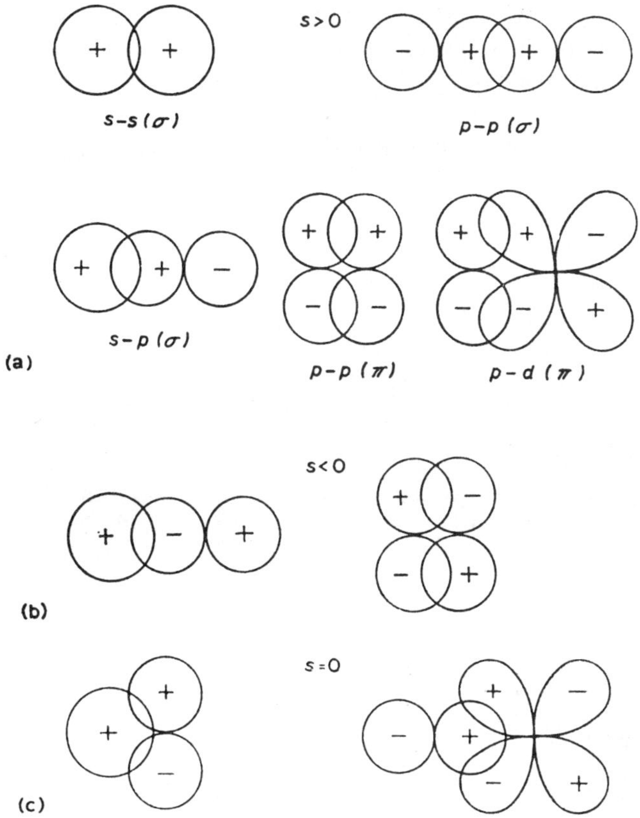

FIGURE 5.4. Overlap of atomic orbitals giving positive (a) negative (b) and zero (c) overlap integrals

orbitals are either completely filled or empty, and if the filled ones overlap with any other orbital containing an electron this will tend to violate the exclusion principle. For the same reason, once a molecule has reached a situation in which all its electrons are paired in low energy orbitals it will not readily add on further atoms.

The main group elements, which have incomplete s and p shells, tend to combine with other atoms until all their electrons are paired. The transition and rare earth elements, on the other hand, form stable molecules in which there are unpaired electrons. This is because the *d* and *f* orbitals are largely buried in the electron density of inner completely filled shells. (e.g. 3*d* is buried in the 3*s* and 3*p* shells, as can be seen from figure 5.5) and the repulsion from these filled shells makes it difficult to obtain a large overlap between *d* or *f* orbitals and orbitals of neighbouring atoms. In chapter 13 the chemistry of the transition elements will be explained in terms of a theory in which the *d* orbitals are only weakly involved in covalent bond formation.

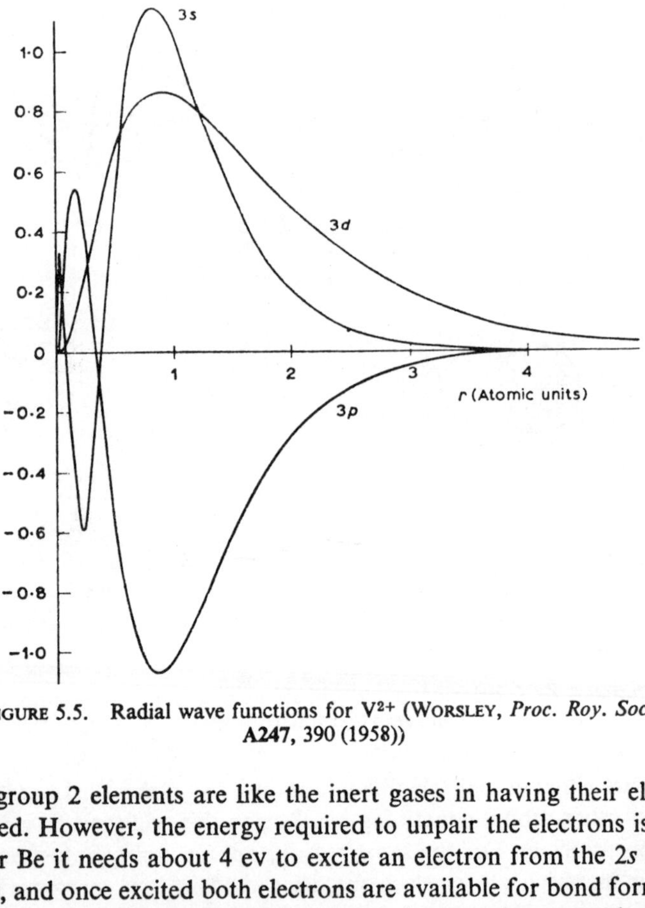

FIGURE 5.5. Radial wave functions for V^{2+} (WORSLEY, *Proc. Roy. Soc.*, **A247**, 390 (1958))

The group 2 elements are like the inert gases in having their electrons all paired. However, the energy required to unpair the electrons is rather low (for Be it needs about 4 ev to excite an electron from the 2*s* to a 2*p* orbital), and once excited both electrons are available for bond formation. The energy gained from forming the bonds is more than enough to compensate for the initial excitation. This is illustrated by the potential energy diagram for BeH shown in figure 5.6. (The non-crossing rule applies as for figure 5.2.) Amongst the first row diatomic hydrides LiH to HF, BeH has an anomalously low dissociation energy because only for Be is it necessary to excite the atom before electrons are available for bond formation.

Our explanation of how Be can form covalent bonds has left us in a rather difficult position, for after excitation we seem to have electrons which can form different kinds of bond. One electron is in a $2p$ orbital, and this can form a bond if a suitably orientated orbital approaches either of its two lobes. The electron in the s orbital, on the other hand, can form a bond with an orbital that comes up from any direction. These ideas are not

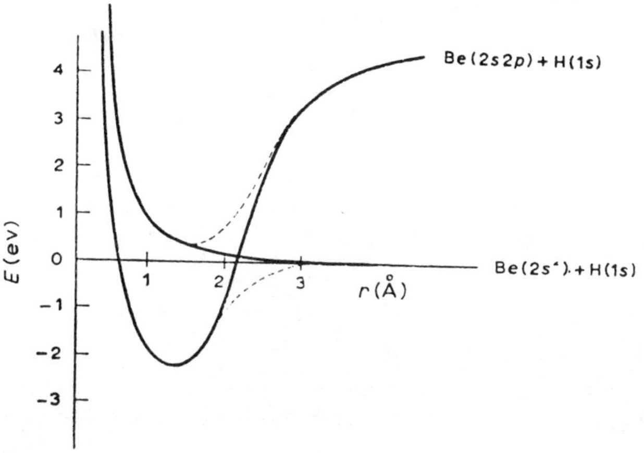

FIGURE 5.6. Potential energy curves for BeH

compatible with the properties of a molecule like BeH_2, which, if it existed, would certainly be linear, with the two BeH bonds being identical†.

It is evident that if we are going to use the concept of the electron-pair bond the two orbitals of Be which are going to form the bonds must be equivalent, and they must have directional properties such that their lobes point to opposite ends of a straight line. Hybrid orbitals which have these properties can be constructed, and they are defined by

$$\sigma_1 = \sqrt{\tfrac{1}{2}}(2s + 2p)$$
$$\sigma_2 = \sqrt{\tfrac{1}{2}}(2s - 2p), \tag{5.4}$$

where the $\sqrt{\tfrac{1}{2}}$ is a normalizing constant. Figure 5.7 gives a rough picture of the shapes of these hybrids, which are called sp hybrids.

† BeH_2 is only known in a polymeric form. The closest example of a molecule showing the structure expected for BeH_2 is $Hg(CH_3)_2$ which has colinear Hg—C bonds.

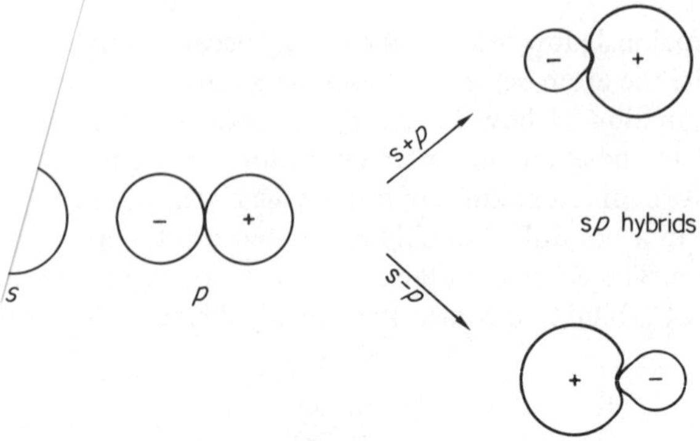

FIGURE 5.7. The formation of *sp* hybrids

It should be noted that they have an even more pronounced directional property than the *p* orbitals, for they will only form a good bond when the hydrogen atom comes up towards their positive lobe (figure 5.8).

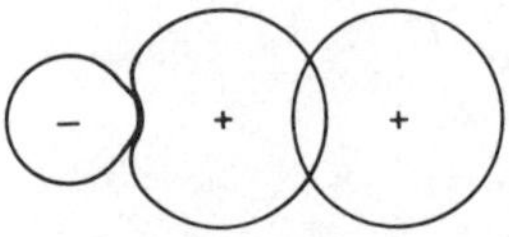

FIGURE 5.8. Bonding through the overlap of an *sp* hybrid and an *s* orbital

We need not have equal contributions from *s* and *p* orbitals in a hybrid. A more general expression for a normalized hybrid wave function is

$$\Psi = a\, s + b\, p \qquad (5.5)$$

with $a^2+b^2 = 1$. Its fractional *s* character is a^2 and its fractional *p* character is b^2 and we can call it an $s^{a^2}\, p^{b^2}$ or $sp^{(b/a)^2}$ hybrid. Reversing the process we can see that an sp^x hybrid has a normalized wave function

$$\Psi = (1+x)^{-\frac{1}{2}}(s+\sqrt{x}\, p) \qquad (5.6)$$

where x may be any positive number.

The direction in which these hybrids point is the same as the direction of the *p* orbital. As the three *p* orbitals p_x, p_y and p_z have an angular variation like unit vectors in the x, y and z directions respectively, we can always form a *p* orbital pointing in a specified direction by the normal rules of vector algebra. If the *p* orbital is to point in the direction of the unit vector (a, b, c) then its wave function is (see problem 3.2)

$$p\,(a, b, c) = a\, p_x + b\, p_y + c\, p_z. \qquad (5.7)$$

Thus an sp^x hybrid pointing in this direction has, from (5.6), the wave function

$$\Psi = (1+x)^{-\frac{1}{2}} \{s+\sqrt{x}(a\, p_x+b\, p_y+c\, p_z)\} \tag{5.8}$$

We are therefore in a position to write down the wave function of any hybrid formed from s and p orbitals pointing in any direction. If we are going to construct two or more hybrids from a given set of orbitals we will usually want these to be orthogonal (see problem 3.3); that is to be independent of each other. Because the s and three p orbitals are initially orthogonal and normalized, the two hybrids

$$\Psi_1 = a_1\, s+b_1\, p_x+c_1\, p_y+d_1\, p_z$$

$$\Psi_2 = a_2\, s+b_2\, p_x+c_2\, p_y+d_2\, p_z \tag{5.9}$$

will be orthogonal if

$$a_1a_2+b_1b_2+c_1c_2+d_1d_2 = 0. \tag{5.10}$$

These rules can be generalized to include d or other orbitals although the transformation properties of these are not as simple as for p orbitals.

Hybridization plays an important role in describing the bonds of all the elements other than hydrogen, but it is perhaps most important for carbon. Carbon in its ground state has the $1s$ and $2s$ orbitals filled and two electrons in $2p$ orbitals (we say it has the electron configuration $1s^22s^22p^2$). As such it might be expected to be divalent and to form bonds, through the $2p$ orbitals, at 90° to one another. But in nearly all of its compounds carbon is tetravalent. In order to explain this tetravalency we must suppose that the state of carbon which is forming the bonds is one having the electron configuration $1s^22s\ 2p^3$, for then the $2s$ and three $2p$ electrons are all available for bond formation.

Methane has four equivalent CH bonds, which point towards the corners of a tetrahedron. To associate these bonds with four equivalent electron pairs we need to construct four equivalent tetrahedrally orientated hybrid orbitals from the one $2s$ and the three $2p$ orbitals of the carbon atom. Since each hybrid must contain the same fraction of s character and the same fraction of p character, we deduce immediately that each contains $(1/4)s$ character and $(3/4)$ p character. From (5.5) it follows that each of these hybrids therefore has a wave function of the form

$$\Psi = \tfrac{1}{2}(s+\sqrt{3}\, p). \tag{5.11}$$

To obtain more detailed expressions for the hybrids we must choose some orientation for the tetrahedron with respect to the cartesian axes. The most convenient orientation is to inscribe the tetrahedron in a cube whose edges are parallel to the cartesian axes as shown in figure 5.9.

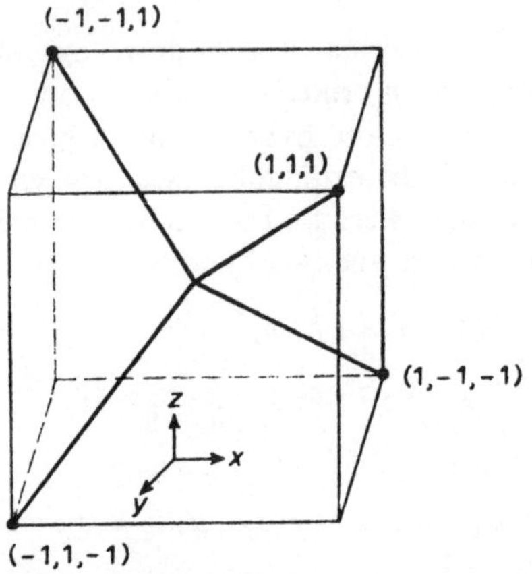

FIGURE. 5.9. **Tetrahedral bonds pointing towards the corners of a cube. The centre of the cube is the origin of the coordinates**

The corners of the tetrahedron can be labelled by their cartesian coordinates as shown. A p orbital pointing to the corner (1,1,1) will from (5.7) have a wave function

$$\sqrt{\tfrac{1}{3}}(p_x + p_y + p_z). \tag{5.12}$$

It follows from (5.11) that the corresponding hybrid must have a wave function

$$\sigma(1, 1, 1) = \tfrac{1}{2}(s + p_x + p_y + p_z). \tag{5.13}$$

the other three hybrids must have the wave functions

$$\begin{aligned}\sigma(1,-1,-1) &= \tfrac{1}{2}(s + p_x - p_y - p_z)\\ \sigma(-1, 1,-1) &= \tfrac{1}{2}(s - p_x + p_y - p_z)\\ \sigma(-1,-1, 1) &= \tfrac{1}{2}(s - p_x - p_y + p_z).\end{aligned} \tag{5.14}$$

It is easy to confirm that the four hybrid orbitals we have constructed are orthogonal.

The four sp^3 hybrids are strongly directional, having a similar shape to the sp hybrids discussed earlier. A contour diagram is shown in figure (5.10) for one of them.

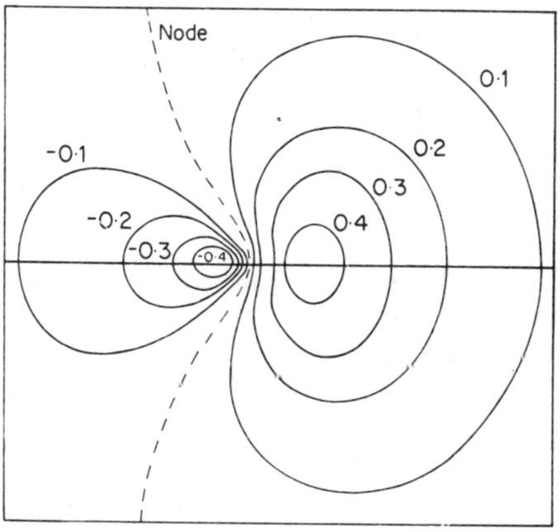

FIGURE 5.10. Contours for an sp^3 hybrid

In olefins and aromatic hydrocarbons a carbon atom is joined to three other atoms which lie in a plane and which subtend angles of 120° at the carbon atom. If we call the plane of these atoms the yz plane, then it is clear that the carbon p_x orbital cannot contribute to hybrid orbitals which point along the three bonds. The three hybrid orbitals required must therefore be made up of s, p_y and p_z, and they are called sp^2 hybrids. If the three hybrids are equivalent, then they must each contain $\frac{1}{3}s$ and $\frac{2}{3}p$ character.

If we choose axes as in figure 5.11, then the hybrid pointing towards atom A cannot contain any p_y, since this has a nodal plane through A. We therefore conclude that

$$\sigma_A = \sqrt{\tfrac{1}{3}}s + \sqrt{\tfrac{2}{3}}p_z. \tag{5.15}$$

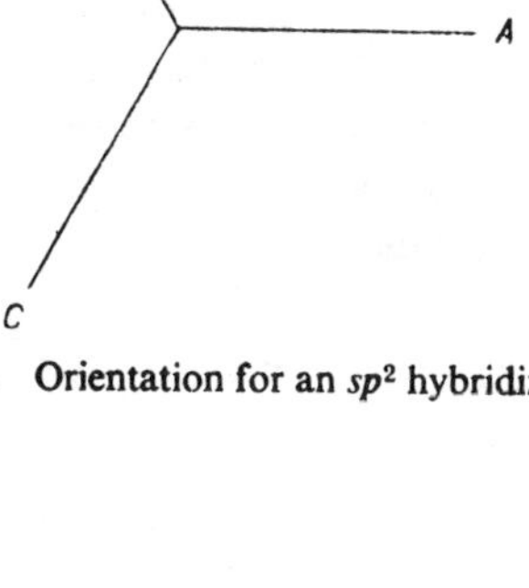

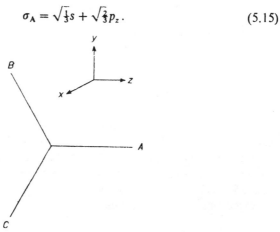

FIGURE 5.11. Orientation for an sp^2 hybridized molecule

To illustrate another approach to the construction of hybrids we consider what happens when we rotate each of the functions in (5.15) by $2\pi/3$. Then $\sigma_A \to \sigma_B$, $s \to s$ and $p_z \to p_z \cos \frac{2}{3}\pi + p_y \sin \frac{2}{3}\pi$. Introducing these new functions into (5.15) gives

$$\sigma_B = \sqrt{\tfrac{1}{3}}s - \sqrt{\tfrac{1}{6}}p_z + \sqrt{\tfrac{1}{2}}p_y. \tag{5.16}$$

Likewise from a rotation by $-2\pi/3$ we have

$$\sigma_C = \sqrt{\tfrac{1}{3}}s - \sqrt{\tfrac{1}{6}}p_z - \sqrt{\tfrac{1}{2}}p_y. \tag{5.17}$$

The hybrid orbitals that we have constructed so far are really determined by the symmetry of the problem. It can also be shown that they have the property of giving the maximum overlap of the orbitals involved in the bonds. If we maximize the overlap then we maximize the bond strength (roughly speaking), hence maximum overlap can be used as a general criterion for defining the best set of hybrid orbitals. Taking methane, for example, we could seek four hybrid orbitals which maximize the sum of the overlap integrals between the four hydrogen 1*s* orbitals and the four hybrids with which they are being paired. This would lead to the same set of orbitals that we obtained by symmetry. A general method exists for constructing the best hybrid orbitals according to the maximum overlap criterion in a case where there is no symmetry†.

The bond angles in a molecule indicate the nature of the hybrids involved in the bond; restricting our attention to *s* and *p* orbitals we have

Bond angle	Hybrids
180°	sp
120°	sp^2
109° 28′	sp^3
90°	p

Bonds with other angles between 180° and 90° have intermediate hybrid character. No hybrids can be formed from *s* and *p* orbitals which satisfy the orthogonality condition, and which have angles less than 90°. For this reason molecules like cyclopropane are referred to as having bent bonds (figure 5.12). This description is in agreement with the fact that these bonds are much weaker than a normal C—C bond.

† See RANDIĆ, *J. Chem. Phys.*, **36**, 3278 (1962), and papers referred to therein.

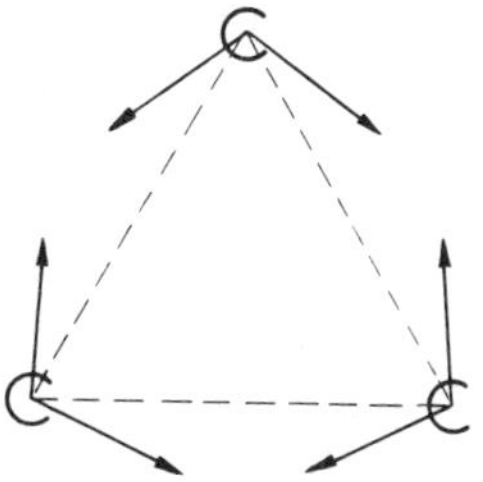

FIGURE 5.12. Bent bonds in cyclopropane. The arrows indicate the directions of the best-bonding hybrids

If *d* orbitals are included in the construction of hybrids then sets of hybrid orbitals can be obtained appropriate to other geometries. For example,

octahedron	sp^3d^2
trigonal bipyramid	sp^3d
square plane	sp^2d

Our discussion of hybridization so far has been purely *deducta a posteriori*. It may be said that hybridization is very useful to describe the wave functions of electrons in bonds—we shall see later that it is a vital step in the VB theory—but it is not, in general, an explanation of the shapes of molecules Strictly speaking, hybridization is a mathematical tool of the theoretical chemist, but it is very useful for a qualitative description of the chemical bond, as will be shown in many examples in this book.

Molecular shape is largely determined by two contributions to the energy: bond energies, which depend on overlap, and the electrostatic repulsion between different pairs of electrons. The latter is often the dominant effect. For example, in $Hg(CH_3)_2$ the electrostatic repulsion between the electron pairs in the Hg—C bonds is minimized when the molecule is linear; in BF_3 the repulsion of the three electron pairs is minimized by the molecule being planar and having bond angles of 120°; in CH_4 the repulsion of the four bond pairs is minimized in a tetrahedral arrangement.

We shall now give some examples which will illustrate the basic points about the covalent bond which have been discussed in this chapter.

1. *Ethane, ethylene and acetylene*

It is conventional to write the carbon–carbon bonds in ethane, ethylene and acetylene as single, double and triple respectively, so that in all these compounds carbon is tetravalent. What does this mean in quantum-mechanical language? The configurations of the three molecules are shown in figure 5.13. Acetylene is a linear molecule, ethylene is planar, in ethane the two CH_3 groups have almost free rotation relative to one another about the C—C bond.

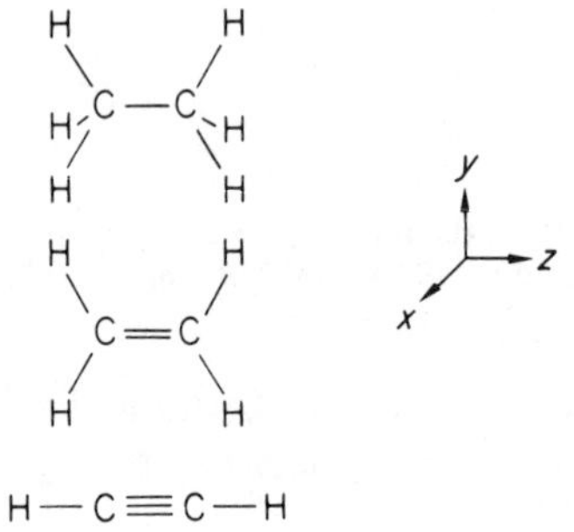

FIGURE 5.13. Defined orientations for some carbon compounds

From the bond angles we deduce that in ethane carbon is sp^3 hybridized, in ethylene sp^2 and in acetylene sp hybridized.

In ethane three of the sp^3 hybrids of each carbon are used to form electron-pair bonds with the hydrogen atoms. The fourth is used to form the C—C bond. The C—C bond is axially symmetrical and is called a σ bond. This is shown schematically in figure 5.14.

In ethylene two of the sp^2 hybrids of each carbon are used to form the C—H bonds, and the third forms a carbon–carbon σ bond. There is one of the carbon $2p$ atomic orbitals which is not used in constructing the sp^2 hybrids, namely p_x, for which the plane of the molecule is a nodal plane.

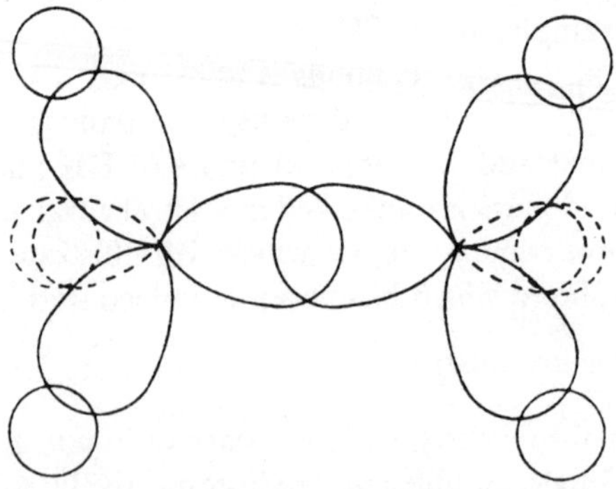

FIGURE 5.14. Bonding through sp^3 hybrids in ethane

The two carbon p_x orbitals can form an electron-pair bond between the two carbon atoms. The electron density of these two electrons is zero in the plane of the molecule, and this type of bond is called a π bond. The σ and π bonds of ethylene are illustrated in figure 5.15.

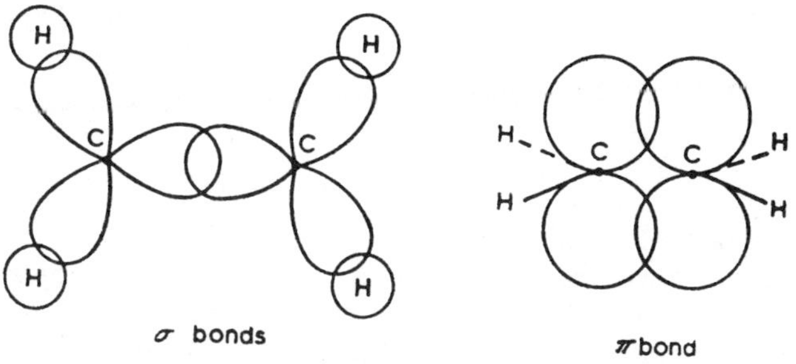

FIGURE 5.15. σ and π bonds in ethylene

In acetylene the carbon *sp* hybrids form the C—H bonds and a carbon–carbon σ bond. There are two *p* orbitals of each carbon atom not used in forming the σ-bond structure, and these can be used to form π bonds. If we make an arbitrary choice of axes in space, as in figure 5.13, we can say that there is a p_x–p_x π bond and a p_y–p_y π bond, but the only thing that is potentially measurable is the total electron density in these π bonds; this is symmetrical around the carbon–carbon bond axis but is zero on the axis†. In chapter 10 we shall go more carefully into the meaning of the symbols σ and π which we have used here.

A π bond is expected to be weaker than a σ bond because the overlap integral between two $p\pi$ atomic orbitals is considerably less than between two σ hybrid atomic orbitals. π bonding reduces the carbon–carbon bond length from 1·54 Å in ethane to 1·35 Å in ethylene, and the presence of the second π bond in acetylene is associated with a further reduction to 1·21 Å. By analysing thermochemical data it has been found possible to give an energy value to each type of carbon–carbon bond which can be taken to be approximately constant in all molecules. Pauling's values for these energies are as follows‡.

	C—C	C=C	C≡C
Bond energy (ev)	2·6	4·3	5·3

† The density $p_x^2+p_y^2$ is axially symmetric about the carbon–carbon axis

‡ PAULING, *The Nature of the Chemical Bond*, Cornell, 1961.

2. *Benzene*

There are two classical valence structures which can be written for benzene; they are known as Kekulé structures.

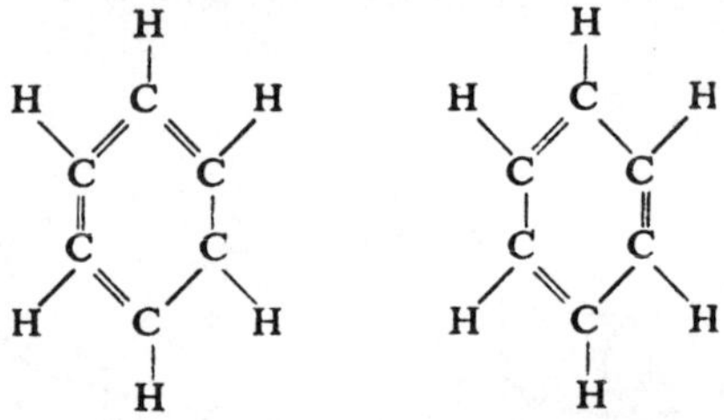

The bond angles of benzene being 120°, it is natural to describe the bonding in this molecule in terms of sp^2 hybrids, as in the case of ethylene. From this we can build up a framework of C—H and C—C single bonds, which we call the σ-bond system. There remains one p orbital on each carbon atom available for π bonding as shown in figure 5.16.

Ample physical and chemical evidence exists to show that benzene has a regular hexagonal structure, and not an alternating set of long and short bonds which change places rapidly according to the equilibrium

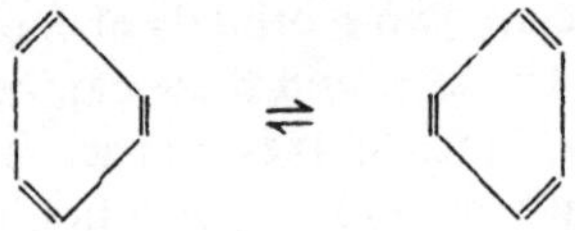

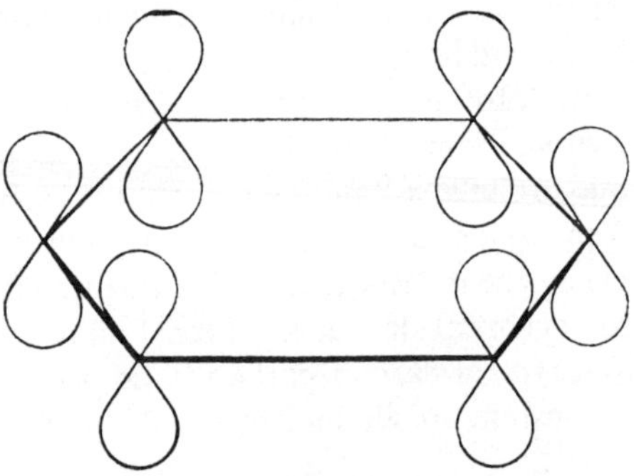

FIGURE 5.16. π bonding orbitals for benzene

For a regular hexagonal structure there is no obvious way of forming electron-pair bonds from the $p\pi$ atomic orbitals. An orbital has just as large an overlap with the orbital on its left as with the one on its right. We picture the π-electron density as being roughly concentrated in two rings, one above the benzene plane and one below, as in figure 5.17, and we refer to the π bonding as being delocalized around the ring. Each carbon–carbon bond can be said to have half a π bond, and their length, 1·40 Å, is intermediate between that of ethane and ethylene.

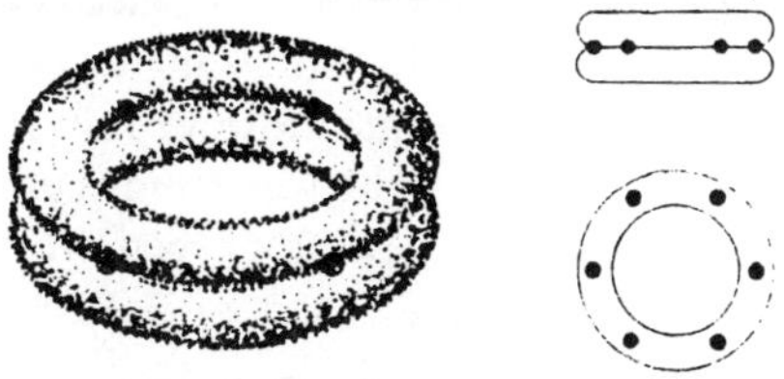

FIGURE 5.17. Delocalized π bonds in benzene

This delocalization of the π electrons is responsible for the great stability of the benzene molecule. The characteristic chemical reactions of benzene are substitution rather than the additions which occur so readily for olefins. We shall leave a detailed study of the π bonding in benzene and related compounds until chapter 15. At this point we shall simply emphasize the chemical importance of delocalized π-electron systems. These occur not only in benzenoid compounds but in all compounds which contain conjugated unsaturated bonds, for example

Butadiene $CH_2{=}CH{-}CH{=}CH_2$

Acrolein $CH_2{=}CH{-}CH{=}O$

Fulvene

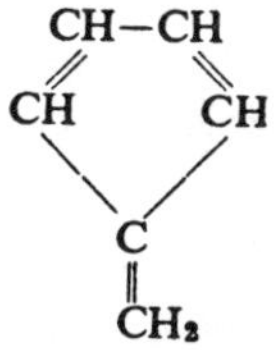

These molecules have delocalized π-electron systems even though they can be represented by single classical valence structures.

3. *The Gillespie-Nyholm approach to molecular structures.*

As we have already mentioned, two factors, bond energies and electrostatic repulsions, may play a significant part in determining the geometry of a molecule. For any one molecule, it is not easy to assess their relative importance, and to illustrate this difficulty we shall consider the water molecule.

A ground state oxygen atom has the electron configuration $1s^2\ 2s^2\ 2p^4$, and if we arrange the electrons one each in $2px$ and $2py$, and two in $2pz$ then we would expect the unpaired electrons to give rise to two OH bonds with an interbond angle of 90°. The bond angles found for H_2S (92°), H_2Se (91°) and H_2Te (89·5°) are quite close to this value, but for water the angle is much larger (104°). For the water molecule, the bond angle is closer to the value expected (109·5°) if the oxygen orbitals involved in the bonding are two sp^3 hybrids rather than pure p orbitals. However, it requires energy to form these hybrids because a $2s$ electron has to be promoted to a $2p$ orbital and, in the case of the water molecule, the only compensating factor could be an increase in bond strength.

Whether one believes that the 'natural' bond angle in the water molecule is 90° or 109·5°, the discrepancy with the observed bond angle may be explained in terms of electrostatic repulsion. In the first case, that of a natural bond angle of 90°, Pauling† has argued that the angle is opened up because the dipolar O—H bonds repel one another (figure 5.18).

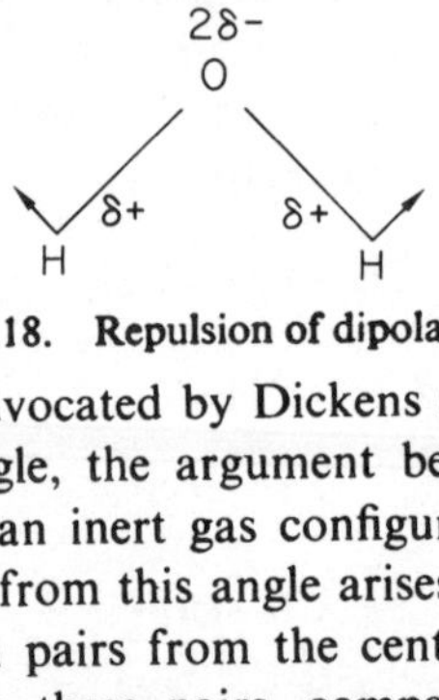

FIGURE 5.18. Repulsion of dipolar OH bonds

The second approach, advocated by Dickens and Linnett‡, assumes that 109·5° is the natural angle, the argument being based on the electron distribution expected in an inert gas configuration (O^{2-} is isoelectronic with Ne). The distortion from this angle arises because the protons draw out the bonding electron pairs from the central atom; this reduces the repulsion energy between these pairs, compared with the non-bonding pairs, and causes the bond angle to decrease. In both theories, the difference between oxygen and the other six group elements is that the latter are less electronegative than oxygen, so that their electrons are pulled out more towards the hydrogen atoms. This will reduce the dipolar repulsion

† PAULING, *The Nature of the Chemical Bond*, Cornell, 1961.
‡ DICKENS and LINNETT, *Quart. Rev. (London)*, **11**, 291 (1957).

invoked in Pauling's theory and will also reduce the repulsion of the electron pairs invoked by Dickens and Linnett.

However, there is another and more pertinent difference between oxygen and, for example, sulphur. This is that for oxygen, the overlap integral between the $2s$ orbital and a hydrogen $1s$ orbital (at the equilibrium distance) is appreciably greater than is the overlap involving the $2p$ orbital. But for sulphur, the $3p$ overlap is greater than the $3s$. There is therefore much more to be gained in overlap by bringing the s orbitals into a hybrid scheme for the first row, than for the second. Therefore a larger bond angle (associated with more s character) is to be expected for oxygen. The relevant numbers are given in table 5.2, where, to show the generality of the approach, we also include data on the NH_3 and PH_3 molecules.

TABLE 5.2. Overlap integrals between s, p and sp^3 hybrids and hydrogen $1s$ orbitals. (Calculated by Murrell and Randić using SCF orbitals)

	$2s$	$2p$	$(2s\ 2p^3)$	Increase in overlap from pure p to sp^3
N–H	0·54	0·41	0·63	45%
O–H	0·50	0·34	0·54	59%
	$3s$	$3p$	$(3s\ 3p^3)$	
P–H	0·50	0·53	0·71	34%
S–H	0·42	0·51	0·65	27%

On the surface it may appear a gross deficiency of quantum theory not to have one clear and acceptable theory of the bond angles in these molecules. In fact we can see that the problem is one of the most difficult that is met. The total electronic energy of H_2O is about 2000 ev. This is the value that would be obtained for the energy if the Schrödinger equation were solved for the ground state of the molecule. The dissociation energy of H_2O is about 10 ev and the energy required to distort the bond angle from 104° to 90° is about 0·1 ev. This means that if we want to show by a calculation starting from first principles that the bond angle of H_2O is 104° and not 90°, we have to obtain a solution of the Schrödinger equation accurate to at least 0·1 percent for several different nuclear configurations. The best calculations on H_2O available at the present time (1969) give energies accurate to about 99·6 percent. That is, the errors are about 10 ev. The bond angles obtained from these calculations are about 108°. It should be evident that it is rather difficult to explain the bond angle of water by an *a priori* calculation.

It was first pointed out by Sidgwick and Powell, that the molecular structure of compounds of the non-transition elements could be rationalized on the assumption that bond angles are determined solely by repulsive forces. This approach has subsequently been elaborated by Gillespie and Nyholm, and the following section is largely based on their work.

It is postulated that the valence electron pairs associated with an atom keep as far apart as possible. For the water molecule, for example, the oxygen atom is associated with four electron pairs, two of which correspond to O—H bonds and two to lone-pair electrons. The requirement that these four electron pairs keep as far apart as possible leads to a prediction that there should be a tetrahedral arrangement of them around the oxygen atom and, consequently, a $H\hat{O}H$ bond angle of 109·5°. Similar arguments lead us to the predictions of molecular geometry given in Table 5.3, in which we also give examples of molecules with the predicted structure. For more than six-electron pairs the predictions are usually not clear-cut, although it is often easy to see that some geometrical arrangement is not favoured. For example, the cubic structure is not the most stable arrangement of eight-valence electron pairs. For each of the molecules given in Table 5.3 all the valence electron pairs are

TABLE 5.3. Predictions of gross molecular structure based on electron pair repulsions

Number of valence electron pairs	Predicted geometrical arrangement	Examples
2	linear	$HgCl_2$
3	equilateral triangular	BCl_3
4	tetrahedral	CCl_4
5	trigonal bipyramidal	PCl_5 (gaseous)
6	octahedral	SF_6

associated with bonds to a halogen and the structures are exactly those predicted. When some of the valence electron pairs are non-bonding whilst others are bonding (as in the water molecule) then this lack of equivalence is reflected in a distortion of the molecular structure. The form of the distortion will be that already discussed, under the Dickens-Linnett approach to the structure of water molecule, leading to the prediction that the inter-electron pair repulsions will vary in the order non-bonding—non-bonding > non-bonding—bonding > bonding—bonding. Further, for any one of these we expect the repulsion to decrease rapidly as the distance between the two centres of gravity of electron density increases; that is, as the 'angle' between the electron pairs increases. It follows, that when there are both bonding and non-bonding electron pairs present on the same atom, the angle between bonding electron pairs should be smaller than that suggested by Table 5.3. For the case of five-electron pairs, where in the expected geometry there are two non-equivalent positions, we would expect non-bonding electron pairs to occupy the equatorial sites rather than the axial. This is because in the latter position there would be three other electron pairs at an angle of 90°, but in the former only two electron pairs at 90°. In Table 5.4 we compare these predictions with the structures of some molecules containing both non-bonding and bonding electrons. Figure 5.19 shows the structures of some of the molecules listed in Table 5.4.

TABLE 5.4. Structures of some molecules with bonding and non-bonding electron pairs

Number of electron pairs		3	4	5	6
Number of non-bonding pairs	1	CH_2†(103°)	NH_3(107°)	$TeCl_4$(~105° and 186°)	BrF_5(170°)
	2	—	H_2O(104·5°)	ClF_3(175°)	XeF_4(planar)
	3	—	—	XeF_2 (linear)	no known example

†In the lowest singlet-spin state

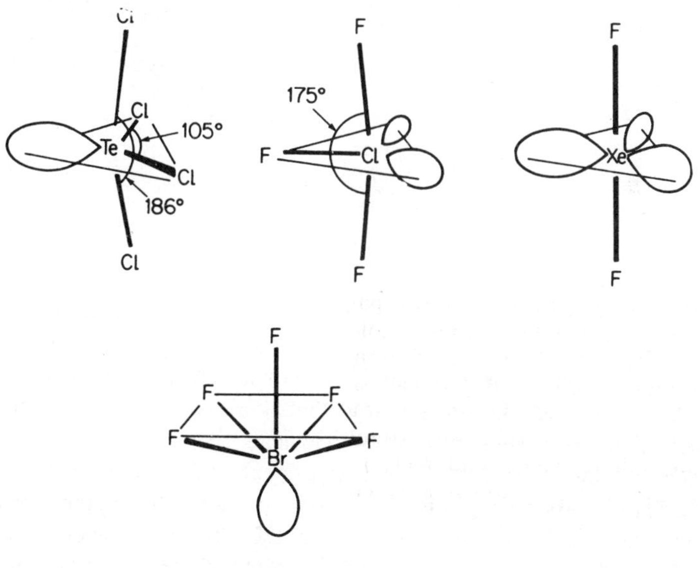

FIGURE 5.19. Geometries of molecules referred to in Table 5.4

Particularly interesting is the case of BrF_5 (IF_5 is probably similar). where the bromine atom is below the basal plane of the square pyramid formed by the fluorine atoms.

In all of the molecules we have given in Tables 5.3 and 5.4 there is no example of a molecule in which π-bonding causes the delocalization of electron density from the central to the surrounding atoms. Evidently, if such π-bonding occurs it has the effect of reducing the electron repulsions involving the delocalized electron pair from which the π-electrons 'originate'. For example, our discussion so far would lead us to predict that the $[C(CN)_3]^-$ anion is approximately tetrahedral, the NC-C-CN bond angle being less than 109·5°. In fact the anion is planar, the angle being 120°. This is understandable if π-bonding occurs as shown in Figure 5.20 (an antibonding orbital of the isolated $C\equiv N$ unit is involved).

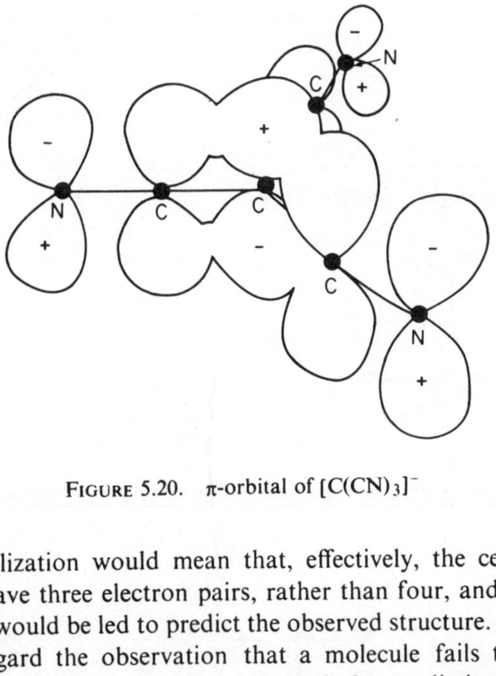

FIGURE 5.20. π-orbital of $[C(CN)_3]^-$

Such delocalization would mean that, effectively, the central carbon atom would have three electron pairs, rather than four, and so, as Table 5.3 shows, we would be led to predict the observed structure. It is common practice to regard the observation that a molecule fails to follow the Gillespie-Nyholm rules but follows instead the prediction based on a smaller number of electron pairs, as evidence of π-bonding. Where it is evident that π-bonding must occur (as, for example, in the nitrite and nitrate anions, NO_2^- and NO_3^-) then the correct molecular geometry may be predicted by applying the Gillespie-Nyholm rules to the σ electrons alone.

The Gillespie-Nyholm approach applies to isolated molecules, not to solids. Thus although $BeCl_2$ is predicted to be a linear molecule in the gas phase, the theory makes no predictions about the solid-state structure which, in fact, consists of $BeCl_4$ tetrahedra, each chlorine being bonded to two beryllium atoms. However, granted that each beryllium is bonded to four chlorine atoms, it is predicted that these will be arranged tetrahedrally.

Exceptions to the Gillespie-Nyholm model are sufficiently rare for them to cause some excitement when they are discovered. Examples include gaseous Li_2O, which appears to be linear, rather than bent, like water. This, presumably, may be explained in terms of $Li^+ - Li^+$ repulsions, in a manner analogous to that used by Pauling to explain why the bond angle in water is greater than 90°. A family of exceptions occur for some group 5 anions of general formula $[MX_6]^{3-}$, in which the atom M has seven valence electron pairs. Sometimes these anions have a regular octahedral arrangement around M and sometimes a seven-coordinate arrangement with, apparently, a 'hole' containing a non-bonding electron pair. No really convincing explanation of this phenomenon has been offered.

In summary, the Nyholm-Gillespie model of electron pair repulsion offers a simple qualitative understanding of a wide range of molecular structures. However, one might criticize it on the grounds that there is no quantitative basis behind the relative repulsions of lone pairs and bond pairs. It is therefore not very useful in tackling difficult cases like ClF_3, where the energy gained by relieving non-bonding—non-bonding repulsions has to be set against that lost by increasing non-bonding–bonding repulsions.

Problems 5

5.1 What is the coulomb attraction energy of the ions in NaCl (bond length 2·5 Å) assuming that the electron clouds of Na^+ and Cl^- are not overlapping at this distance?

5.2 The ionization potential of Li is 5·27 ev. The electron affinity of Cl is 3·83 ev. At what distance would the potential energy curve for the ionic state cross that of the covalent state?

5.3 It is often convenient to have an algebraic form for the potential energy curve. The most commonly used is the Morse curve expressed as

$$V = D_e\,(1 - e^{-\beta(r-r_0)})^2.$$

Is this a satisfactory function for all values of r? The vibration frequency of a diatomic molecule AB is given by

$$\nu = (1/2\pi)(k/\mu)^{\frac{1}{2}},$$

where μ is the reduced mass ($\mu = M_A M_B/(M_A + M_B)$) and k is the force constant which is the curvature of the potential energy curve at $r = r_0$; $k = (d^2V/dr^2)_{r_0}$. Show that the constant β in the Morse function is given by

$$\beta = \nu(2\pi^2\mu/D_e)^{\frac{1}{2}}.$$

5.4 A table of bond dipole moments has been compiled on the assumption that the total dipole moment of a molecule is a vector sum of bond dipoles. Examine the correlation between bond dipole moment and electronegativity from the following data and table 5.1.

	μ (D)		μ (D)
HF	1·91	NH	1·32
HCl	1·03	PH	0·34
FCl	0·88	NF	0·18
OH	1·51	PF	0·77

Molecular dipole moments are measured in Debye units ($1\,\text{D} = 10^{-18}$ esu). For example, an electron and a proton separated by 1Å have a dipole moment of 4·8 D.

5.5 There are several ways in which d orbitals can overlap to give good chemical bonds. Give sketches of these in which the z axis is the internuclear axis.

5.6 The lowest molecular orbital for H_2 can be approximated by the function

$$\Psi = N(1s_a + 1s_b),$$

where $1s_a$ and $1s_b$ are the two hydrogen $1s$ atomic orbitals. Show that N the normalizing constant, is given by

$$N = (2 + 2S)^{-\frac{1}{2}},$$

where S is the overlap integral. Show that two electrons in this molecular orbital have a larger density between the two nuclei than if one electron were in $1s_a$ and the other in $1s_b$.

5.7 Write down the wave function for an sp^3 hybrid pointing along the z axis.

5.8 Comment on the hybridization of the carbon atoms and the delocalization of π electrons in the following compounds.

$$CH_2{=}CH{-}CH{=}CH_2$$
$$CH{\equiv}C{-}CH{=}CH_2$$
$$CH_3{-}CH{=}C{=}CH_2$$

5.9 The overlap integral between a hydrogen $1s$ orbital and a carbon $2s$ orbital separated by the C—H bond length is 0·57, and that between the hydrogen $1s$ and carbon $2p_z$ (take the C—H axis as z axis) is 0·46. What are the overlap integrals between the hydrogen $1s$ and carbon sp, sp^2 and sp^3 hybrid orbitals directed along the z axis? How do these results correlate with the C—H stretching force constants (see problem 5.3) in the following molecules

	$CH{\equiv}CH$	$CH_2{=}CH_2$	CH_4	
$k =$	5·85	5·1	4·79	$\times 10^5$ dyn cm^{-1}.

Chapter 6

The mathematical foundations of quantum mechanics

6.1 Postulates†

In chapter 1 the Schrödinger equation for an atomic particle was derived from the classical equation for a harmonic standing wave and the de Broglie relationship. For systems containing many particles, with perhaps external electric and magnetic fields, a more general approach to the equations of quantum mechanics is needed.

The foundations of quantum mechanics are best considered to be a set of postulates from which the equations of motion can be derived. The original postulates are then justified by the fact that the solutions of these equations agree with experiment. Consider a system of n particles which could be described classically by specifying the values of the $3n$ coordinates (q) and $3n$ momenta (p) at a given time. To describe such a system in quantum mechanics the following postulates are made.

Postulate 1. The system of particles can be described by a function $\Psi(q_1 \ldots q_{3n}, t)$, called the wave function, which determines all measurable quantities of the system. Ψ is interpreted physically by $\Psi^*\Psi \, dq_1 \ldots dq_{3n}$ being the probability of finding the particles with coordinates lying between $q_1 \ldots q_{3n}$ and $q_1 + dq_1 \ldots q_{3n} + dq_{3n}$.

Since each particle must be somewhere in space, the integrated probability density must be unity. This is the normalization condition

$$\int \Psi^*\Psi \, dv = 1, \tag{6.1}$$

† This chapter covers much of the mathematics needed for the remainder of the book. There are, however, large sections of the remaining chapters which can be followed without a detailed understanding of this mathematics so if you find it difficult do not despair.

where $dv = dq_1 \dots dq_{3n}$, and the integral is over all of the $3n$-dimensional space.

Postulate 2. Every observable can be characterized in quantum mechanics by a linear operator, $\mathscr{B}$, say. The average value of this observable is given by†

$$\bar{b} = \int \Psi^* \mathscr{B} \Psi \, dv. \tag{6.2}$$

The rule for constructing the quantum-mechanical operators is as follows: the classical expression for the observable is expressed in terms of the q's and p's, and the quantum-mechanical operator is obtained from this by leaving the q's alone but replacing p_k by $(h/2\pi i)(\partial/\partial q_k)$.

As examples of expression (6.2) we have:

(a) The mean value of the x coordinate for a single particle

$$\bar{x} = \iiint \Psi^* x \Psi \, dx \, dy \, dz. \tag{6.3}$$

(b) The mean value of the x component of the momentum of a single particle

$$\bar{p}_x = \iiint \Psi^* \left(\frac{h}{2\pi i}\frac{\partial}{\partial x}\right) \Psi \, dx \, dy \, dz. \tag{6.4}$$

It should be noted that if $\mathscr{B}$ is an algebraic function of the coordinates as in (a), it does not matter where it is placed in the integral, but if $\mathscr{B}$ is a differential operator, then it must be sandwiched between Ψ^* and Ψ such that it only operates on Ψ.

† If we want to convert one function $f(x)$, say, into another $g(x)$ then this is expressed algebraically by

$$\mathscr{B} f(x) = g(x)$$

and $\mathscr{B}$ is called an operator. For example

$$[+2]\, x^3 = 2 + x^3 \quad \text{(a)}$$
$$[x]\, x^3 = x^4 \quad \text{(b)}$$
$$[\sqrt{\ }]\, x^3 = x^{3/2} \quad \text{(c)}$$
$$\left[\frac{d}{dx}\right] x^3 = 3x^2. \quad \text{(d)}$$

In all these equations the operator is in parenthesis. Operators operate on functions to the right of them. A linear operator is one which fulfils the conditions

$\mathscr{B}(f(x) + g(x)) = \mathscr{B} f(x) + \mathscr{B} g(x)$ and $\mathscr{B} k f(x) = k\, \mathscr{B} f(x)$ (k being a constant).

Only (b) and (d) in the above examples are linear operators.

Postulate 3. For a system whose total energy does not vary with time (a conservative system), the classical expression for the energy written in terms of the q's and p's is known as Hamilton's function. The corresponding operator in quantum mechanics (the operator corresponding to the energy as observable) is called the Hamiltonian and is given the symbol $\mathscr{H}$. In such a system the wave function satisfies the operator equation

$$\mathscr{H}\Psi(q, t) = E\Psi(q, t), \tag{6.5}$$

where E is the energy, which is a constant independent of the coordinates and the time (t).

It should be noted that in equation (6.5) the same function Ψ appears on both sides of the equation. An equation such as (6.5) is commonly called an eigenvalue equation. E is an eigenvalue of $\mathscr{H}$ associated with the eigenfunction Ψ.

As a familiar example of an equation of the type (6.5) we have

$$\frac{\mathrm{d}}{\mathrm{d}x}\mathrm{e}^{kx} = k\mathrm{e}^{kx}. \tag{6.6}$$

The eigenfunctions of $\mathrm{d}/\mathrm{d}x$ are e^{kx} and their eigenvalues are k. It is mathematical nonsense to cancel out the functions e^{kx} from both sides of (6.6) (or Ψ from both sides of (6.5)).

Postulate 4. More generally the wave function satisfies the equation

$$\mathscr{H}\Psi(q, t) = \frac{-h}{2\pi i}\frac{\partial\Psi(q, t)}{\partial t}. \tag{6.7}$$

This is Schrödinger's time-dependent equation, which is valid (unlike 6.5) even if the Hamiltonian contains the time.

Once we know Ψ at one time this equation allows us to deduce Ψ at all subsequent times. However, since we are not concerned with time-dependent phenomena in this book, we shall not meet this equation in later chapters.

For a conservative system Ψ satisfies both (6.5) and (6.7), hence

$$\frac{-h}{2\pi i}\frac{\partial\Psi}{\partial t}(q, t) = E\Psi(q, t). \tag{6.8}$$

This equation has a general solution

$$\Psi(q, t) = \Psi(q)\exp\left(\frac{-2\pi iEt}{h}\right), \tag{6.9}$$

and, since for a conservative system the Hamiltonian does not contain the time, we can substitute (6.9) into (6.5), cancel the exponential functions and obtain

$$\mathscr{H}\Psi(q) = E\Psi(q). \tag{6.10}$$

Equation (6.10) is the general form of Schrödinger's equation for a so-called *stationary state* of the system, that is, one whose energy does not vary with time. For a stationary state we can obtain the value of any observable by using the time-independent wave functions $\Psi(q)$ rather than the more complete functions $\Psi(q, t)$. This is because expression (6.2) becomes

$$\int \Psi^*(q)\exp\left(\frac{2\pi iEt}{h}\right)\mathscr{B}\Psi(q)\exp\left(\frac{-2\pi iEt}{h}\right) \mathrm{d}v = \int \Psi^*(q)\mathscr{B}\Psi(q)\,\mathrm{d}v \tag{6.11}$$

if $\mathscr{B}$ is not a function of time.

Hamilton's function for an electron having a potential energy V is

$$T + V = \frac{1}{2m}(p_x{}^2 + p_y{}^2 + p_z{}^2) + V. \tag{6.12}$$

The Hamiltonian for this system is then (using the rule given in postulate 2)

$$\mathscr{H} = \frac{-h^2}{8\pi^2 m}\left\{\frac{\partial^2}{\partial x^2} + \frac{\partial^2}{\partial y^2} + \frac{\partial^2}{\partial z^2}\right\} + V \tag{6.13}$$

and equation (6.10) becomes, after rearranging,

$$\nabla^2\Psi + \frac{8\pi^2 m}{h^2}(E - V)\Psi = 0, \tag{6.14}$$

which is the form of Schrödinger's equation given in chapter 1.

Suppose we have two solutions of equation (6.10)

$$\mathscr{H}\Psi_a = E_a\Psi_a, \qquad \mathscr{H}\Psi_b = E_b\Psi_b. \tag{6.15}$$

If the first equation is multiplied by a constant λ and the second by a constant μ, and the two are added, the result is

$$\mathscr{H}(\lambda\Psi_a + \mu\Psi_b) = \lambda E_a\Psi_a + \mu E_b\Psi_b. \tag{6.16}$$

If the expression on the right-hand side could be factorized as $k(\lambda\Psi_a + \mu\Psi_b)$, where k is a constant, then $\lambda\Psi_a + \mu\Psi_b$ would be an eigenfunction of $\mathscr{H}$. But in general this is not the case, so that *linear combinations of eigenfunctions are not themselves eigenfunctions.* The only exception is when $E_a = E_b$, for then

$$\mathscr{H}(\lambda\Psi_a + \mu\Psi_b) = E_a(\lambda\Psi_a + \mu\Psi_b). \tag{6.17}$$

If two or more eigenfunctions have the same eigenvalue they are said to be degenerate. In this case any linear combination of these degenerate eigenfunctions is also an eigenfunction of the Hamiltonian. We made use of this theorem in chapter 3 when the wave functions of *p* and *d* atomic orbitals were given in both real and complex forms.

The observable quantities of atomic systems are of two kinds: those whose values are well defined, for example, the energy which has only discrete (quantized) values in any bounded system, and those for which any measurement can give only the average value of a probability distribution. If an observable characterized by the operator $\mathscr{B}$ is well defined then this means that the wave functions of the system, which are eigenfunctions of the Hamiltonian, are also eigenfunctions of $\mathscr{B}$. That is,

$$\mathscr{H}\Psi = E\Psi,$$

and

$$\mathscr{B}\Psi = b\Psi. \tag{6.18}$$

If the observable is not well defined then

$$\mathscr{B}\Psi \neq b\Psi, \tag{6.19}$$

although there may be another set of functions which are eigenfunctions of $\mathscr{B}$. If the observable is not well defined, the average value of the observable can still be calculated from expression (6.2).

The criterion for Ψ satisfying equation (6.18) is that the operators $\mathscr{H}$ and $\mathscr{B}$ shall commute, that is

$$\mathscr{B}\mathscr{H} = \mathscr{H}\mathscr{B}. \tag{6.20}$$

In general, operators do not commute. For example, if $\mathscr{A} = (\mathrm{d}/\mathrm{d}x)$ and $\mathscr{B} = x$, then

$$\mathscr{A}\mathscr{B}f = \frac{\mathrm{d}}{\mathrm{d}x}xf = f + x\frac{\mathrm{d}f}{\mathrm{d}x}$$

$$\mathscr{B}\mathscr{A}f = x\frac{\mathrm{d}f}{\mathrm{d}x}$$

so that

$$\mathscr{A}\mathscr{B} - \mathscr{B}\mathscr{A} = 1. \tag{6.21}$$

We will now prove that if two operators commute a set of functions can be obtained which are simultaneously eigenfunctions for both operators. Suppose the eigenfunctions of $\mathscr{A}$ are θ and of $\mathscr{B}$ are χ; then

$$\mathscr{A}\theta_i = a_i\theta_i, \tag{6.22}$$

$$\mathscr{B}\chi_j = b_j\chi_j. \tag{6.23}$$

Multiplying (6.23) from the left by $\mathscr{A}$ gives

$$\mathscr{A}\mathscr{B}\chi_j = \mathscr{A}b_j\chi_j = b_j\mathscr{A}\chi_j. \tag{6.24}$$

But if $\mathscr{A}\mathscr{B} = \mathscr{B}\mathscr{A}$ (6.24) becomes

$$\mathscr{B}(\mathscr{A}\chi_j) = b_j(\mathscr{A}\chi_j). \tag{6.25}$$

Equation (6.25) shows that $\mathscr{A}\chi_j$ is an eigenfunction of $\mathscr{B}$ with eigenvalue b_j. But χ_j is an eigenfunction of $\mathscr{B}$ with eigenvalue b_j. Therefore either $\mathscr{A}\chi_j$ differs from χ_j by some multiplying constant

$$\mathscr{A}\chi_j = k\chi_j, \tag{6.26}$$

or, if χ_j is a member of a degenerate set of eigenfunctions, then $\mathscr{A}\chi_j$ can be a linear combination of members of this set

$$\mathscr{A}\chi_j = k\chi_j + k'\chi_{j'} + k''\chi_{j''} + \ldots$$

In the non-degenerate case expression (6.26) shows that χ_j is an eigenfunction of $\mathscr{A}$, that is, it is one of the set of functions θ. In the degenerate case one can always choose some linear combinations of the degenerate functions χ_j which will be eigenfunctions of $\mathscr{A}$ (and of course of $\mathscr{B}$). For example, suppose we have a case of double degeneracy, and

$$\mathscr{A}\chi_j = a\chi_j + b\chi_{j'}$$

$$\mathscr{A}\chi_{j'} = c\chi_j + d\chi_{j'}.$$

Then if new constants λ, μ, k and k' are defined through the four equations

$$k\lambda = \lambda a + \mu c, \qquad k\mu = \lambda b + \mu d,$$

$$k'\mu = \mu a - \lambda c, \qquad k'\lambda = \lambda d - \mu b,$$

we find that

$$\mathscr{A}(\lambda\chi_j + \mu\chi_{j'}) = k(\lambda\chi_j + \mu\chi_{j'}),$$

$$\mathscr{A}(\mu\chi_j - \lambda\chi_{j'}) = k'(\mu\chi_j - \lambda\chi_{j'})$$

and these equations define eigenfunctions of $\mathscr{A}$.

The commutation relationships between operators form the basis of many of the most fundamental results of quantum mechanics. For example, if two operators do not commute then there will not be a set of functions which are simultaneously eigenfunctions for both operators, and, as a result, there can be no experiment in which it is possible to have the observables of both these operators well defined. The Heisenberg uncertainty principle as stated in chapter 1 is a special case of this. Since x and $(h/2\pi i)\partial/\partial x$ do not commute (6.21) then a particle cannot have a precise value of both x and p_x.

In quantum mechanics the eigenfunctions are always restricted to a class of functions which are single-valued, continuous and normalizable (these have been called functions of class Q). These conditions are necessary in order that the probability density shall be a well behaved function. A restriction must also be imposed on the operators because when we make a measurement the result is a real number. This means that the operators of quantum mechanics must give calculated average values (from (6.2)) which are real numbers.

If

$$\bar{b} = \int \Psi^* \mathscr{B} \Psi \, dv, \tag{6.27}$$

then taking complex conjugates

$$(\bar{b})^* = \int \Psi \mathscr{B}^* \Psi^* \, dv. \tag{6.28}$$

But if $\bar{b} = (\bar{b})^*$, which is true for a real number, then

$$\int \Psi^* \mathscr{B} \Psi \, dv = \int \Psi \mathscr{B}^* \Psi^* \, dv. \tag{6.29}$$

More generally it can be shown that the operator must satisfy

$$\int \Psi_1^* \mathscr{B} \Psi_2 \, dv = \int \Psi_2 \mathscr{B}^* \Psi_1^* \, dv, \tag{6.30}$$

where Ψ_1 and Ψ_2 are arbitrary functions of class Q.

An operator which satisfies (6.30) for functions of class Q is said to be Hermitian. If we are constructing a quantum-mechanical operator from the classical expression for the observable, using postulate 2, then it is necessary to arrange the terms in the operator such that it is Hermitian. For example, if the classical expression is xp_x, then the quantum-mechanical operator is not $(h/2\pi i)\, x\partial/\partial x$, which is not Hermitian, but $(h/4\pi i)$ $(x(\partial/\partial x) + (\partial/\partial x)x)$ which is. In other words, we start from the symmetrical classical expression $\frac{1}{2}(xp_x + p_x x)$. An alternative would be to start from the expression $x^{\frac{1}{2}}p_x x^{\frac{1}{2}}$; only experiment would show which gives the correct quantum-mechanical operator.

There are three important properties of the eigenfunctions and eigenvalues of Hermitian operators.

1. The eigenvalues of Hermitian operators are real. This follows from (6.27)–(6.29) if Ψ is an eigenfunction of $\mathscr{B}$.

2. If two eigenfunctions of a Hermitian operator have different eigenvalues, they are orthogonal. That is if

$$\mathscr{B}\Psi_1 = b_1\Psi_1 \tag{6.31}$$

and

$$\mathscr{B}\Psi_2 = b_2\Psi_2, \tag{6.32}$$

then

$$\int \Psi_2{}^*\Psi_1 \, dv = 0. \tag{6.33}$$

To prove this take the complex conjugate of (6.32)

$$\mathscr{B}^*\Psi_2{}^* = b_2\Psi_2{}^*. \tag{6.34}$$

Multiply (6.31) from the left by $\Psi_2{}^*$ and integrate over all space; likewise multiply (6.34) on the left by Ψ_1 and integrate. Subtract the two equations

$$\int \Psi_2{}^*\mathscr{B}\Psi_1 \, dv - \int \Psi_1\mathscr{B}^*\Psi_2{}^* \, dv = (b_1 - b_2)\int \Psi_2{}^* \Psi_1 \, dv. \tag{6.35}$$

But by the definition of a Hermitian operator (6.30) the left-hand side is zero. It follows that if $b_2 \neq b_1$, then (6.33) holds.

The term orthogonality occurs in vector algebra; if two vectors **a**, **b** make an angle of 90° with one another then the scalar product of the vectors is zero, $\mathbf{a}.\mathbf{b} = 0$, and the vectors are said to be orthogonal. This means that if we try to compound **a** in terms of other vectors in the space, then **a** contains no component of **b**; in other words, **a** and **b** are completely independent of one another. Likewise, when eigenfunctions are orthogonal this means that they are completely independent functions; one contains no component of the other.

Suppose we do try to write one eigenfunction as a combination of all others of the set, say

$$\Psi_1 = \sum_{i \neq 1} c_{i1}\Psi_i. \tag{6.36}$$

Then multiplying by $\Psi_j{}^*$ ($j \neq 1$) and integrating we have

$$\int \Psi_j^*\Psi_1 \, dv = \sum_{i \neq 1} c_{i1} \int \Psi_j{}^*\Psi_i \, dv. \tag{6.37}$$

But using the orthogonality condition, the left-hand side is zero and the only integral on the right which is non-zero is when $i = j$. It follows that $c_{j1} = 0$ thus establishing the linear independence of Ψ_1, on all other eigenfunctions; and this is true for all j.

We can combine the orthogonality and normalization condition (6.1) for eigenfunctions into the expression

$$\int \Psi_i^* \Psi_j \, dv = \delta_{ij}, \tag{6.38}$$

where δ_{ij} is called the Kronecker delta, which is zero unless $i = j$, in which case it is unity. A set of functions which satisfies (6.38) is said to be *orthonormal.*

3. The eigenfunctions Θ_i of a Hermitian operator form a complete set of functions which can be used as an expansion set for any function having the same boundary conditions as the eigenfunctions.
That is,

$$\Psi = \sum_i c_i \Theta_i \tag{6.39}$$

is exact if the summation is taken over all the eigenfunctions (which will be an infinite summation). The proof of this statement has not been established in general but is found to be valid for the Hermitian operators encountered in quantum mechanics. As will be seen in the next section and in other parts of the book, the expansion method is the most common approach to obtaining approximate solutions of the Schrödinger equation.

6.2 Variation and perturbation theory

In quantum mechanics we are faced with the problem that for all except the simplest systems the Schrödinger equation cannot be solved exactly. We have to be content with obtaining approximate solutions of the equation and try to make these approximate solutions as close to the exact solutions as possible. The most common approach is to write the solutions as a summation over a known set of functions, according to expression (6.39), and then to vary the coefficients c_i until the best wave function is obtained. There are two common methods of obtaining approximate solutions to the wave equation, the perturbation method and the variation method.

The perturbation method starts from a set of eigenfunctions of some Hamiltonian $\mathscr{H}^0$ which is a close approximation to the true Hamiltonian $\mathscr{H}$ of the system. The eigenfunctions of $\mathscr{H}$ are then expanded in terms of the complete set of eigenfunctions of $\mathscr{H}^0$, and successive approximations to the coefficients in this expansion are found. For example, to obtain the wave functions and energies of the hydrogen atom in the presence of an electric field (the Stark effect), the eigenfunctions of the hydrogen atom in the absence of a field can be used as an expansion set (this set is only complete if the eigenfunctions for both bound and unbound states are included).

Suppose

$$\mathscr{H}^0\Psi_i^0 = E_i^0\Psi_i^0, \tag{6.40}$$

and the corresponding eigenfunction and eigenvalue of $\mathscr{H}$ are required.

$$\mathscr{H}\Psi_i = E_i\Psi_i. \tag{6.41}$$

We write

$$\Psi_i = \Psi_i^0 + \sum_{j \neq i} c_{ij}\Psi_j^0, \tag{6.42}$$

(this function can be normalized at a later stage) and substituting (6.42) into (6.41) gives

$$(\mathscr{H} - E_i)\Psi_i^0 + \sum_{j \neq i} c_{ij}(\mathscr{H} - E_i)\Psi_j^0 = 0. \tag{6.43}$$

The Hamiltonian is now split up into $\mathscr{H}^0$ and a perturbation term $\mathscr{H}'$

$$\mathscr{H} = \mathscr{H}^0 + \mathscr{H}'. \tag{6.44}$$

Substituting (6.44) into (6.43) and making use of (6.40) gives

$$(\mathscr{H}' + E_i^0 - E_i)\Psi_i^0 + \sum_{j \neq i} c_{ij}(\mathscr{H}' + E_j^0 - E_i)\Psi_j^0 = 0. \tag{6.45}$$

Multiplying (6.45) from the left by Ψ_i^{0*}, integrating over all space and making use of the orthonormality of the functions Ψ^0, this gives

$$(\mathscr{H}_{ii}' + E_i^0 - E_i) + \sum_{j \neq i} c_{ij}\mathscr{H}_{ij}' = 0, \tag{6.46}$$

where

$$\mathscr{H}_{ij}' = \int \Psi_i^{0*}\mathscr{H}'\Psi_j^0 \, dv. \tag{6.47}$$

Likewise, multiplying (6.45) from the left by Ψ_k^{0*} $(k \neq i)$ gives

$$\mathscr{H}_{ki}' + c_{ik}(\mathscr{H}_{kk}' + E_k^0 - E_i) + \sum_{j \neq i,k} c_{ij}\mathscr{H}_{kj}' = 0. \tag{6.48}$$

The energy and coefficients are now expanded as follows

$$E_i = E_i^0 + E_i' + E_i'' + \dots$$

$$c_{ij} = c_{ij}' + c_{ij}'' + \dots \tag{6.49}$$

where the prime represents the contribution of first-order in $\mathscr{H}'$, the double prime that of second-order, etc. We now pick out terms from (6.46) and (6.48) which are of the same order in $\mathscr{H}'$.

The first-order terms from (6.46) give

$$E_i' = \mathscr{H}_{ii}' = \int \Psi_i^{0*} \mathscr{H}' \Psi_i^0 \, dv. \tag{6.50}$$

That is, the *first-order* correction to the energy is just calculated from the zeroth-order wave functions and the perturbation to the Hamiltonian. The first-order terms in (6.48) are

$$\mathscr{H}_{ki}' + c_{ik}'(E_k^0 - E_i^0) = 0, \tag{6.51}$$

hence we have the first-order correction to the coefficients

$$c_{ik}' = \frac{\mathscr{H}_{ki}'}{E_i^0 - E_k^0}. \tag{6.52}$$

The second-order terms in (6.46) are

$$- E_i'' + \sum_{j \neq i} c_{ij}' \mathscr{H}_{ij}' = 0, \tag{6.53}$$

and, making use of (6.52), this gives us the second-order correction to the energy

$$E_i'' = \sum_{j \neq i} \frac{\mathscr{H}_{ji}' \mathscr{H}_{ij}'}{E_i^0 - E_j^0}. \tag{6.54}$$

If Ψ^0 are real functions $\mathscr{H}_{ji}' \mathscr{H}_{ij}' = (\mathscr{H}_{ij}')^2$, but in any case, from the Hermitian property of $\mathscr{H}'$ it follows that $\mathscr{H}_{ji}' \mathscr{H}_{ij}'$ is real and positive.

It is seen from (6.52) and (6.54) that the first-order correction to the coefficients and the second-order correction to the energy involve the energy difference between the states Ψ^0 in the denominator; the states which have the largest effect on Ψ_i^0 are those which are closest in energy. Because $\mathscr{H}_{ji}' \mathscr{H}_{ij}'$ is always real and positive, states of energy greater than E_i^0 decrease the ith state and those of lower energy increase its energy.

Expressions (6.50), (6.52) and (6.54) are the most important of perturbation theory; it is rarely necessary to go to higher-order terms. The criterion for validity of the perturbation expansion is that

$$|\mathscr{H}_{ki}'| \ll |E_i^0 - E_k^0|. \tag{6.55}$$

It is clear that the theory breaks down if there are degenerate states (say $E_i^0 = E_k^0$), since the terms in c_{ik}' and E_i'' may go to infinity. However, these cases can be treated using a modification of the variation method, which will now be described.

The most powerful criterion for the accuracy of an approximate wave function is provided by the *variation theorem*. This states that the energy evaluated from an approximate wave function will always be greater than the lowest eigenvalue of the Hamiltonian. That is, if Θ is an approximate wave function, not necessarily normalized, then

$$\left[\int \Theta^* \mathscr{H} \Theta \, \mathrm{d}v \Big/ \int \Theta^* \Theta \, \mathrm{d}v\right] > E_0. \tag{6.56}$$

To prove this theorem we expand Θ in terms of the set of eigenfunctions of $\mathscr{H}$,

$$\Theta = \sum_i c_i \Psi_i. \tag{6.57}$$

Substituting this in (6.56) and making use of the orthonormality of Ψ_i it is then necessary to prove†

$$\sum_i c_i^2 E_i > E_0 \sum_i c_i^2. \tag{6.58}$$

But this is necessarily so if E_0 is the lowest eigenvalue of $\mathscr{H}$ since

$$\sum_i c_i^2 (E_i - E_0) > 0. \tag{6.59}$$

Thus if we choose a wave function which contains variable parameters, evaluate the energy of the function in terms of these parameters and then minimize this energy with respect to the parameters, we shall obtain the best approximation to the ground state energy using a wave function of the chosen form. For example, if we take a wave function

$$\phi = \mathrm{e}^{-\zeta r} \tag{6.60}$$

for the 1*s* orbital of He, then, as will be shown in section 9.3, the energy of the ground state of He is calculated to be

$$E = \zeta^2 - (27/8)\zeta. \tag{6.61}$$

Differentiating this expression with respect to ζ and equating the result to zero gives $\zeta = 27/16$. Substituting this value into (6.60) gives the best form for the 1*s* orbital of He written as a single exponential term.

The most common way of applying the variation principle is to expand the wave function as a linear combination of a known set of functions, which may or may not be orthonormal, and to vary the coefficients in this set until the energy is minimized.

† We are assuming here that the coefficients c are real, if not then c_i^2 should be replaced by $c_i^* c_i$. This assumption holds for the rest of this chapter.

Let the wave function be written as a linear combination of n functions $\Theta_1 \ldots \Theta_n$.

$$\Psi = \sum_{i=1}^{n} c_i \Theta_i, \tag{6.62}$$

where the coefficients c_i are variation parameters. Then

$$E = \int \Psi^* \mathscr{H} \Psi \, \mathrm{d}v \Big/ \int \Psi^* \Psi \, \mathrm{d}v, \tag{6.63}$$

and substituting (6.62) into this gives

$$\sum_{i=1}^{n} \sum_{j=1}^{n} c_i c_j (\mathscr{H}_{ij} - E S_{ij}) = 0, \tag{6.64}$$

where

$$\mathscr{H}_{ij} = \int \Theta_i^* \mathscr{H} \Theta_j \, \mathrm{d}v \quad \text{and} \quad S_{ij} = \int \Theta_i^* \Theta_j \, \mathrm{d}v. \tag{6.65}$$

Differentiating (6.64) with respect to c_k, keeping the other coefficients constant (remembering that $\mathscr{H}_{ij}$ and S_{ij} are constants) gives

$$2 \sum_{i=1}^{n} c_i (\mathscr{H}_{ik} - E S_{ik}) - \sum_{i=1}^{n} \sum_{j=1}^{n} c_i c_j S_{ij} \frac{\partial E}{\partial c_k} = 0. \tag{6.66}$$

For an energy minimum, $(\partial E/\partial c_k) = 0$, we have

$$\sum_{i=1}^{n} c_i (\mathscr{H}_{ik} - E S_{ik}) = 0. \tag{6.67}$$

There is one equation of this type for every coefficient c_k that can be varied; n equations in all. Now n linear equations in n unknown coefficients of the type (6.67) are consistent only if the determinant of the factors multiplying the coefficients is zero; that is†

$$|\mathscr{H}_{ik} - E S_{ik}| = 0. \tag{6.68}$$

† For example, the equations

$$(1 - E)c_1 + (2 - 0{\cdot}5E)\, c_2 = 0 \tag{a}$$
$$(2 - 0{\cdot}5E)\, c_1 + (1 - E)\, c_2 = 0 \tag{b}$$

are only consistent if

$$\begin{vmatrix} 1 - E & 2 - 0{\cdot}5E \\ 2 - 0{\cdot}5E & 1 - E \end{vmatrix} = 0.$$

That is, if

$$(1 - E)^2 - (2 - 0{\cdot}5E)^2 = 0$$

or $E = +2$ or -2.

It can be seen that any other value of E substituted into (a) and (b) does not give the same ratio of c_1/c_2 from the two equations.

The determinant (6.68), which is called the *secular determinant*, gives on expansion a polynomial of the nth degree in E. The n values of E obtained by solving the resulting equation all correspond to turning points of the function (6.63). The lowest value of E gives the best value for the ground state energy that can be obtained from a function of the type (6.62). The coefficients of this function are obtained by substituting this energy back into the set of simultaneous equations (6.67), and solving in the usual way; this will give the ratio of the coefficients and their absolute value will be obtained by normalizing the function. An example of this technique will be given in chapter 12 so it will not be elaborated further here.

It can be shown that the solutions of (6.67) give not only the best ground state function but also the best excited state functions that can be obtained from the limited expansion set (6.62) (subject to the restriction that an excited state function is orthogonal to all states of lower energy).

If the functions Θ_i of our expansion set are eigenfunctions of some Hamiltonian $\mathscr{H}^0$, then they form an orthonormal set, and we can write $S_{ik} = \delta_{ik}$, $\mathscr{H}_{ik} = \mathscr{H}_{ik}' + \delta_{ik}E_k^0$ (using the nomenclature followed in the discussion on perturbation theory). The secular equations (6.67) then become

$$\sum_{i=1}^{n} c_i(\mathscr{H}_{ik}' + (E_k^0 - E)\delta_{ik}) = 0, \tag{6.69}$$

which is analogous to the perturbation theory expression (6.46). The perturbation expressions for the energy (6.50) and (6.54) can then be considered as approximations to the exact solutions of the secular determinant

$$|\mathscr{H}_{ik}' + (E_k^0 - E)\delta_{ik}| = 0 \tag{6.70}$$

if we take the summations not to infinity, but only over the states included in the expansion set.

When the perturbation expressions given earlier cannot be applied, because of degeneracies or near degeneracies in the zeroth-order functions then it is necessary to go to the secular determinant to determine the energies. This method of handling degeneracies in perturbation theory is fully discussed in most advanced texts in quantum mechanics†.

The labour involved in using the expansion technique to obtain approximate wave functions depends primarily on the accuracy to which the wave functions and their energies are required. It almost goes without saying that it is worth while to choose an expansion set which gives a rapid convergence to the exact eigenfunction, rather than one which will give a slow convergence.

† For example, EYRING, WALTER and KIMBALL, *Quantum Chemistry*, pp. 96–99, Wiley, 1944.

A considerable saving in labour can result if the functions in the expansion set are eigenfunctions of an operator which commutes with the Hamiltonian. Thus if Θ_1 and Θ_2 are eigenfunctions of an operator $\mathscr{B}$ with different eigenvalues, and $\mathscr{B}$ commutes with $\mathscr{H}$, then

$$\int \Theta_1{}^*\mathscr{H}\Theta_2 \, dv = 0. \tag{6.71}$$

To prove this consider the integral

$$\int \Theta_1{}^*\mathscr{H}\mathscr{B}\Theta_2 \, dv. \tag{6.72}$$

If $\mathscr{B}\Theta_2 = b_2\Theta_2$, then this integral is equal to

$$b_2 \int \Theta_1{}^*\mathscr{H}\Theta_2 \, dv. \tag{6.73}$$

However, if $\mathscr{H}$ and $\mathscr{B}$ commute, and $\mathscr{B}$ is Hermitian, expression (6.72) becomes

$$\int \Theta_1{}^*\mathscr{B}(\mathscr{H}\Theta_2) \, dv = \int \mathscr{H}\Theta_2\mathscr{B}^*\Theta_1{}^* dv = b_1 \int (\mathscr{H}\Theta_2)\Theta_1{}^* \, dv$$

$$= b_1 \int \Theta_1{}^*\mathscr{H}\Theta_2 \, dv. \tag{6.74}$$

Expressions (6.73) and (6.74) must be equal, but if $b_1 \neq b_2$ this is only true if the integral (6.71) is zero.

This result is extremely important, and we illustrate its use by the following example. Suppose wave functions are required for the atomic orbitals of a many-electron atom. A suitable expansion for these functions will be

$$\varphi = \sum_i c_i\Theta_i, \tag{6.75}$$

where the functions Θ_i have the general form

$$\Theta_i = r^n \, e^{-\zeta r} Y_{lm}(\vartheta, \varphi), \tag{6.76}$$

with n, and ζ as constants which can be chosen to be different for each function i. The functions $Y_{lm}(\vartheta, \varphi)$ are the spherical harmonics (3.10), and we shall see in chapter 9 that these functions are also eigenfunctions of the orbital angular momentum operators; l and m characterizing their eigenvalues. Now the Hamiltonian commutes with the angular momentum operators (if spin terms are not included) so there will be no Hamiltonian integrals involving functions which have different eigenvalues of angular momentum. That is, integrals of the type (6.71) will be zero if Θ_1 and Θ_2 have different l or m quantum numbers. It follows that the secular determinant will split up into blocks on the principal diagonal, each block involving functions having the same quantum numbers l and m, and there will be no off-diagonal terms linking blocks of different l or m. The secular determinant can be factorized into smaller determinants each having a characteristic l and m, and the wave functions obtained from these determinants are also characterized by these quantum numbers. It is because the angular momentum operators commute with the Hamiltonian that the atomic-orbital wave functions can be written in the form (4.3), where the spherical harmonic is a factor of the whole wave function.

Further illustrations of this rule will be found in other parts of the book. In particular chapter 8 will show how the symmetry of functions may be used to simplify the construction of wave functions using the expansion technique.

Problems 6

6.1 Show that if Ψ_i and Ψ_j are eigenfunctions of $\mathscr{B}$ with different eigenvalues then

$$\int \Psi_i^* \mathscr{B} \Psi_j \, dv = 0.$$

6.2 Show that if b is an eigenvalue of $\mathscr{B}$ then b^n is an eigenvalue of $\mathscr{B}^n$.

6.3 What are the eigenfunctions of the momentum operator?

6.4 If Ψ is an eigenfunction of the Hamiltonian for a single electron having a potential energy V, prove, by considering the commutation relationship between $\mathscr{H}$ and x, that

$$\int \Psi_i^* \left(\frac{\partial}{\partial x}\right) \Psi_j dv = -\left(\frac{4\pi^2 m}{h^2}\right)(E_i - E_j) \int \Psi_i^* \, x \Psi_j dv.$$

This formula can be generalized to relate an integral involving $\partial^n/\partial x^n$ to one involving $\partial^{n-1}/\partial x^{n-1}$.

6.5 Prove that if ϑ is a complete orthonormal set of functions then

$$\sum_i \int \vartheta_a^* \mathscr{A} \vartheta_i \, d\tau \, . \int \vartheta_i^* \mathscr{B} \vartheta_b \, d\tau = \int \vartheta_a^* \mathscr{A} \mathscr{B} \vartheta_b \, d\tau$$

or

$$\sum_i \mathscr{A}_{ai} \mathscr{B}_{ib} = (\mathscr{A} \mathscr{B})_{ab}.$$

This is an important theorem of quantum mechanics known as the matrix sum rule.

Chapter 7

The wave functions of many-electron systems

7.1 The antisymmetry rule

For a many-electron system the electron repulsion terms e^2/r_{ij} must be included as part of the potential energy in the wave equation. The wave function of the electrons is then a function of all the electron coordinates. If it were not for the electron repulsion terms, the wave equation could be solved by writing the many-electron wave function as a product of one-electron functions

$$\Psi = \psi_a(1)\psi_b(2) \ldots \psi_k(n), \tag{7.1}$$

and it would then be found that the wave equation would be separable into a set of equations each involving the coordinates of only one electron, and the solutions of these equations would give the functions ψ. Because of the electron repulsion terms this is not possible, and so an exact solution of the wave equation for a many-electron system has not been obtained. This is not solely a limitation of quantum mechanics, the many-particle problem has not been solved classically either.

Because of the difficulty of manipulating and tabulating functions of many variables, there is little option when dealing with many-electron systems but to work with functions of the type (7.1) and then to make these as close as possible to the true solutions of the wave equation, for example, by using the variation theorem. It is only when working with functions like (7.1) that an orbital can be defined, for *an orbital is a one-electron wave function.*

The method of obtaining the best functions for atomic orbitals by the SCF method, was outlined in chapter 4. Similar techniques can be used to obtain molecular orbitals, and these will be discussed in chapter 10.

Let us consider a set of atomic or molecular orbitals ψ. We feed electrons into these orbitals and associate an electron spin with each orbital. We have seen in chapter 4 that there are two different spin states for an electron

($s = \pm\frac{1}{2}$). We shall represent the spin wave functions for these two states by α and β, corresponding to $s = +\frac{1}{2}$ and $s = -\frac{1}{2}$ respectively. The combined space and spin function of an electron is called a *spin-orbital.* For example, we can have $1s\alpha$ or $1s\beta$ as the complete wave function of an electron in a $1s$ orbital. For brevity we sometimes shall write these spin-orbitals $1s$ and $\overline{1s}$ respectively if it is important to label the spin of the electron.

As a first guess we might write a wave function for a many-electron system as

$$\psi_a(1)\psi_b(2) \ldots \psi_k(n), \tag{7.2}$$

where we have assigned electron 1 to spin-orbital ψ_a electron 2 to ψ_b etc†. But electrons are indistinguishable from one another. An equally good wave function to (7.2) would be

$$\psi_a(2)\psi_b(1) \ldots \psi_k(n) \tag{7.3}$$

or any other of the $n!$ functions that can be obtained by permuting the n electrons amongst these spin-orbitals. The most general wave function will be a linear combination of these $n!$ functions which gives them equal weight. It turns out, however, that only one combination gives results which agree with experiment; this is the combination which changes sign on exchanging the coordinates of any two electrons. The wave function is said to be antisymmetric to the exchange of any two electrons. For two electrons, for example, the function

$$\psi_a(1)\psi_b(2) - \psi_a(2)\psi_b(1) \tag{7.4}$$

satisfies this condition. In general, if the wave function is written as a determinant

$$\sqrt{\frac{1}{n!}}\begin{vmatrix} \psi_a(1) & \psi_b(1) & \ldots & \psi_k(1) \\ \psi_a(2) & \psi_b(2) & \ldots & \psi_k(2) \\ \cdots & \cdots & \cdots & \cdots \\ \psi_a(n) & \psi_b(n) & \ldots & \psi_k(n) \end{vmatrix} \tag{7.5}$$

the antisymmetry property is satisfied, since exchanging the coordinates of two electrons is equivalent to exchanging two rows of the determinant, and this does change the sign of the function. On normalizing the function by dividing by $(n!)^{\frac{1}{2}}$, as in (7.5), we have what is called a *Slater determinant*. Since a function like (7.5) is rather bulky to write out in full we shall abbreviate it thus:

$$|\psi_a\psi_b \ldots \psi_k|. \tag{7.6}$$

† For example, the ground state of Be would be $1s(1)\,\overline{1s}(2)\,2s(3)\,\overline{2s}(4)$.

The Pauli exclusion principle follows from the above restriction on the form of the wave function. If two electrons in the same atom have the same four quantum numbers then there must be two spin-orbitals in the wave function which are the same. In this case two rows of the determinant are the same, which makes the determinant zero—in other words this type of function cannot exist.

It has been found that the rule of antisymmetry of wave functions is held not only by electrons but also by many other elementary particles, notably protons and neutrons. For example, it is the antisymmetry restriction on protons that leads to two kinds of H_2, ortho and para; but this is a topic best discussed elsewhere.

7.2 The Born–Oppenheimer approximation

The Hamiltonian for a system of nuclei (μ, ν ...) of coordinates X, and electrons (i, j ...) of coordinates x, is as follows

$$\mathscr{H}(x, X) = -\sum_{\mu} \frac{h^2}{8\pi^2 M_\mu} \nabla_\mu^2 - \sum_i \frac{h^2}{8\pi^2 m} \nabla_i^2 + V_{\text{ne}}(x, X) + V_{\text{ee}}(x) + V_{\text{nn}}(X), \tag{7.7}$$

where, for example, V_{ee} represents the electron–electron repulsion terms,

$$V_{\text{ee}}(x) = \sum_{i>j} \frac{e^2}{r_{ij}}. \tag{7.8}$$

If we separate off the nuclear kinetic energy terms

$$\mathscr{H}_{\text{n}}(X) = -\sum_{\mu} \frac{h^2}{8\pi^2 M_\mu} \nabla_\mu^2, \tag{7.9}$$

then

$$\mathscr{H}(x, X) - \mathscr{H}_{\text{n}}(X) = \mathscr{H}_{\text{e}}(x, X) \tag{7.10}$$

is the Hamiltonian which describes the motion of the electrons for fixed positions of the nuclei. It should be noted that $\mathscr{H}_{\text{e}}$ depends on the position but not on the momentum of the nuclei.

We now assume that the complete wave function, which is a solution of

$$\mathscr{H}\Psi = E\Psi \tag{7.11}$$

can be written as

$$\Psi = \Psi_{\text{e}}(x, X)\chi_{\text{ne}}(X), \tag{7.12}$$

where the 'electronic' wave function is defined by

$$\mathscr{H}_{\text{e}}(x, X)\Psi_{\text{e}}(x, X) = E_{\text{e}}(X)\Psi_{\text{e}}(x, X), \tag{7.13}$$

and the 'nuclear' wave function by

$$[\mathscr{H}_n(X) + E_e(X)]\chi_{ne}(X) = E\chi_{ne}(X). \tag{7.14}$$

These equations describe the Born–Oppenheimer approximation.

What do these equations mean? They mean that the total wave function of a molecule can be separated into an electronic and a nuclear part. The electronic wave function is obtained for various fixed positions of the nuclei by solving (7.13). This gives us an electronic energy E_e, which if plotted as a function of X for say a diatomic molecule (X being in this case the internuclear distance) may give a curve as in figure 5.1. This electronic energy is then taken as the potential energy determining the motion of the nuclei so that the Schrödinger equation for the nuclei has the form (7.14). The solutions χ_{ne} are independent of the electron coordinates, but depend on the nature of the electronic state, because the potential curve depends on the nature of the electronic state. Each electronic state Ψ_e has its own set of nuclear wave functions χ_{ne}.

The electronic potential energy is not an observable quantity, although it can be calculated. It is usual to write the total energy as a sum of an electronic energy, evaluated at the equilibrium configuration, plus a nuclear energy.

$$E = E_e(X_0) + E_n. \tag{7.15}$$

In this way an electronic and a nuclear energy may be separated from the total energy.

There is clear evidence from spectroscopy that in general the Born–Oppenheimer approximation is a good one. In infrared spectroscopy transitions are observed between different vibrational levels of the lowest electronic state and this vibrational energy pattern can be interpreted in terms of the wave functions χ_{ne} and energies E_n defined above. In the visible and ultraviolet regions of the spectrum we observe transitions between different electronic energy levels, and the resulting spectra have structures arising from the different vibrational components of the two electronic levels. The spacing of this vibrational fine structure and the relative intensities of the components can again be well understood on the basis of the above equations.

We now discuss the conditions under which the Born–Oppenheimer approximation is valid. On substituting (7.10) and (7.12) into (7.11), one obtains

$$[\mathscr{H}_e(x, X) + \mathscr{H}_n(X)]\Psi_e(x, X)\chi_{ne}(X) = E\Psi_e(x, X)\chi_{ne}(X). \tag{7.16}$$

Because the only differential terms in $\mathscr{H}_e$ are functions of x, we can write, using (7.13),

$$\mathscr{H}_e(x, X)\Psi_e(x, X)\chi_{ne}(X) = \chi_{ne}(X)\mathscr{H}_e(x, X)\Psi_e(x, X)$$
$$= \chi_{ne}(X)E_e(X)\Psi_e(x, X). \qquad (7.17)$$

But $\mathscr{H}_n$ is a differential function of X, and both Ψ_e and χ_{ne} are functions of X, hence (using 7.9)

$$\mathscr{H}_n(X)\Psi_e(x, X)\chi_{ne}(X) = -\sum_\mu \frac{h^2}{8\pi^2 M_\mu}\{\Psi_e(x, X)\nabla_\mu{}^2\chi_{ne}(X)$$
$$+ 2\nabla_\mu\Psi_e(x, X)\nabla_\mu\chi_{ne}(X) + \chi_{ne}(X)\nabla_\mu^2\Psi_e(x, X)\}. \qquad (7.18)$$

On substituting (7.17) and (7.18) into (7.16) we arrive at expression (7.14) which determines the nuclear wave function, providing the terms in $\nabla_\mu\Psi_e$ and $\nabla_\mu{}^2\Psi_e$ in (7.18) can be neglected. In other words, the Born–Oppenheimer approximation is valid providing the electronic wave function Ψ_e is a slowly varying function of the nuclear coordinates. The only situation in which the Born–Oppenheimer approximation is likely to break down is when we have degeneracy or near degeneracy between two electronic wave functions. In these cases the terms involving $\nabla_\mu\Psi_e$, which are neglected in the Born–Oppenheimer approximation, will cause an appreciable interaction between the two degenerate or near-degenerate electronic states, so that a single wave function of the type (7.12) will not be a good representation of the system.

In this book we shall be concerned with methods of calculating the electronic wave function Ψ_e and the potential energy E_e. For this reason any future references to the total wave function, energy or Hamiltonian will mean the electronic terms although we will not bother to write in the suffix e in each case.

Bibliography

A description of situations in which the Born–Oppenheimer approximation breaks down is given by LONGUET-HIGGINS, *Advances in Spectroscopy*, II, p. 429, Interscience, 1962.

Chapter 8

Symmetry

8.1 Introduction to symmetry groups

If a molecule has symmetry, then we can say something about the form of its wave functions without going to the trouble of solving the wave equation. We shall illustrate this by taking formaldehyde as an example.

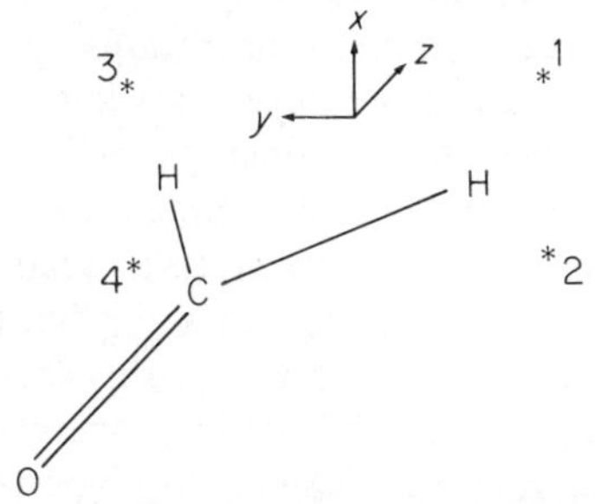

FIGURE 8.1. Equivalent points with respect to the nuclei of formaldehyde. 1 and 3 are above the plane of the molecule, 2 and 4 are below

Suppose we are interested in finding molecular orbitals for formaldehyde with the nuclei in their equilibrium configuration. If an orbital ψ is occupied by an electron, then the probability of finding the electron at a point in space is given by the value of ψ^2 at that point. Let us consider an arbitrary point (labelled 1 in figure 8.1) defined by the vector $\mathbf{r}_1$ and let this be somewhere in front of the nuclear plane of the molecule. The probability density for the electron at that point is $\psi^2(\mathbf{r}_1)$. But there are three other points in space, indicated in the figure, which are entirely equivalent to the point 1. For example, point 2 lies at the back of the nuclear plane, but it is the same distance from all the nuclei as is the point 1. It is clear that the electron density at 1 and 2 must be the same; that is, $\psi^2(\mathbf{r}_1) = \psi^2(\mathbf{r}_2)$. There are two ways of satisfying this condition: either $\psi(\mathbf{r}_1) = \psi(\mathbf{r}_2)$ or $\psi(\mathbf{r}_1) = -\psi(\mathbf{r}_2)$. That is, we can either have a wave function which is symmetric to the interchange of points 1 and 2, or one which is antisymmetric to such an interchange.

The points 1 and 2 are related through the action of symmetry operations which leave the nuclear configuration of the molecule unchanged. More generally, a symmetry operation is a movement of a body which transforms it into an equivalent orientation, so that after the movement every point of the body is coincident with an equivalent point (or the same point) of the body in its original orientation. In other words, if an observer saw the body before and after but not during the symmetry operation he would not be able to tell whether the symmetry operation had been carried out or not.

If we rotate the figure representing formaldehyde by 180° about the z axis then we interchange points 1 and 4, 2 and 3, but the final orientation of the molecule is equivalent to that with which we started. Similarly, if we reflect in the yz plane, which is the plane containing the nuclei, then we interchange the points 1 and 2, 3 and 4, but we get an equivalent orientation. Lastly, if we reflect in the xz plane then we interchange 1 and 3, 2 and 4, but again get an equivalent orientation.

We therefore have three symmetry operations which convert figure 8.1 into an equivalent orientation: two reflections and a rotation. These symmetry operations are subject to a mathematical discipline known as group theory. In any system of mathematics something is needed which plays the role of the number 1 in ordinary algebra. That is, 1 multiplied by a function is equal to that function. In matrix algebra, for example, this role is taken by the unit matrix. In group theory this unit element is called the identity. For symmetry groups the identity operation is that operation which leaves the figure unmoved. The identity operation and the rotation and two reflection operations make up what is called the point group of symmetry operations for figure 8.1.

The general definition of a group is that it is a collection of elements which are interrelated according to certain rules. To examine the formal mathematics of group theory we do not have to specify what these elements are nor need we attribute any physical significance to them. Although in this book we shall be concerned with the groups formed by symmetry operations, it is convenient to start with some basic definitions and theorems which are generally applicable to any group.

1. The number of elements in a group is called the *order* (it is given the conventional symbol h).

2. The product of any two elements of the group, and the square of each element, is another element in the group. That is $AB = C$ where A, B and C are elements of the group. However, elements of a group do not necessarily commute (page 65) that is, AB and BA are not necessarily equivalent. We do not define more rigorously what we mean by the *product* of two elements until we specify the nature of the elements of the group.

3. One element of the group is the identity element. We give this the symbol I although E is also frequently used. $IA = AI = A$ for any element A of the group.

4. The association law of algebra holds, that is, in evaluating the product ABC it does not matter if we first combine BC and then combine their product with A, or first combine AB and then combine their product with C: $A(BC) = (AB)C$.

5. Every element has a reciprocal which is also an element of the group. Thus B is the reciprocal of A if $BA = AB = I$. We can also write $B = A^{-1}$.

6. Within any group there will be smaller sub-groups which satisfy requirements 2–5 (as a trivial example, I is the group of order 1). It can be shown that the order of a sub-group is an integral factor of the order of the full group. For example, a group of order 6 can only have sub-groups of order 1, 2 or 3.

7. If A and X are two elements of the group then $X^{-1}AX$ is also an element of the group, B say.

$$X^{-1}AX = B. \tag{8.1}$$

B is said to be the similarity transform of A by X, and A and B are said to be conjugate. It should be noted that for the above expression, we have, after multiplying from the left by X and from the right by X^{-1},

$$XX^{-1}AXX^{-1} = XBX^{-1}. \tag{8.2}$$

But $XX^{-1} = I$, and if the reciprocal of X is Y, then

$$A = Y^{-1}BY. \tag{8.3}$$

In other words, the conjugate property of A and B is mutual. A complete set of elements which are mutually conjugate is called a *class*. That is, if $X^{-1}AX$, $X^{-1}BX$ and $X^{-1}CX$ all give either A, B or C for any operation X, then A, B and C are said to form a class. The number of elements in a class must be an integral factor of the order of the group, but there need not be a class for every integral factor. Thus in a group of order six the classes can contain 1, 2 or 3 elements.

We turn now to a discussion of symmetry groups, and in the first place to an examination of the different types of symmetry operation. We have already met three: the identity operation, rotation about an axis, and reflection in a plane. There are two more that we shall meet: inversion in a centre of symmetry (a point with coordinates x, y, z is moved to the point $-x, -y, -z$), and rotation about an axis followed by reflection in a plane perpendicular to that axis. The conventional symbols for these operations are listed in the following table.

TABLE 8.1. The symbols for symmetry group operations[a]

Symbol	Meaning
I	The identity.
C_n	A rotation by $2\pi/n$. The axis of highest n is called the principal axis. If there are two-fold axes perpendicular to the principal axis the appropriate operations are labelled C_2'.
σ	A reflection. σ_h represents a reflection in a plane perpendicular to the principal axis (h for horizontal). σ_v represents a reflection in a plane containing the principal axis (v for vertical). σ_d represents reflection in a plane containing the principal axis and bisecting two C_2' axes (d for dihedral).
S_n	A rotation by $2\pi/n$ followed by reflection in a plane perpendicular to the axis of rotation.
i	Inversion in a centre of symmetry.

[a] The only symmetry groups with which we shall be concerned are the so-called point groups, which have the elements listed in this table. There are also symmetry groups which contain elements of translation, and these are called space groups. They are useful in the study of crystals.

For formaldehyde the symmetry operations are I, $C_2(z)$, $\sigma_v(xz)$, $\sigma_v'(yz)$ as shown in figure 8.2. This set of operations is called the $\mathbf{C}_{2v}$ symmetry group. We shall first confirm that they satisfy the basic requirements for a group which we have already stated.

We first check that the product of two group operations is itself a group operation. The results are shown in table 8.2, which is called the group multiplication table

TABLE 8.2. The group multiplication table for $\mathbf{C}_{2v}$

	I	$C_2(z)$	$\sigma_v(xz)$	$\sigma_v'(yz)$
I	I	C_2	σ_v	σ_v'
C_2	C_2	I	σ_v'	σ_v
σ_v	σ_v	σ_v'	I	C_2
σ_v'	σ_v'	σ_v	C_2	I

For example (returning to figure 8.2), operating on the figure first with $\sigma_v(xz)$ and then with $\sigma_v'(yz)$ is equivalent to a single operation of C_2. We write this in algebraic form $\sigma_v'\sigma_v = C_2$ and insert a C_2 as the element of column σ_v and row σ_v' in table 8.2. For this particular group the result is

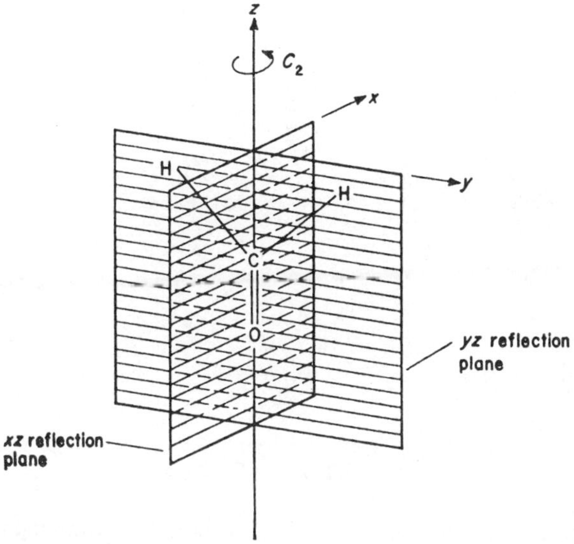

FIGURE 8.2. The symmetry operations of formaldehyde

independent of the order in which the two operations are carried out. The elements of the group commute, and the group is said to be Abelian. Molecules belong to Abelian symmetry groups providing they have no axes of rotation higher than two-fold. It is seen also that each element is its own reciprocal (the identity operator is on the principal diagonal). The operations I, C_2 are a sub-group of order 2, but I, σ_v, σ_v' is not a sub-group since $\sigma_v . \sigma_v' = C_2$.

Since all the elements of the group commute we have

$$X^{-1}AX = AX^{-1}X = A,$$

so that each element is in a class of one. This is a general property of Abelian symmetry groups.

We will now examine a non-Abelian group, the $\mathbf{C}_{3v}$ group, to which molecules like NH_3 belong. As illustrated in figure 8.3 the symmetry operations are the identity, a rotation by $2\pi/3$ (C_3) or by $4\pi/3$, which we can write C_3^2, and three vertical symmetry planes each of which contains one of the hydrogen nuclei.

Let us identify the three hydrogen atoms by symbols a, b and c and the three reflection operations by σ_{va}, σ_{vb} and σ_{vc}. Figure 8.3 shows the effect of the successive operations C_3 and σ_{va} on the figure. The final positions of atoms a, b and c depend on the order in which these operations are

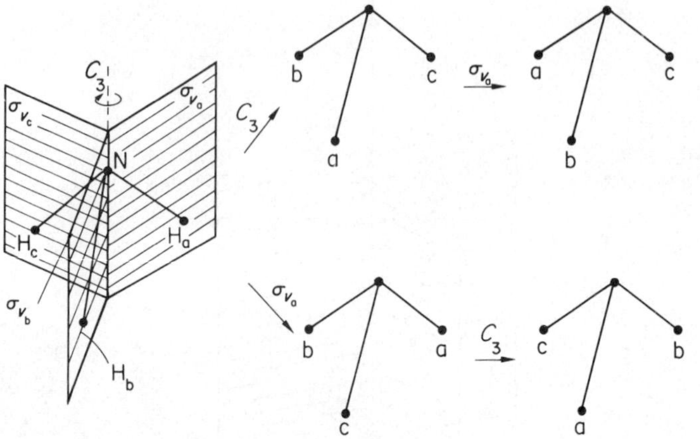

FIGURE 8.3. The effect of successive operations of rotation and reflection on a molecule having C_{3v} symmetry

carried out: $\sigma_{va}C_3 = \sigma_{vb}$; $C_3\sigma_{va} = \sigma_{vc}$, The group multiplication table for the C_{3v} group is given in table 8.3.

TABLE 8.3. The group multiplication table for C_{3v}

2nd operation	1st operation					
	I	C_3	C_3^2	σ_{va}	σ_{vb}	σ_{vc}
I	I	C_3	C_3^2	σ_{va}	σ_{vb}	σ_{vc}
C_3	C_3	C_3^2	I	σ_{vc}	σ_{va}	σ_{vb}
C_3^2	C_3^2	I	C_3	σ_{vb}	σ_{vc}	σ_{va}
σ_{va}	σ_{va}	σ_{vb}	σ_{vc}	I	C_3	C_3^2
σ_{vb}	σ_{vb}	σ_{vc}	σ_{va}	C_3^2	I	C_3
σ_{vc}	σ_{vc}	σ_{va}	σ_{vb}	C_3	C_3^2	I

It is seen that each reflection operation is its own reciprocal, but C_3^2 is the reciprocal of C_3 and vice versa. I, C_3 and C_3^2 are a sub-group of order 3.

It can be shown that the C_{3v} group has three classes of operation. The identity is a class, C_3 and C_3^2 are another class and σ_{va}, σ_{vb} and σ_{vc} a third class. We shall prove that the two rotations are a class by examining their similarity transformations (8.1) with all the elements of the group.

$$I^{-1}C_3I = IC_3I = C_3, \qquad \sigma_{va}^{-1}C_3\sigma_{va} = \sigma_{va}C_3\sigma_{va} = C_3^2$$

$$C_3^{-1}C_3C_3 = C_3^2C_3C_3 = C_3, \qquad \sigma_{vb}^{-1}C_3\sigma_{vb} = \sigma_{vb}C_3\sigma_{vb} = C_3^2$$

$$(C_3^2)^{-1}C_3C_3^2 = C_3C_3C_3^2 = C_3, \qquad \sigma_{vc}^{-1}C_3\sigma_{vc} = \sigma_{vc}C_3\sigma_{vc} = C_3^2.$$

Likewise, the similarity transformation of C_3^2 by I, C_3 and C_3^2 gives C_3^2, and by σ_{va}, σ_{vb} and σ_{vc} gives C_3. Thus C_3 and C_3^2 are members of the same class. In exactly the same manner the three reflections can be shown to belong to the same class. The three classes of the $\mathbf{C}_{3v}$ symmetry group are thus the three different types of operation that can be performed on the figure, and this is a general rule. *Two operations belong to the same class if one may be replaced by the other in a new coordinate system which is accessible by a symmetry operation.*

The first stage in using symmetry group theory in a molecular problem is to establish the symmetry group to which the molecule belongs, or more precisely to which its symmetry operations belong.

The following rules will enable this to be done.

1. There are three special groups for molecules which do not have a unique axis of high symmetry. These are the group of operations of the tetrahedron, $\mathbf{T}_d$ (e.g. CH_4), the octahedron, $\mathbf{O}_h$ (e.g. SF_6), and the icosahedron, $\mathbf{I}_h$ (which occurs in some boron compounds).

2. If a molecule does not belong to any of the special groups, we look for an axis of symmetry. If there is none but the molecule has a plane of symmetry then it belongs to the group $\mathbf{C}_s$ (e.g. quinoline).

3. If the molecule possesses one or more axes of symmetry, the group is labelled by the order n of the highest symmetry axis (the principal axis). There may be no unique principal axis but instead three mutually perpendicular C_2 axes (e.g. in naphthalene). In this case one of these axes is arbitrarily chosen to be the principal axis.

4. If the molecule has the symmetry operation S_n (see table 8.1), where n is twice the order of the principal axis, but there are no other symmetry elements except perhaps an inversion, then the group is $\mathbf{S}_n$ (e.g. $\mathbf{S}_4$: tetraphenylmethane).

5. We next look for n two-fold symmetry axes perpendicular to the principal n-fold axis. If these are present then the group is a **D** group, if not it is a **C** group.

6. The **C** groups are of three types: $\mathbf{C}_n$ has no symmetry planes ($\mathbf{C}_1$ is any molecule with no symmetry), $\mathbf{C}_{nv}$ has vertical symmetry planes ($\mathbf{C}_{3v}$: ammonia), $\mathbf{C}_{nh}$ has a horizontal plane ($\mathbf{C}_{2h}$: *trans*-dibromoethylene).

7. The **D** groups are also of three kinds: $\mathbf{D}_n$ has no symmetry planes ($\mathbf{D}_2$: twisted ethylene), $\mathbf{D}_{nh}$ has a horizontal symmetry plane ($\mathbf{D}_{6h}$: benzene), $\mathbf{D}_{nd}$ has no horizontal plane but n dihedral planes ($\mathbf{D}_{2d}$: allene).

8. For linear molecules the nuclear axis is an infinite rotation axis. These molecules belong to infinite groups which are of two kinds: $\mathbf{C}_{\infty v}$ those without a horizontal symmetry plane (e.g. HCl), and $\mathbf{D}_{\infty h}$ those with a horizontal symmetry plane (e.g. CO_2).

One example of the use of these rules is shown in figure 8.4, and problem 8.1 gives some practice in identifying other molecular symmetry groups.

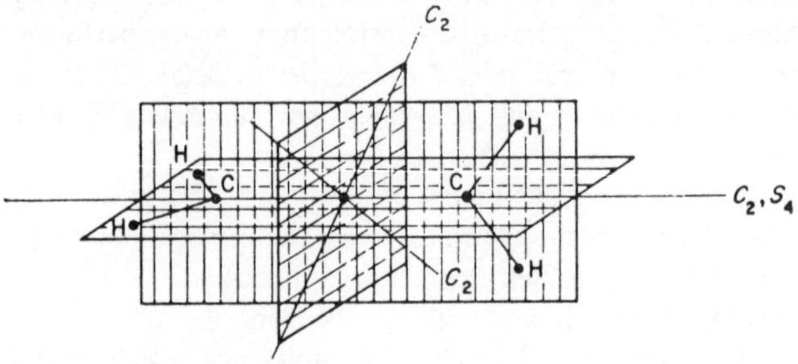

FIGURE 8.4. The symmetry operations for allene: the D_{2d} group

8.2 Representations of groups

We saw, at the beginning of this chapter, that a molecular orbital of formaldehyde must be unchanged or changed in sign by a symmetry operation of the C_{2v} group. Using the convention of operator algebra we can write, for example (see 6.5),

$$C_2\psi = +1\psi \quad \text{or} \quad C_2\psi = -1\psi$$

so that ψ is an eigenfunction of C_2 with eigenvalue $+1$ or -1. If we have a molecular orbital for which

$$I\psi = +1\psi \qquad \sigma_v\psi = +1\psi$$

$$C_2\psi = +1\psi \qquad \sigma_v'\psi = +1\psi$$

then the set of four numbers $+1$, $+1$, $+1$, $+1$ characterizes the symmetry of the orbital; in this case the orbital is said to be totally symmetric.

There are other sets of numbers which would characterize the symmetry of other orbitals. However, the group multiplication rules (table 8.2) do place some restriction on which sets are allowed. For example, since $\sigma_v'.\sigma_v = C_2$, then if an orbital is symmetric under σ_v and antisymmetric under σ_v' it must be antisymmetric under C_2. There are only four sets of numbers which satisfy the requirements of the C_{2v} group multiplication table, with the limitation that the number $+1$ is the eigenvalue of the identity operator; these are shown in table 8.4.

TABLE 8.4. Sets of numbers satisfying the C_{2v} multiplication table

I	C_2	σ_v	σ_v'
1	1	1	1
1	1	−1	−1
1	−1	1	−1
1	−1	−1	1

Although there are only four sets of numbers which satisfy the group multiplication table, there are also sets of matrices which satisfy it†. Two such sets are shown below

$$\begin{array}{lcccc} & I & C_2 & \sigma_v & \sigma_v' \\ \text{(a)} & \begin{pmatrix}1 & 0\\0 & 1\end{pmatrix} & \begin{pmatrix}-1 & 0\\0 & -1\end{pmatrix} & \begin{pmatrix}1 & 0\\0 & -1\end{pmatrix} & \begin{pmatrix}-1 & 0\\0 & 1\end{pmatrix}; \\ \text{(b)} & \begin{pmatrix}1 & 0\\0 & 1\end{pmatrix} & \begin{pmatrix}-1 & 0\\0 & -1\end{pmatrix} & \begin{pmatrix}0 & 1\\1 & 0\end{pmatrix} & \begin{pmatrix}0 & -1\\-1 & 0\end{pmatrix}. \end{array}$$

It is easy to check that these satisfy the group multiplication table. For example,

$$\begin{pmatrix}0 & 1\\1 & 0\end{pmatrix}\begin{pmatrix}0 & -1\\-1 & 0\end{pmatrix} = \begin{pmatrix}-1 & 0\\0 & -1\end{pmatrix},$$

which satisfies the requirement $\sigma_v . \sigma_v' = C_2$.

Any set of matrices which multiply together in such a way as to satisfy the group multiplication table is said to be a *representation* of a group. The sets of numbers given in table 8.4 show the special case where the matrices are of order 1.

The representation of a group is the sets of matrices which express mathematically the operations of a group on a function or set of functions. For example, the operations of a group on the totally-symmetric wave function are represented by $+1$ for all operations. Again, consider the two functions x and y (coordinates as in figure 8.1): under the identity operation the point (x, y) remains (x, y), under C_2 it becomes $(-x, -y)$, under σ_v

† Some elementary knowledge of matrix algebra is necessary for a full understanding of this chapter. The most essential requirement is to know how to multiply two matrices. If $\mathbf{C} = \mathbf{AB}$, then the element in the row r column s of matrix $\mathbf{C}$ is given by $c_{rs} = \sum_t a_{rt} b_{ts}$ where the summation is over each column t of $\mathbf{A}$ row t of $\mathbf{B}$.

it becomes $(x, -y)$ and under σ_v' it becomes $(-x, y)$. We can express this in matrix form as follows

$$I\begin{pmatrix} x \\ y \end{pmatrix} = \begin{pmatrix} 1 & 0 \\ 0 & 1 \end{pmatrix}\begin{pmatrix} x \\ y \end{pmatrix} = \begin{pmatrix} x \\ y \end{pmatrix}$$

$$C_2\begin{pmatrix} x \\ y \end{pmatrix} = \begin{pmatrix} -1 & 0 \\ 0 & -1 \end{pmatrix}\begin{pmatrix} x \\ y \end{pmatrix} = \begin{pmatrix} -x \\ -y \end{pmatrix}$$

$$\sigma_v\begin{pmatrix} x \\ y \end{pmatrix} = \begin{pmatrix} 1 & 0 \\ 0 & -1 \end{pmatrix}\begin{pmatrix} x \\ y \end{pmatrix} = \begin{pmatrix} x \\ -y \end{pmatrix}$$

$$\sigma_v'\begin{pmatrix} x \\ y \end{pmatrix} = \begin{pmatrix} -1 & 0 \\ 0 & 1 \end{pmatrix}\begin{pmatrix} x \\ y \end{pmatrix} = \begin{pmatrix} -x \\ y \end{pmatrix}.$$

The set of 2×2 matrices which we have here are then representations of the group (the representation (a) of the previous page). We say that the functions x and y are a *basis for a representation of the group* or *the functions transform according to the representation* (a).

Clearly the number of representations of the group that we can construct is limited only by our ingenuity in thinking up sets of functions which generate them. Any set of functions $f_1 \ldots f_n$, which is transformed by the group operations into other sets $f_1' \ldots f_n'$, where the functions f' can be written as linear combinations of the functions f_1, forms a basis for a representation of the group. This is an important theorem to which we will return in the next section.

Although each group has an infinite set of representations, there is a finite number which have a special significance and these are called *irreducible representations.* This is a term which we use rather often in this book and we think it convenient to give it the abbreviation *I.R.* (the italics should prevent any confusion with the abbreviation for infrared, which in any case we do not use in this book). It is useful, as we shall see later, to draw an analogy between a representation and a vector. In a finite space there are an infinite number of vectors but a finite number of orthogonal unit vectors; for a finite group there are an infinite number of representations but a finite number of *I.R.*'s. Just as one can break any vector down into component orthogonal unit vectors, so a representation can be broken down into component *I.R.*'s.

The method of breaking down a representation is by means of a similarity transformation (cf. 8.1). Suppose we have a set of matrices **A, B, C** ... which are a representation of the group; if there is a similarity transformation by a matrix **X** which converts these into matrices which have smaller sub-matrices on the diagonals and zeros elsewhere, for example

$$\mathbf{X}^{-1}\mathbf{A}\mathbf{X} = \mathbf{A}' = \begin{pmatrix} \mathbf{A}_1' & & & \\ & \mathbf{A}_2' & & \\ & & \mathbf{A}_3' & \\ & & & \mathbf{A}_4' \end{pmatrix}, \tag{8.4}$$

and the sub-matrices $\mathbf{A}_k'$, $\mathbf{B}_k'$, $\mathbf{C}_k'$ ... are of the same order (but $\mathbf{A}_1'$, $\mathbf{A}_2'$ ... are not necessarily of the same order), then the representation is said to be reducible.

Suppose a requirement of the group multiplication table is

$$\mathbf{A}.\mathbf{B}. = \mathbf{C}, \tag{8.5}$$

then

$$\mathbf{X}^{-1}\mathbf{A}\mathbf{X}\mathbf{X}^{-1}\mathbf{B}\mathbf{X} = \mathbf{X}^{-1}\mathbf{C}\mathbf{X}, \tag{8.6}$$

hence

$$\mathbf{A}'.\mathbf{B}' = \mathbf{C}'. \tag{8.7}$$

But from the rules of matrix multiplication

$$\mathbf{A}_k'.\mathbf{B}_k' = \mathbf{C}_k', \tag{8.8}$$

hence each set of sub-matrices $\mathbf{A}_k'$, $\mathbf{B}_k'$, $\mathbf{C}_k'$ ... satisfies the requirements of the group multiplication table and is therefore a representation of the group.

If it is not possible to find a similarity transformation which reduces all of the matrices in a representation in the above manner, then the representation is said to be irreducible. It follows that any representation can be reduced to the *I.R.*'s of the group.

We shall now list, without proof, some important properties of the *I.R.*'s of a group†.

1. The sum of the squares of the dimensions of the *I.R.*'s of a group is equal to the order of the group h.

2. The sum of the squares of the characters of the matrices in any *I.R.* is equal to h. The character (or trace) of a matrix is the sum of its diagonal elements, and is given the symbol χ. Thus

$$\sum_R [\chi_i(R)]^2 = h, \tag{8.9}$$

† For complete proofs the reader must go to a fairly advanced text on group theory: Cotton (see bibliography) gives a good account of the properties but does not give complete proofs.

where i labels the *I.R.* and $\chi(R)$ is the character of the matrix appropriate to the symmetry operation R.

3. The characters of the *I.R.*'s can be taken to be components of vectors in a space of dimensions h. The vectors associated with two different *I.R.*'s are orthogonal†. Combining this property with (8.9) we have

$$\sum_R \chi_i(R)\chi_j(R) = h\delta_{ij}. \tag{8.10}$$

4. The characters of matrices associated with operations of the same class are the same if these matrices belong to the same representation.

5. The number of *I.R.*'s of a group is equal to the number of classes in the group.

Let us start with the C_{2v} group as our first example. It has four elements and four classes (page 91), hence it must have four *I.R.*'s of dimension 1. These are the sets of numbers given in table 8.4. These numbers, which are also the characters of the *I.R.*'s if we consider them as being 1 × 1 matrices, are set out in what is called the character table of the group.

TABLE 8.5. The C_{2v} character table

C_{2v}	I	C_2	σ_v	σ_v'
A_1	1	1	1	1
A_2	1	1	−1	−1
B_1	1	−1	1	−1
B_2	1	−1	−1	1

Each *I.R.* is given a symbol, indicated on the left of the table. These symbols are standard and the following are general rules on nomenclature.

1. A and B are used to label *I.R.*'s of dimension 1, E those of dimension 2, and T those of dimension 3 (sometimes F has been used instead of T).

2. A is used when the character of the highest rotation operation is +1, and B when it is −1.

3. If the molecule has a centre of symmetry a subscript g is used if the character of the inversion operation is + 1, and a subscript u if it is − 1.

4. There is always one *I.R.* which consists of +1 for each operation. This is the totally-symmetric *I.R.* and is labelled A (A_g if there is a centre of symmetry); or if there is more than one A species it is A_1, A' or A_{1g}.

† More generally if $\Gamma_i(R)_{mn}$ is the element in the mth row nth column of a matrix in the ith *I.R.*, then the sets $\Gamma_i(R)_{mn}$, one element for each operation R, behave as the components of a complete set of orthogonal vectors in a space of dimensions h. We shall not, however, need to use this more general result.

One frequently finds in the character table a note of the simple functions which form a basis for the *I.R.*'s. Thus in the $\mathbf{C}_{2v}$ group z forms a basis for A_1, x for B_1, y for B_2, xy for A_2 etc. (the axes being defined as in figure 8.1).

We return now to the $\mathbf{C}_{3v}$ group whose multiplication rules were given in table 8.3. This group is of order six and has three classes. It follows that there must be three *I.R.*'s, two one-dimensional and one two-dimensional: $1^2 + 1^2 + 2^2 = 6$.

These three *I.R.*'s are shown below (check the result stated in the footnote on page 98).

I	C_3	$C_3^{\,2}$	σ_{va}	σ_{vb}	σ_{vc}
1	1	1	1	1	1
1	1	1	−1	−1	−1
$\begin{pmatrix}1 & 0\\ 0 & 1\end{pmatrix}$	$\begin{pmatrix}-\frac{1}{2} & \frac{\sqrt{3}}{2}\\ -\frac{\sqrt{3}}{2} & -\frac{1}{2}\end{pmatrix}$	$\begin{pmatrix}-\frac{1}{2} & -\frac{\sqrt{3}}{2}\\ \frac{\sqrt{3}}{2} & -\frac{1}{2}\end{pmatrix}$	$\begin{pmatrix}-1 & 0\\ 0 & 1\end{pmatrix}$	$\begin{pmatrix}\frac{1}{2} & -\frac{\sqrt{3}}{2}\\ -\frac{\sqrt{3}}{2} & -\frac{1}{2}\end{pmatrix}$	$\begin{pmatrix}\frac{1}{2} & \frac{\sqrt{3}}{2}\\ \frac{\sqrt{3}}{2} & -\frac{1}{2}\end{pmatrix}$

The characters of the two-dimensional *I.R.* are

$$2 \quad -1 \quad -1 \quad 0 \quad 0 \quad 0$$

It should be noted that rules 3 and 4 (page 98) hold for these three sets of characters. The character table of the group only lists the appropriate character for each class of operation, as illustrated in table 8.6.

TABLE 8.6. The character table for the group $\mathbf{C}_{3v}$

$\mathbf{C}_{3v}$	I	$2C_3$	$3\sigma_v$
A_1	1	1	1
A_2	1	1	−1
E	2	−1	0

We shall not list the character tables for symmetry groups unless they are used in this book. Complete lists may be found in the books mentioned in the bibliography at the end of this chapter.

8.3 Uses of group theory in valence problems

At the beginning of this chapter we showed that the molecular orbitals of formaldehyde must obey certain conditions when subjected to the symmetry operations of the nuclear framework, in order that the electron density at equivalent points in the molecule should be identical. We shall now look at the symmetry properties of wave functions in more detail and show how the results we have obtained in the previous sections may be used to simplify problems of valency.

Firstly, what is meant by the symmetry operations of a molecule? We have so far discussed symmetry operations using terms which are appropriate to the molecule being a solid figure, or even a set of points in space. But a molecule is neither of these things, but is an assembly of electrons and nuclei whose positions in space are described only in terms of probability densities. The symmetry of a wave function must clearly be examined in the light of the symmetry of the Hamiltonian of which the wave function is an eigenfunction. In other words, the symmetry of a wave function is determined by those symmetry operations for which the Hamiltonian is invariant.

Let us again illustrate this by the example of formaldehyde. Electronic wave functions are defined within the Born–Oppenheimer approximation (section 7.2), that is, in terms of fixed positions of the nuclei. If we take the nuclei in their equilibrium configuration then an interchange of the two protons will not change the potential field acting on the electrons, so that such an interchange will leave the electronic Hamiltonian invariant. In general, any symmetry operation acting on the figure defined by the position of the nuclei will leave the Hamiltonian invariant. If $\mathscr{H}$ is invariant under a symmetry operation R then $\mathscr{H}$ and R commute. It follows that if we have an eigenfunction Ψ of $\mathscr{H}$

$$\mathscr{H}\Psi_k = E_k\Psi_k, \tag{8.11}$$

then

$$R\mathscr{H}\Psi_k = RE_k\Psi_k, \tag{8.12}$$

so that

$$\mathscr{H}(R\Psi_k) = E_k(R\Psi_k). \tag{8.13}$$

In other words, the function $\Psi_k' = R\Psi_k$ is an eigenfunction of $\mathscr{H}$ with the same eigenvalue as Ψ_k. But we know that this is only possible if Ψ_k' is a linear combination of a degenerate set of functions all with energy E_k (page 69). If the set is non-degenerate then $\Psi_k' = \pm\Psi_k$ for normalized functions. The operation of R on any member of a set of degenerate eigenfunctions will therefore send it into a linear combination of members of the set, or, the set of degenerate eigenfunctions is a basis for a representation of the group (page 96). More specifically, however, the set is a basis for an *I.R.* of the group. If this were not so then we could take linear combinations of the set and subdivide these into smaller sub-sets which would be bases for *I.R.*'s of the group. But there would then be no symmetry requirement indicating why members of different sub-sets should be degenerate. In other words, providing that there are no degeneracies other than those required by the symmetry properties of the Hamiltonian (no accidental degeneracies), then a degenerate set of eigenfunctions must form a basis for an *I.R.* of the group. An alternative way of saying this is that a set of eigenfunctions, each having the same energy, transforms like an *I.R.* of the group.

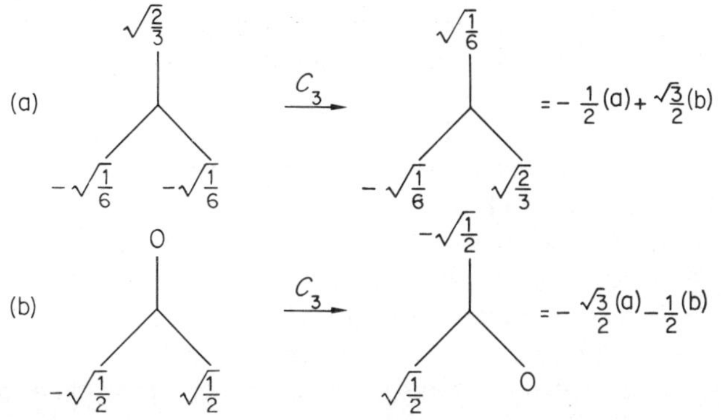

FIGURE 8.5. The effect of a rotation on a pair of functions of species *E* of the C_{3v} group

As an example of this we take two wave functions for a molecule having C_{3v} symmetry, as shown in figure 8.5. The coefficients represent the weighting of the wave functions about three equivalent points in the molecule. It is seen that rotations by $2\pi/3$ or $4\pi/3$ send these functions into mixtures of (a) and (b). If (a) and (b) are degenerate eigenfunctions then these linear combinations are also eigenfunctions, and they satisfy the symmetry requirements of the C_{3v} group as far as the rotation operators are concerned. It can be shown that they satisfy the requirements for any other operation, that is, they form a basis for a representation of the group. From the characters of this representation it can be shown that it is an *I.R.* of species *E*; we shall return to this point later.

This result has important implications for the construction of molecular wave functions. We pointed out in chapter 6, that when an exact solution of the Schrödinger equation is not possible, and this is the case for all many-particle systems, one usually seeks an approximate solution by taking a linear combination of a set of functions which is complete for the expansion of the wave function. Because eigenfunctions of the Hamiltonian transform like *I.R.*'s of the group, it is convenient to take the basis set of expansion functions in such a form that they also transform like *I.R.*'s of the group. The eigenfunctions of the Hamiltonian will then be a linear combination of the expansion functions which transform as the same *I.R.* as the eigenfunction. This result will be used when we come to consider the form of the molecular orbitals of H_2O (chapter 12), and of an octahedral complex (chapter 13).

In order to use the technique described above one has to be able to identify the *I.R.*'s to which the individual members of the basis set belong, or, if they do not transform individually as an *I.R.* of the group, then they must be combined together to give new functions which do. The latter is not always a trivial matter, and the technique by which it is done will now be described.

We first start with a set of functions which are transformed into one another by the operations of the group. These form a basis for a representation of the group, and we must find the component *I.R.*'s of this representation. To find these we need to know only the characters of the representation, and not (thankfully) the full matrix form of the representation, or the form of the matrix which leads to its reduction. The first step then is to find the characters of the representation.

Suppose we have two functions f_1 and f_2, and an operation of the group R transforms these into new functions f_1' and f_2', where

$$f_1' = af_1 + bf_2, \qquad f_2' = cf_1 + df_2,$$

then we can write the effect of this operation in matrix form

$$\begin{pmatrix} a & b \\ c & d \end{pmatrix}\begin{pmatrix} f_1 \\ f_2 \end{pmatrix} = \begin{pmatrix} f_1' \\ f_2' \end{pmatrix},$$

and the matrix

$$\begin{pmatrix} a & b \\ c & d \end{pmatrix}$$

is the component of the representation of the group associated with the operation R which has the functions f_1 and f_2 as basis.

The character of this matrix is $a + d$; a is the amount of f_1 in f_1' that is left after the operation, d is the amount of f_2 in f_2' that is left after the operation. To find the character of the representation associated with a particular operation we therefore ask how much of each original function is left after the operation and sum up these residues.

Let us take an example from the $\mathbf{C}_{3v}$ group. Suppose we are forming molecular orbitals of NH_3. A first step is to take linear combinations of the hydrogen 1*s* atomic orbitals which transform like *I.R.*'s of the group. We label these wave functions ϕ_a, ϕ_b and ϕ_c as in figure 8.3. To find the characters of the representation having ϕ_a, ϕ_b and ϕ_c as basis we ask how many of these functions are unchanged by each symmetry operation (since an operation either leaves a function unchanged or changes it completely into another function). For the identity operator all three are unchanged; for a C_3 rotation all three are changed completely into another function; for each reflection there is one function that is unchanged (e.g. σ_{va} leaves ϕ_a unchanged but interchanges ϕ_b and ϕ_c.) The character of the representation, which we label Γ, is then

I	$2C_3$	$3\sigma_v$
$\Gamma = 3$	0	1

In a case for which the functions are not just unchanged or completely changed into another function by the operation of the group it is sometimes difficult to evaluate even the characters of the representation. In chapter 13 a useful formula will be derived which gives the characters of any rotation operator for a basis set of spherical harmonic functions. Using this formula one can easily evaluate the character of any representation derived from a basis set of atomic orbitals centred on a point which is invariant under any operation of the group.

We now want to find the component *I.R.*'s of this representation. The reason why we only need the characters of the representation to find the component *I.R.*'s is that the character of a matrix is unchanged by a similarity transformation (8.4)†. From this it follows that the character of a representation associated with a particular symmetry operation R is equal to the sum of the characters of the component *I.R.*'s of that operation. That is

$$\chi(R) = \sum_i a_i \chi_i(R), \tag{8.14}$$

where i labels a component *I.R.* of the representation and a_i is the number of times that component contributes to the total representation; a_i is always a positive integer.

Let us now return to the example of NH_3 which we are considering. By inspection of table 8.6 one can see that the only combination of the characters of the *I.R.*'s which give the set Γ is $A_1 + E$.

In a more complicated case it may be difficult to see by inspection how to compound the characters of the *I.R.*'s to get the characters of the representation. There is, however, a simple method for doing this which is based on the fact that the characters of the *I.R.*'s can be considered as a set of orthogonal vectors whose length is equal to $h^{\frac{1}{2}}$ (8.10). Taking the vector defined by the characters of the representation, it is a straightforward

† This is easily proved by the rules of matrix multiplication.

matter to find the components of the orthogonal vectors which are the characters of the *I.R.*'s, using the scalar product rule. That is, if

$$\chi(R) = \sum_i a_i \chi_i(R), \tag{8.15}$$

then multiplying each side by $\chi_j(R)$ and summing over R we have

$$\sum_R \chi(R)\chi_j(R) = \sum_i \sum_R a_i \chi_i(R)\chi_j(R). \tag{8.16}$$

But from (8.10)

$$\sum_R \chi_i(R)\chi_j(R) = h\delta_{ij}, \tag{8.17}$$

hence

$$\sum_R \chi(R)\chi_j(R) = ha_j. \tag{8.18}$$

In other words, if the set of characters $\chi(R)$ is taken to define a vector **P**, and the set $\chi_j(R)$ a vector $\mathbf{Q}_j$, then

$$a_j = \frac{1}{h}\mathbf{P}.\mathbf{Q}_j. \tag{8.19}$$

For the example we are considering we have

$$\begin{aligned} \Gamma.A_1 &= 3 \times 1 + 2(0 \times 1) + 3\,(1 \times 1) = 6 \\ \Gamma.A_2 &= 3 \times 1 + 2(0 \times 1) - 3\,(1 \times 1) = 0 \\ \Gamma.E \ &= 3 \times 2 - 2(0 \times 1) + 3\,(1 \times 0) = 6. \end{aligned}$$

This establishes the result, obtained previously by inspection, that $\Gamma = A_1 + E$.

The next step in the analysis is to find the form of the linear combinations of basis functions which transform like *I.R.*'s of the group. This may, in a simple case, be obvious by inspection of the character table. For example, what combination of the hydrogen 1*s* orbitals of formaldehyde transfers like *I.R.*'s of the $\mathbf{C}_{2v}$ group? If ϕ_1 and ϕ_2 are the two atomic orbitals then by the method described above it can easily be shown that these two functions are a basis for a representation which has components $A_1 + B_2$. Now A_1 is a totally-symmetric function and it follows that the combination transforming as the A_1 *I.R.* is $\phi_1 + \phi_2$. For all operations of the group we have $R\,(\phi_1 + \phi_2) = +1(\phi_1 + \phi_2)$. The combination transforming as B_2 must be antisymmetric under C_2 and σ_v, and therefore is $\phi_1 - \phi_2$. These two combinations can be called symmetry-adapted functions.

The general method of generating functions which transform as *I.R.*'s of the group is as follows. Take one function ϕ in the basis set and operate on it with each of the operations of the group; multiply the result of each operation by the appropriate character of the *I.R.* and sum the result. That is,

$$\psi_t = \sum_R \chi_t(R)(R\phi), \tag{8.20}$$

where ψ_t is the symmetry-adapted function which transforms like the *I.R.* labelled t. This method gives un-normalized functions.

We shall prove this result for a non-degenerate *I.R.* We can always write any of the basis functions ϕ in terms of the set of symmetry-adapted functions ψ to which they give rise

$$\phi = \sum_s b_s\psi_s. \tag{8.21}$$

It follows that

$$R\phi = \sum_s b_s R\psi_s = \sum_s b_s\chi_s(R)\psi_s. \tag{8.22}$$

Multiplying both sides by $\chi_t(R)$ and summing over all R gives

$$\sum_R \chi_t(R)(R\phi) = \sum_R \sum_s b_s\chi_t(R)\chi_s(R)\psi_s. \tag{8.23}$$

But from the orthogonality of the characters of different *I.R.*'s the only term retained from the summation over s is that for which $s = t$, hence

$$\sum_R \chi_t(R)(R\phi) = \sum_R b_t\chi_t(R)\chi_t(R)\psi_t = b_t h\psi_t, \tag{8.24}$$

which gives expression (8.20) apart from a normalizing constant.

Expression (8.20) holds also for the components of a degenerate *I.R.*†, but in this case, to get the components of a k-fold degenerate set of functions, it is necessary to carry out the procedure k times with different starting functions ϕ. The k functions which are generated in this way are not, however, generally orthogonal to one another, so that a further stage of orthogonalization is usually necessary.

To illustrate the technique we return to the NH_3 problem that we have already considered.

	I	C_3	C_3^2	σ_{va}	σ_{vb}	σ_{vc}
(1) operating on ϕ_a:	ϕ_a	ϕ_b	ϕ_c	ϕ_a	ϕ_c	ϕ_b
(2) operating on ϕ_b:	ϕ_b	ϕ_c	ϕ_a	ϕ_c	ϕ_b	ϕ_a.

Multiplying the first row by the characters of the A_1 *I.R.* and adding, we have

$$\phi_a + \phi_b + \phi_c + \phi_a + \phi_c + \phi_b.$$

† For a proof see EYRING, WALTER and KIMBALL, *Quantum Chemistry*, p.189, Wiley, 1944.

After normalization (neglecting overlap) this gives the symmetry-adapted function

$$\psi(A_1) = \sqrt{\tfrac{1}{3}}(\phi_a + \phi_b + \phi_c).$$

It should be noted that the same result is obtained if one works with the second row. For the symmetry-adapted function of type E we need two components. Multiplying the first row by the characters of the E *I.R.* and adding, we get

$$2\phi_a - \phi_b - \phi_c,$$

which after normalization gives

$$\psi_1(E) = \sqrt{\tfrac{1}{6}}(2\phi_a - \phi_b - \phi_c).$$

Carrying out the same procedure for the second row we get a similar function

$$\psi_2(E) = \sqrt{\tfrac{1}{6}}(2\phi_b - \phi_c - \phi_a).$$

However, ψ_1 and ψ_2 are not orthogonal. One method of obtaining orthogonal functions (known as the Schmidt procedure) is to take new functions,

$$\psi_1 \quad \text{and} \quad \psi_2' = \psi_2 + c\psi_1, \tag{8.25}$$

such that ψ_1 and ψ_2' are orthogonal. That is,

$$\int \psi_1\psi_2'\,dv = \int \psi_1\psi_2\,dv + c\int \psi_1\psi_1\,dv = 0. \tag{8.26}$$

This condition is fulfilled if

$$c = -\int \psi_1\psi_2\,dv. \tag{8.27}$$

In the example we are considering

$$\int \psi_1\psi_2\,dv = \frac{1}{6}\int (\phi_c^2 - 2\phi_a^2 - 2\phi_b^2)\,dv = -\tfrac{1}{2},$$

so that

$$\psi_2'(E) = \sqrt{\tfrac{1}{6}}(2\phi_b - \phi_c - \phi_a) + (\tfrac{1}{2})\sqrt{\tfrac{1}{6}}(2\phi_a - \phi_b - \phi_c).$$

After normalization

$$\psi_2'(E) = \sqrt{\tfrac{1}{2}}(\phi_b - \phi_c).$$

This pair of functions, $\psi_1(E)$ and $\psi_2'(E)$, were shown by the discussion following figure 8.5 to satisfy the symmetry requirements of the $\mathbf{C}_{3v}$ group.

Another problem that arises frequently in quantum mechanics is the following. Suppose we have two sets of functions $f_1 \ldots f_m$ and $g_1 \ldots g_n$, where the functions f are a basis for a representation of the group with characters $\chi_f(R)$ and the functions g are a basis for a representation with characters $\chi_g(R)$. The set of functions $f_i g_j$ ($m \times n$ functions in all) must also form a basis for a representation of the group; what are its characters? Now

$$Rf_i = \sum_k a_{ik} f_k; \qquad Rg_j = \sum_l b_{jl} g_l, \tag{8.28}$$

hence

$$R(f_i g_j) = \sum_k \sum_l a_{ik} b_{jl} f_k g_l = \sum_k \sum_l c_{ij,kl} f_k g_l. \tag{8.29}$$

It follows that

$$\chi_{fg}(R) = \sum_i \sum_j c_{ij,ij} = \sum_i a_{ii} \sum_j b_{jj} = \chi_f(R)\chi_g(R). \tag{8.30}$$

The set of functions fg is called the direct product of the sets f and g. The characters of the representation of the direct product are equal to the product of the corresponding characters of the component representations.

If two sets of functions transform like *I.R.*'s of the group then their direct product will, in general, be the basis of a reducible representation, but this can be easily split up into its component *I.R.*'s according to the method given earlier. Let us take some examples from the $\mathbf{C}_{3v}$ group

$\mathbf{C}_{3v}$	I	$2C_3$	$3\sigma_v$	
A_1	1	1	1	
A_2	1	1	-1	
E	2	-1	0	
$A_1.A_2$	1	1	-1	$= A_2$
$A_2.A_2$	1	1	1	$= A_1$
$A_2.E$	2	-1	0	$= E$
$E.E$	4	1	0	$= E + A_1 + A_2$

Note that A_1, the totally-symmetric *I.R.*, only appears from the direct product of functions transforming as the same *I.R.* This is a general result for the symmetry point groups.

Consider the integral over all space

$$\int F \, dv. \tag{8.31}$$

Now the function F can be compounded from functions which transform as *I.R.*'s of the group. This integral will be zero unless there is a component of F which transforms as the totally-symmetric *I.R.*; for all other components the positive and negative regions will cancel in the integration over all space.

We now come to two integrals which are of importance in quantum mechanics

$$\int \Theta_a \Theta_b \, dv, \qquad \int \Theta_a \mathscr{B} \Theta_b \, dv. \tag{8.32}$$

The first will be zero unless Θ_a and Θ_b transform as the same *I.R.* of the group, otherwise their direct product cannot give rise to a representation which contains the totally-symmetric *I.R.* Similarly, for the second integral it is necessary that the direct product of Θ_a and Θ_b gives a representation which contains the component *I.R.* having the same transformation properties as the operator $\mathscr{B}$. For example, if in a molecule having C_{3v} symmetry we introduce a perturbation transforming as A_2, then this perturbation will cause an interaction between two states of the unperturbed molecule if one state transforms as A_1 and the other as A_2, or if both transform as E.

In the remaining chapters of this book we shall make use of group theory in two ways. Firstly, we shall use group theory symbols to label wave functions where this is illuminating. Secondly, group theory will be used to simplify the mathematics involved in calculating wave functions and predicting how these functions will behave under the influence of a perturbation. It is important to realize that the solution to these problems can be obtained without using group theory, but in a situation where there is high symmetry, such as an octahedral transition metal complex, the mathematical simplification that is obtained by using group theory can be very great.

Bibliography

EYRING, WALTER and KIMBALL, *Quantum Chemistry*, chapter 10 and appendix 7, Wiley, 1944.

COTTON, *Chemical Applications of Group Theory*, Interscience, 1963.

WILSON, DECIUS and CROSS, *Molecular Vibrations*, McGraw-Hill, 1955.

Problems 8

8.1 Determine the symmetry groups of the following molecules: CH_2Cl_2, *trans*-CHCl=CHCl, C_2H_4, IF_7 (pentagonal bipyramid), C_6H_5Cl, HCN, CO_2.

8.2 Construct the group multiplication table for the C_{2h} group.

8.3 Construct the character table for the C_{2h} group and label the *I.R.*'s.

8.4 What are the component *I.R.*'s of the following reducible representations of the C_{2v} group?

	I	C_2	σ_v	σ_v'
(a)	4	0	0	0
(b)	5	−1	−3	−1

8.5 What are the characters of the reducible representations which have as basis
(a) The hydrogen 1*s* orbitals
(b) The chlorine 2*p* orbitals
of *trans*-dichloroethylene (see answer to problem 8.2)?
Resolve these into their component *I.R.*'s.

8.6 What linear combinations of hydrogen 1*s* orbitals transform like *I.R.*'s of the groups of the following molecules?
C_2H_4 (see table 14.1)
CH_4 (see table 13.11).

8.7 What are the direct products of the *I.R.*'s of the group C_{2h} (see answer to problem 8.3)? What states will interact with each other under the influence of an operator transforming like the A_u *I.R.*?

Chapter 9

Angular momentum and atomic energy levels

9.1 Angular momentum operators

We shall now examine the relationship between the quantum numbers l and m of atomic orbitals and the orbital angular momentum of the electron, and we shall see how electron spin is treated by quantum mechanics. This will lead on to a more detailed discussion of atomic energy levels than was given in chapters 3 and 4.

Classically, angular momentum about a point O is defined by the vector product

$$\mathbf{L} = \mathbf{r} \times \mathbf{p} \qquad \begin{aligned} L_x &= yp_z - zp_y \\ L_y &= zp_x - xp_z \\ L_z &= xp_y - yp_x, \end{aligned} \tag{9.1}$$

where $\mathbf{p}$ is the linear momentum of the particle..To obtain the observable values of angular momentum in quantum mechanics we need to know the appropriate angular momentum operators. We construct these using the rule given in chapter 6 (postulate 2), namely by replacing $\mathbf{p}$ by $(h/2\pi i)\partial/\partial q$. This gives

$$\mathbf{L}_x = \frac{h}{2\pi i}\left(y\frac{\partial}{\partial z} - z\frac{\partial}{\partial y}\right)$$

$$\mathbf{L}_y = \frac{h}{2\pi i}\left(z\frac{\partial}{\partial x} - x\frac{\partial}{\partial z}\right)$$

$$\mathbf{L}_z = \frac{h}{2\pi i}\left(x\frac{\partial}{\partial y} - y\frac{\partial}{\partial x}\right) \tag{9.2}$$

If we multiply these operators together we find that they do not commute, but that

$$\mathbf{L}_x\mathbf{L}_y - \mathbf{L}_y\mathbf{L}_x = \frac{ih}{2\pi}\mathbf{L}_z$$

$$\mathbf{L}_y\mathbf{L}_z - \mathbf{L}_z\mathbf{L}_y = \frac{ih}{2\pi}\mathbf{L}_x$$

$$\mathbf{L}_z\mathbf{L}_x - \mathbf{L}_x\mathbf{L}_z = \frac{ih}{2\pi}\mathbf{L}_y. \tag{9.3}$$

We shall also be interested in the operator which will give us the square of angular momentum

$$\mathbf{L}^2 = \mathbf{L}.\mathbf{L} = \mathbf{L}_x{}^2 + \mathbf{L}_y{}^2 + \mathbf{L}_z{}^2, \tag{9.4}$$

and we now find, using the definition (9.2), that $\mathbf{L}^2$ commutes with all the components of $\mathbf{L}$

$$\mathbf{L}^2\mathbf{L}_x - \mathbf{L}_x\mathbf{L}^2 = 0$$

$$\mathbf{L}^2\mathbf{L}_y - \mathbf{L}_y\mathbf{L}^2 = 0$$

$$\mathbf{L}^2\mathbf{L}_z - \mathbf{L}_z\mathbf{L}^2 = 0. \tag{9.5}$$

Since the components of orbital angular momentum do not commute, it is never possible to set up an experiment in which the values of $\mathbf{L}_x$, $\mathbf{L}_y$ and $\mathbf{L}_z$ are all well defined, that is, have definite values rather than just average values (see page 70). But $\mathbf{L}^2$ commutes with all the components of $\mathbf{L}$, so it is possible that the observable value of $\mathbf{L}^2$ and any *one* component can be well defined. This is the best that we can do. By convention it is usual to choose wave functions for an atom such that $\mathbf{L}_z$ is this component. If we perform an experiment to measure the value of $\mathbf{L}^2$ and $\mathbf{L}_z$, for example by applying a magnetic field in the z direction and looking at the absorption of light by the atom, then in our experiment the values of $\mathbf{L}_y$ and $\mathbf{L}_x$ are not definitely fixed. We can always calculate the average values of $\mathbf{L}_x$ and $\mathbf{L}_y$, but as the vector $\mathbf{L}$ is just as likely to lie along the $+x$ direction as along the $-x$ (and similarly for y) these average values will be zero.

Since $\mathbf{L}^2$ and $\mathbf{L}_z$ commute, the eigenfunctions of both operators can be chosen to be the same. These eigenfunctions F_{lm} will be defined by the equations

$$\mathbf{L}^2 F_{lm} = k_l F_{lm}$$

$$\mathbf{L}_z F_{lm} = k_m F_{lm}, \tag{9.6}$$

where k_l and k_m are constants, the eigenvalues of these operators. If these equations hold then $\mathbf{L}_x F_{lm} \neq k' F_{lm}$ and $\mathbf{L}_y F_{lm} \neq k' F_{lm}$, where k' is any constant (except for the special case $k_l = 0$).

If we now return to the specific expressions for the orbital angular momentum operators (9.2), and solve equations (9.6), it is found that the functions F_{lm} are the spherical harmonics Y_{lm}, defined in chapter 3, which give the angular variation of atomic orbitals. In addition it is found that the constants k_l and k_m are given by

$$k_l = l(l+1)\frac{h^2}{4\pi^2}$$

$$k_m = m\frac{h}{2\pi}, \tag{9.7}$$

where $l = 0, 1, 2 \ldots$ and for a given l, m can have the $2l+1$ values $l, l-1, \ldots 0, \ldots -(l-1), -l$. We therefore identify the integers l and m with the quantum numbers introduced in chapter 3, so that if an atomic orbital is written as a complex function (3.10) then an electron in this orbital has definite values of $\mathbf{L}^2$ and $\mathbf{L}_z$. If the real functions (3.12) are used then $\mathbf{L}^2$ has a definite value but $\mathbf{L}_z$ does not except for the cases where $m = 0$.

If $\mathbf{L}^2$ has the definite value $l(l+1)\,h^2/4\pi^2$, the modulus of the orbital angular momentum vector has the value $(h/2\pi)[l(l+1)]^{\frac{1}{2}}$. However, the direction of this vector is not fixed in space since this would require that the three components of $\mathbf{L}$ were simultaneously quantized.

It is possible to arrive at the eigenvalues of angular momentum without using the definitions (9.2) and solving equations (9.6). If the commutation relationships (9.3) and (9.5) are taken as a general definition of angular momentum operators in quantum mechanics, it can be shown from these alone that k_l and k_m are given by expressions (9.7). However, more generally, the constant l can be a member either of the set of integers

$$l = 0, 1, 2, \ldots,$$

or the set of half integers

$$l = \tfrac{1}{2}, \tfrac{3}{2}, \tfrac{5}{2}, \ldots.$$

For a given value of l, m can again run from l to $-l$ in unit steps. These results are true for any operators which satisfy the same commutation relationships as $\mathbf{L}^2$ and the components of $\mathbf{L}$, but for *orbital* angular momentum operators only *integral* values of l are allowed†.

We saw in chapter 4 that the electron behaves as a spinning particle and has an associated spin angular momentum. If it is assumed that the spin angular momentum operators obey the commutation relationships (9.3) and (9.5), then agreement with experiment is found if, for a single electron,

† These statements are proved in EYRING, WALTER and KIMBALL, *Quantum Chemistry*, chapter 3, Wiley, 1944.

$l = \frac{1}{2}$ is taken. If $l = \frac{1}{2}$ then $m = \frac{1}{2}$ or $-\frac{1}{2}$; that is, there are two spin states for the electron. It is conventional to label these two eigenfunctions α and β (page 83). Using the symbol **S** for the spin angular momentum operators we have, from (9.6) and (9.7),

$$\mathbf{S}^2\alpha = \tfrac{1}{2}(1+\tfrac{1}{2})\frac{h^2}{4\pi^2}\alpha; \qquad \mathbf{S}^2\beta = \tfrac{1}{2}(1+\tfrac{1}{2})\frac{h^2}{4\pi^2}\beta;$$

$$\mathbf{S}_z\alpha = \frac{1}{2}\frac{h}{2\pi}\alpha; \qquad \mathbf{S}_z\beta = -\frac{1}{2}\frac{h}{2\pi}\beta. \tag{9.8}$$

These spin functions are assumed to be orthonormal, that is,

$$\int \alpha^2\, ds = \int \beta^2\, ds = 1, \qquad \int \alpha\beta\, ds = 0. \tag{9.9}$$

Classically, for an electron in an orbit the magnetic moment μ is given by

$$\mu = \frac{-e}{2mc} \times \text{angular momentum}, \tag{9.10}$$

$(e/2mc)$ is called the magnetogyric ratio. This equation holds in quantum mechanics for orbital angular momentum; e.g. $\boldsymbol{\mu}_z(\text{orbital}) = (e/2mc)\mathbf{L}_z$. However, putting the electron spin angular momentum, $h/4\pi$, into this expression gives $\mu = eh/8\pi mc$, which is half the value deduced from the Stern–Gerlach experiments (and deduced also by Uhlenbeck and Goudsmit from the fine structure of spectral lines). For spin angular momentum we therefore have a magnetogyric ratio of e/mc, that is, the operators of spin magnetic moment are

$$\boldsymbol{\mu}_z\,(\text{spin}) = \frac{-e}{mc}\mathbf{S}_z, \quad \text{etc.} \tag{9.11}$$

9.2 *Atomic energy levels*

For each electron i in an atom we can define spin and orbital angular momentum operators,

$$\mathbf{L}_{ix}, \mathbf{L}_{iy}, \mathbf{L}_{iz}, \mathbf{L}_i^2; \qquad \mathbf{S}_{ix}, \mathbf{S}_{iy}, \mathbf{S}_{iz}, \mathbf{S}_i^2.$$

But for a many-electron atom it is the total angular momentum of all the electrons which is an observable, since the electrons cannot be distinguished one from the other. These total angular momentum operators are defined by

$$\mathbf{L}_x = \sum \mathbf{L}_{xi}, \qquad \mathbf{S}_x = \sum \mathbf{S}_{xi}, \quad \text{etc.} \tag{9.12}$$

and

$$\mathbf{L}^2 = \mathbf{L}_x{}^2 + \mathbf{L}_y{}^2 + \mathbf{L}_z{}^2, \quad \text{etc.} \tag{9.13}$$

In addition we can define an operator of the total spin and orbital angular momentum of the whole atom.

$$\mathbf{J} = \mathbf{L} + \mathbf{S} = \sum_i \mathbf{J}_i = \sum_i \mathbf{L}_i + \sum_i \mathbf{S}_i,$$

$$\mathbf{J}^2 = \mathbf{J}_x{}^2 + \mathbf{J}_y{}^2 + \mathbf{J}_z{}^2, \quad \text{etc.} \tag{9.14}$$

Each of these angular momenta must obey the quantum-mechanical laws given by (9.6) and (9.8). We shall use the symbols L, M_L, S, M_S, J and M_J for the quantum numbers associated with the eigenvalues of $\mathbf{L}^2$, $\mathbf{L}_z$, $\mathbf{S}^2$, $\mathbf{S}_z$, $\mathbf{J}^2$ and $\mathbf{J}_z$ respectively†. For example, since $\mathbf{J}^2$ and $\mathbf{J}_z$ commute, one can find the functions F_{JM_J} which satisfy

$$\mathbf{J}^2 F_{JM_J} = J(J+1)\frac{h^2}{4\pi^2} F_{JM_J}$$

$$\mathbf{J}_z F_{JM_J} = M_J \frac{h}{2\pi} F_{JM_J}, \tag{9.15}$$

where J is either integral or half integral and $M_J = J, J-1, \ldots, -J$.

The importance of these angular momentum operators in atomic spectroscopy rests in their commutation relationships with one another and with the Hamiltonian. If an operator commutes with the Hamiltonian then the wave functions of the system (the eigenfunctions of the Hamiltonian) can be chosen to be eigenfunctions of this operator. For example, if $\mathbf{L}^2$ commutes with $\mathscr{H}$ then the quantum number L can be used to label the wave functions; each wave function is associated with a definite value of L. If $\mathbf{L}^2$ does not commute with $\mathscr{H}$ then the wave functions do not have a definite value of L, and only the average value of the orbital angular momentum can be measured. We can say some useful things about the properties of a wave function if we know the values of its angular momentum quantum numbers, even though we may not know the exact form of the wave function.

The commutation relationships between the angular momentum operators and the Hamiltonian are shown in table 9.1. The table is built up using the following rules.

† The quantum number of $\mathbf{J}_z$ is usually just given the symbol M, but we use M_J just to emphasize its connection with the total angular momentum.

1. The spin and orbital operators operate on different parts of the wave function, hence all the L's commute with all the S's.

2. Because $\mathbf{J}_z = \mathbf{L}_z + \mathbf{S}_z$, $\mathbf{J}_z$ does not commute with $\mathbf{L}_x$, $\mathbf{S}_y$, etc.

3. Because $\mathbf{J}^2 = (\mathbf{L} + \mathbf{S})^2 = \mathbf{L}^2 + \mathbf{S}^2 + 2\mathbf{L}.\mathbf{S} = \mathbf{L}^2 + \mathbf{S}^2 + 2(\mathbf{L}_x\mathbf{S}_x + \mathbf{L}_y\mathbf{S}_y + \mathbf{L}_z\mathbf{S}_z)$ then $\mathbf{J}^2$ does not commute with the components of **L** and **S**, but it does commute with $\mathbf{L}^2$ and $\mathbf{S}^2$.

4. If the Hamiltonian does not contain terms which represent the coupling between the spin and orbital motion of the electrons (these terms will be discussed later), then the Hamiltonian commutes with all the total spin and total orbital angular momentum operators†.

TABLE 9.1. The commutation relationships between the angular momentum operators and a Hamiltonian which does not include spin-orbit coupling. A tick indicates that the operators in that row and column commute

	$\mathbf{L}_x$	$\mathbf{L}_y$	$\mathbf{L}_z$	$\mathbf{L}^2$	$\mathbf{S}_x$	$\mathbf{S}_y$	$\mathbf{S}_z$	$\mathbf{S}^2$	$\mathbf{J}_x$	$\mathbf{J}_y$	$\mathbf{J}_z$	$\mathbf{J}^2$	$\mathscr{H}$
$\mathbf{L}_x$	✓	·	·	✓	✓	✓	✓	✓	✓	·	·	·	✓
$\mathbf{L}_y$	·	✓	·	✓	✓	✓	✓	✓	·	✓	·	·	✓
$\mathbf{L}_z$	·	·	✓	✓	✓	✓	✓	✓	·	·	✓	·	✓
$\mathbf{L}^2$	✓	✓	✓	✓	✓	✓	✓	✓	✓	✓	✓	✓	✓
$\mathbf{S}_x$	✓	✓	✓	✓	✓	·	·	✓	✓	·	·	·	✓
$\mathbf{S}_y$	✓	✓	✓	✓	·	✓	·	✓	·	✓	·	·	✓
$\mathbf{S}_z$	✓	✓	✓	✓	·	·	✓	✓	·	·	✓	·	✓
$\mathbf{S}^2$	✓	✓	✓	✓	✓	✓	✓	✓	✓	✓	✓	✓	✓
$\mathbf{J}_x$	✓	·	·	✓	✓	·	·	✓	✓	·	·	✓	✓
$\mathbf{J}_y$	·	✓	·	✓	·	✓	·	✓	·	✓	·	✓	✓
$\mathbf{J}_z$	·	·	✓	✓	·	·	✓	✓	·	·	✓	✓	✓
$\mathbf{J}^2$	·	·	·	✓	·	·	·	✓	✓	✓	✓	✓	✓
$\mathscr{H}$	✓	✓	✓	✓	✓	✓	✓	✓	✓	✓	✓	✓	✓

The wave functions can be chosen to be eigenfunctions of any mutually commuting set of operators. For example, $\mathscr{H}$, $\mathbf{L}^2$, $\mathbf{L}_z$, $\mathbf{S}^2$, $\mathbf{S}_z$ and $\mathbf{J}_z$ all commute with one another and we can therefore choose wave functions which are simultaneously eigenfunctions of all these operators; L, M_L, S, M_S and M_J will all be quantum numbers which can be used to label the wave function. Alternatively we can choose eigenfunctions of $\mathscr{H}$, $\mathbf{L}^2$, $\mathbf{S}^2$, $\mathbf{J}^2$, and $\mathbf{J}_z$; L, S, J and M_J will then be quantum numbers. In practice the first of these so-called coupling schemes is easier to work with than the second, and is the one most commonly used. It is called the L–S or *Russell–Saunders scheme*.

† For a proof of this statement see EYRING, WALTER and KIMBALL, *Quantum Chemistry*, p. 134, Wiley, 1944.

Knowing the quantum numbers for the angular momenta of individual electrons, the question now arises of how to find the possible quantum numbers of the total angular momenta? The general rule is that if two angular momentum operators $\mathbf{M}_1^2$ and $\mathbf{M}_2^2$ commute ($\mathbf{M}$ stands for $\mathbf{L}$, $\mathbf{S}$ or $\mathbf{J}$), and if the eigenvalue of $\mathbf{M}_1^2$ is $K_1(K_1+1)h^2/4\pi^2$ and of $\mathbf{M}_2^2$ is $K_2(K_2+1)$ $h^2/4\pi^2$, then $\mathbf{M}^2(\mathbf{M}=\mathbf{M}_1+\mathbf{M}_2)$ has eigenvalues $K(K+1)h^2/4\pi^2$ where $K = K_1 + K_2, K_1 + K_2 - 1, \ldots$ down to $|K_1 - K_2|$. Following the vector addition rule, we project the smaller of the two K's on to the larger, as shown in figure 9.1 for the case $K_1 = 3$, $K_2 = 3/2$.

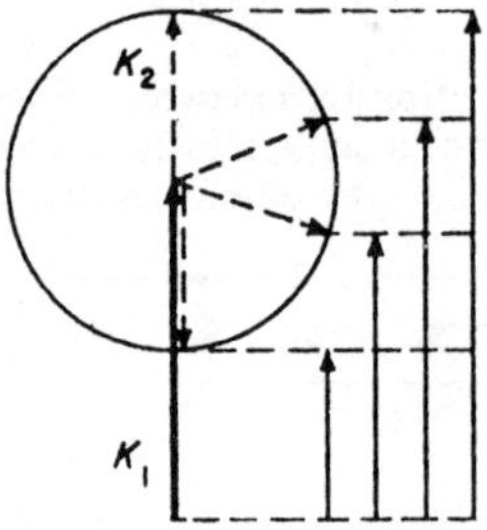

$K_1 = 3, K_2 = 3/2, K = 9/2, 7/2, 5/2, 3/2$

FIGURE 9.1. The vector rule for addition of angular momentum

Suppose we have two electrons, one in a $2p$ orbital and one in a $3p$ orbital. What possible values has the total orbital angular momentum? Electron 1 has $l_1 = 1$, electron 2 has $l_2 = 1$, the values of L are

$$l_1 + l_2 = 2, \qquad l_1 + l_2 - 1 = 1, \qquad l_1 + l_2 - 2 = 0\,(= l_1 - l_2).$$

By analogy with atomic orbitals we label an atomic state $S, P, D, F, \ldots$ if its total L value is $0, 1, 2, 3 \ldots$ respectively. In the above problem we therefore have D, P and S states from the two electrons.

Let us apply the rule to two electron spins. Electron 1 has $s_1 = \frac{1}{2}$, electron 2 has $s_2 = \frac{1}{2}$; the allowed values of S are $s_1 + s_2 = 1$, $s_1 + s_2 - 1 = 0$ $(= s_1 - s_2)$. If $S = 1$, then $M_S = 1, 0 -1$; this is called a *triplet spin state* because it has three spin components. If $S = 0$, then $M_S = 0$; this is a *singlet spin state*. The total *spin multiplicity*, $2S + 1$, of a state is placed as a superior prefix to the orbital angular momentum symbol. From the electron configuration $2p3p$ we therefore get the following

$${}^1D, {}^3D, {}^1P, {}^3P, {}^1S, {}^3S.$$

It is conventional to call these *terms*. Each term has $(2L + 1)$ $(2S + 1)$

components, each of which has a different pair of quantum numbers M_L, M_S.

These terms all have different energies because the repulsion between the electron in the 2*p* orbital and that in the 3*p* orbital is different in each case. There is a rule, Hund's rule, for predicting the most stable term in the case of the ground (lowest energy) configuration

1. It has the highest spin multiplicity.
2. If two terms have the same spin, the one with the larger value of L has the lower energy.

There is no exception to this rule for the ground configuration, but there are frequent exceptions for higher energy configurations. The energies of Russell–Saunders terms will be discussed more fully in section 9.6.

A complete shell of electrons (e.g. s^2, p^6, d^{10}) can only give a ^1S term. The terms for a configuration are therefore determined solely by the incomplete shells. If a shell can take a maximum of n electrons, then the terms arising from k electrons in this shell are the same as those arising from k holes ($n - k$ electrons). Thus p^4 gives rise to the same terms as p^2, d^3 the same as d^7, etc. (see problem 9.4).

We can define the possible values of J for each term. For example, the term ^3D has $S = 1$, L = 2. Possible J values are then $J = 2 + 1, 2 + 1 - 1, 2 - 1$, (3, 2, 1). If there is more than one J value for a given L and S, (which is the case unless either L or S are zero) then the J value is given as an inferior suffix thus: $^3D_3$, $^3D_2$, $^3D_1$. These are usually called *levels*. Each level has $2J + 1$ components with different values of M_J, and in the absence of an external field these are degenerate.

We will summarize the position so far. An atom having an open-shell configuration gives rise to a number of terms which differ in the way that the individual electron angular momenta are coupled together. These terms have different energies. If, for example, we excite Mg from its ground state configuration $^1S(1s^22s^22p^63s^2)$ to the first excited configuration $(1s^22s^22p^63s3p)$ there are not one but two regions of absorption in the spectrum, because the excited configuration gives the terms ^1P and ^3P. The transition $^1P \leftarrow {}^1S$ occurs at 35050 cm^{-1} and $^3P \leftarrow {}^1S$ in the region of 21900 cm^{-1}.

If the $^3P \leftarrow {}^1S$ band is examined with moderately high resolution it is seen to be in fact a triplet with components at 21850, 21870 and 21911 cm^{-1}. These three components correspond to the transitions to the three different levels of the ^3P term, $^3P_0 \leftarrow {}^1S$, $^3P_1 \leftarrow {}^1S$ and $^3P_2 \leftarrow {}^1S$ respectively. The splitting of the ^3P term is due to an interaction between the spin magnetic moment of the electrons and the magnetic field set up by the motion of the electrons around the nucleus. This *spin-orbit* interaction is represented by a term in the Hamiltonian proportional to $\mathbf{L}_i.\mathbf{S}_i$ for each electron

$$\mathscr{H}' = \sum_i \xi_i \mathbf{L}_i.\mathbf{S}_i. \tag{9.16}$$

For a single electron in a central field with potential $V(r)$ it is known that

$$\xi = \frac{1}{2m^2c^2}\left(\frac{1}{r}\frac{\partial V(r)}{\partial r}\right), \tag{9.17}$$

and it is assumed that for a many-electron atom ξ_i has this general form.

When the term $\mathscr{H}'$ is included in the Hamiltonian then $\mathbf{L}^2$, $\mathbf{L}_z$, $\mathbf{S}^2$ and $\mathbf{S}_z$ no longer commute with the Hamiltonian so that L, M_L, S and M_S cease to be good quantum numbers. Only the total orbital angular momentum operators now commute with the Hamiltonian so that strictly speaking only J and M_J are good quantum numbers. However, if the spin-orbit coupling energies are calculated using expression (9.17) then it is found that they are proportional to Z^4†. For light atoms the splitting up of a configuration into different terms through electron repulsion is usually very much greater than the further splitting of these terms into levels through spin-orbit interaction. For this reason one can say that even when spin-orbit coupling is allowed for, L and S are almost good quantum numbers (although M_L and M_S are not). This is illustrated by the results for Mg quoted above.

Suppose we start with a set of terms arising from a configuration each characterized by quantum numbers L and S. We can then treat spin-orbit coupling as a small perturbation, and to first order (see 6.50) we find that a given L–S term is split so that there is one level for each J value ($2L+1$ if $L<S$, $2S+1$ if $S<L$). The relative separation of the levels can be obtained from the rule that the effect of a perturbation term $\sum \mathbf{L}_i.\mathbf{S}_i$ is proportional to the effect of a perturbation

$$\mathbf{L}.\mathbf{S} = \sum_i\sum_j \mathbf{L}_i.\mathbf{S}_j‡.$$

Because

$$(\mathbf{L}+\mathbf{S})^2 = \mathbf{L}^2 + \mathbf{S}^2 + 2\mathbf{L}.\mathbf{S} \tag{9.18}$$

then

$$\mathbf{L}.\mathbf{S} = \tfrac{1}{2}(\mathbf{J}^2 - \mathbf{L}^2 - \mathbf{S}^2), \tag{9.19}$$

† The potential is proportional to Z and hydrogen-like wave functions have a factor $Z^{3/2}$ in their normalizing constant (see table 3.2). An integral of the form (6.27) where $\mathscr{B}$ is $\mathscr{H}'$ will therefore be proportional to Z^4.

‡ For a proof of this see GRIFFITHS, *The Theory of Transition Metal Ions*, chapter 5.1, C.U.P., 1961.

so that if a level has a definite J, L and S, its first-order spin-orbit coupling energy will be proportional to

$$\tfrac{1}{2}\{J(J+1) - L(L+1) - S(S+1)\}\,\frac{h^2}{4\pi^2}. \qquad (9.20)$$

From this formula one can obtain the Landé interval rule, that the interval between two J levels is proportional to the larger J value of the pair (see problem 9.6).

The lowest energy configuration of Fe is $4s^2 3d^6$, and, of the various terms that can arise from this, ^5D has the lowest energy. Spin-orbit coupling then splits this up as shown in table 9.2. The last column would be a constant if the Landé interval rule held exactly; the small deviation from constancy shows the validity of treating spin-orbit coupling by first-order perturbation theory.

TABLE 9.2. The splitting of the ^5D term of Fe

Level	Energy (cm^{-1})	Interval	Interval/J_{greater}
$^5D_0$	978		
		−90	−90
$^5D_1$	888		
		−184	−92
$^5D_2$	704		
		−288	−96
$^5D_3$	416		
		−416	−104
$^5D_4$	0		

If a term arises from a configuration with less than a half-filled shell (e.g. p^2), then after spin-orbit coupling the level with the lowest value of J has the lowest energy (another rule attributed to Hund). This is called a *normal multiplet*. If the configuration has more than a half-filled shell then the level of highest J has the lowest energy; this is called an *inverted multiplet* (cf. table 9.2), and the intervals are given negative signs. From a half-filled shell there is no first-order spin-orbit splitting of the terms, although experimentally very small splittings are observed which are due to second-order effects.

For heavy atoms the spin-orbit coupling energies become quite large, but, providing the electrons are in orbitals of comparable size, the Russell–Saunders coupling scheme is still a good starting point. For example,

$$\text{Ba}(6s6p);\ {}^3P_0 = 12266,\ {}^3P_1 = 12636,\ {}^3P_2 = 13515,\ {}^1P = 18060\ \text{cm}^{-1}.$$

However, if the electron interaction is much smaller than the spin-orbit interaction, and this arises when the electrons are in orbitals of very different sizes, then we have the so-called *j–j* coupling case. Each electron has a characteristic l and s and these can give two or one j values (only one j is possible if $l = 0$). The j values for each electron can then be coupled together according to the standard rule for combining angular momentum, to obtain resultant J values. For example (compare with the energies given earlier for Mg and Ba),

$$\mathrm{Si}(3s^2 3p6s);\ {}^3P_0 = 59221,\ {}^3P_1 = 59273,\ {}^3P_2 = 59506,\ {}^1P_1 = 59636\ \mathrm{cm}^{-1}.$$

The $3p$ electron can give $j = 3/2$ and $1/2$, the $6s$ only has $j = 1/2$. Combining $j = 3/2$ with $j = 1/2$ gives $J = 2, 1$; combining $j = 1/2$ and $j = 1/2$ gives $J = 1, 0$. Figure 9.2 sho·'s the tránsition from the pure Russell–Saunders

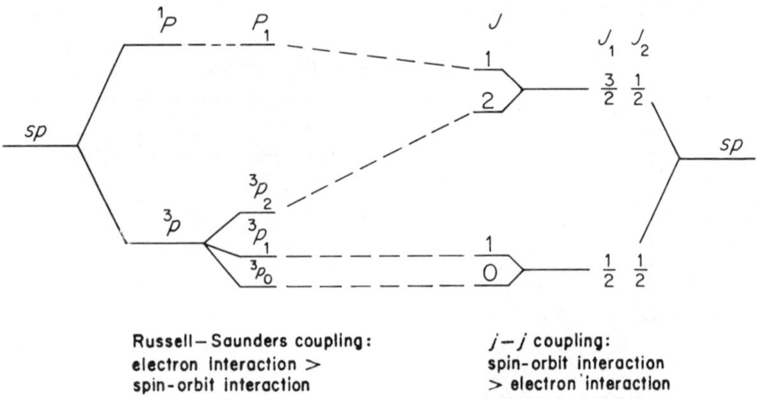

FIGURE 9.2. The transition from Russell–Saunders coupling to *j–j* coupling

coupling scheme to pure *j–j* coupling for the configuration *sp*. Going from left to right there is an increase in the ratio of spin-orbit coupling to electron interaction energies.

9.3 The effect of the Pauli exclusion principle

The vector rule for arriving at the eigenvalues of the total angular momenta does not allow for any restriction which may arise from the Pauli exclusion principle. For example, we have seen that the total spin of two electrons may be 1 or 0, but if the two electrons occupy the same orbit then the term with $S = 1$ is forbidden by the exclusion principle, since the two electrons must occupy the orbital with opposite spins. As a further example we consider the terms which can arise from the configuration p^2. We have seen that the configuration $p.p$ (e.g. $2p3p$) gives rise to the six terms ${}^3D$, ${}^1D$, ${}^3P$, ${}^1P$, ${}^3S$, ${}^1S$. The Pauli exclusion principle makes three of these impossible as we can see in the following manner. Suppose we write down all possible ways of allocating two electrons to a p sub-shell. These are shown in table 9.3. An arrow up will indicate an α-spin electron, an arrow down a β-spin. We choose the p orbitals in complex form so that they have definite m values.

TABLE 9.3. Possible ways of allocating two electrons to a p sub-shell

	$m = 1$	$m = 0$	$m = -1$	$M_L = \Sigma m$	$M_S = \Sigma m_s$	Designation
1	↑↓			2	0	${}^1D$
2	↑	↑		1	1	${}^3P$
3	↑	↓		1	0	${}^1D, {}^3P$ (3, 4)
4	↓	↑		1	0	
5	↓	↓		1	−1	${}^3P$
6	↑		↑	0	1	${}^3P$
7	↑		↓	0	0	${}^1D, {}^3P, {}^1S$ (7, 8, 9)
8		↑↓		0	0	
9	↓		↑	0	0	
10	↓		↓	0	−1	${}^3P$
11		↑	↑	−1	1	${}^3P$
12		↓	↑	−1	0	${}^1D, {}^3P$ (12, 13)
13		↑	↓	−1	0	
14		↓	↓	−1	−1	${}^3P$
15			↑↓	−2	0	${}^1D$

We first look at the arrangement with highest values of M_L. There is just one allocation (1) which has $M_L = 2$, and none with a higher value of M_L; since this has $M_S = 0$, it must be a component of a ${}^1D$ term. A ${}^1D$ term has five components corresponding to $M_L = 2, 1, 0, -1, -2$, and to pick out these five components we simply look for these M_L values with the restriction $M_S = 0$. All five components have $L = 2$, $S = 0$. The ${}^1D$ term uses up (1), (15), one from (3) and (4), one from (12) and (13), and one from (7), (8) and (9).

We next look at the remaining arrangements and again pick out the one with the highest value of M_L, and, if there is more than one, the one with the highest value of M_S. This is the arrangement (2). Since it has $M_L = 1$ and $M_S = 1$, and as there are no higher M_L or M_S values unaccounted for, it must be a component of a ${}^3P$ term. A ${}^3P$ term has $L = 1$, $S = 1$, and nine components corresponding to $M_L = 1, 0$ or -1 and $M_S = 1, 0$ or -1. We can pick out the remaining eight components, as shown in the table.

Of the 15 possible arrangements 14 have now been used up. The only arrangement not accounted for is one from (7), (8), and (9). These all have

$M_L = 0$, $M_S = 0$ so that the remaining term must be ^1S, which has only one component.

The Pauli exclusion principle therefore only allows ^1D, ^3P and ^1S terms from the configuration p^2, which is only half the number that arise from the configuration *p.p.* Spectroscopy confirms that only ^1D, ^3P and ^1S terms occur and presents further conclusive evidence in favour of the Pauli exclusion principle.

The configuration p^2 gives three functions with $M_L = 0$, $M_S = 0$. One is a component of ^1D, one a component of ^3P, and one is ^1S. To find the precise form of the wave functions for these terms we need a few more equations.

From the commutation relationships for the general angular momentum operators, it can be proved that if we have an eigenfunction of $\mathbf{L}^2$ and $\mathbf{L}_z$ such that

$$\mathbf{L}^2 F_{lm} = l(l+1)\frac{h^2}{4\pi^2} F_{lm}, \quad \text{and} \quad \mathbf{L}_z F_{lm} = \frac{mh}{2\pi} F_{lm},$$

then the following expressions hold†

$$\mathbf{L}_+ F_{lm} \equiv (\mathbf{L}_x + \mathrm{i}\mathbf{L}_y) F_{lm} = \frac{h}{2\pi}[(l+m+1)(l-m)]^{\frac{1}{2}} F_{lm+1}.$$

$$\mathbf{L}_- F_{lm} \equiv (\mathbf{L}_x - \mathrm{i}\mathbf{L}_y) F_{lm} = \frac{h}{2\pi}[(l+m)(l-m+1)]^{\frac{1}{2}} F_{lm-1}. \quad (9.21)$$

Expressions (9.21) relate the signs (phases) of the wave functions F_{lm-1}, F_{lm}, F_{lm+1}. For the functions of orbital angular momentum these expressions only give the correct signs if one takes

$$F_{lm} = Y_{lm},\ (m \leqslant 0);\ F_{lm} = (-1)^m\ Y_{lm},\ (m > 0),$$

where Y_{lm} are the spherical harmonics defined by (3.8).
The operator $\mathbf{L}_+$ acting on the function F_{lm} converts it into the eigenfunction with the same l and one higher m value; the operator $\mathbf{L}_-$ steps down the m value by one. These operators are known as *shift operators* or *step operators*. They are useful because knowing one eigenfunction of $\mathbf{L}^2$ and $\mathbf{L}_z$ with a given l and m, we can find the corresponding eigenfunction having the same l and either $m+1$ or $m-1$. Expressions similar to (9.21) hold for **S** and **J**.

Let us take the spin states for two electrons in different orbitals as an example. We can write down four spin functions for two electrons (labelled 1 and 2) as follows:

	$\alpha(1)\alpha(2)$,	$\alpha(1)\beta(2)$,	$\beta(1)\alpha(2)$,	$\beta(1)\beta(2)$.
$M_S =$	1	0	0	−1

† Eyring, Walter and Kimball, *Quantum Chemistry*, p. 46, Wiley, 1944.

Operating on the function $\alpha(1)$ with $\mathbf{S}_-(1)$ gives (putting $l = \frac{1}{2}\, m = \frac{1}{2}$ in (9.21))

$$\mathbf{S}_-(1)\alpha(1) = \frac{h}{2\pi}\beta(1). \tag{9.22}$$

Hence

$$\mathbf{S}_-(1)\alpha(1)\alpha(2) = \frac{h}{2\pi}\beta(1)\alpha(2).$$

Likewise $\mathbf{S}_-(2)$ operating on $\alpha(1)\alpha(2)$ transforms it into $\alpha(1)\beta(2)$. Now the function $\alpha(1)\alpha(2)$ must be the component F_{11} of the triplet spin state ($S = 1$, $M_S = 1$). If we want the component of the triplet with $M_S = 0$, we operate on this with $\mathbf{S}_- = \mathbf{S}_-(1) + \mathbf{S}_-(2)$. It follows from (9.21) that

$$\mathbf{S}_- F_{11} = \frac{h}{2\pi}\sqrt{2}F_{10}.$$

But from (9.22)

$$\mathbf{S}_-\alpha(1)\alpha(2) = \frac{h}{2\pi}\{\alpha(1)\beta(2) + \beta(1)\alpha(2)\}.$$

Hence, the triplet spin function with $M_S = 0$ is

$$F_{10} = \sqrt{\tfrac{1}{2}}\{\alpha(1)\beta(2) + \beta(1)\alpha(2)\}. \tag{9.23}$$

To confirm this we now operate on this function with $\mathbf{S}_-$, and we find

$$\mathbf{S}_-\sqrt{\tfrac{1}{2}}\{\alpha(1)\beta(2) + \beta(1)\alpha(2)\} = \frac{h}{2\pi}\sqrt{2}\beta(1)\beta(2)$$

and $\beta(1)\beta(2)$ is, of course, the component of the triplet F_{1-1} with $M_S = -1$.

To find the wave function of the singlet spin state we just take that combination of the functions having $M_S = 0$ which is orthogonal to the triplet state (9.23). Suppose

$$F_{00} = a\alpha(1)\beta(2) + b\beta(1)\alpha(2).$$

Then from (9.9)

$$\int F_{10}F_{00}\,\mathrm{d}s = 0$$

providing $a = -b$. The normalized singlet function is therefore

$$F_{00} = \sqrt{\tfrac{1}{2}}\{\alpha(1)\beta(2) - \beta(1)\alpha(2)\}. \tag{9.24}$$

If we want to find the eigenfunctions of $\mathbf{L}^2$ and $\mathbf{S}^2$ arising from the configuration p^2 we proceed in the following way. The function (1) (table 9.3) is a component of the D term with $M_L = 2$. Its wave function is easy to write down† $|\psi_1\alpha\,\psi_1\beta|$, where ψ_1 is the p orbital with $m = 1$. If we operate on this with $\mathbf{L}_-$ we will successively generate the functions with $M_L = 1, 0, -1$ and -2. The function (2) is a component of the ^3P term with $M_L = 1$, $M_S = 1$, and again there is no difficulty in writing down its wave function. Operating on this with $\mathbf{S}_-$ will generate the functions with $M_L = 1$, $M_S = 0, -1$. Operating on these with $\mathbf{L}_-$ will give the remaining six components. Lastly, we obtain the wave function of the ^1S term by making it orthogonal to the ^1D and ^3P components which have been obtained.

9.4 Atomic spectra

We shall deal only very briefly with spectroscopy in this book, restricting our attention to those topics which have a direct bearing on the electronic structure of atoms and molecules. Atomic spectra not only give us the energy levels of electrons in atoms, but the changes in spectra produced by the ligands in metal complex tell us something of the strength of the binding between the ligands and the metal, and of the geometry of the ligands.

Monochromatic light can be absorbed by an atom providing that energy is conserved, that is, if the atom is in a state of energy E_0 there must be a state of energy E_1 such that (1.2)

$$E_1 - E_0 = h\nu. \tag{9.25}$$

However, there are other conditions, called selection rules, which must also be satisfied for the light to be absorbed. For an atom these are that the changes in the total angular momentum quantum numbers accompanying the transition must be

$$\Delta J = 0 \quad \text{or} \quad \pm 1 \qquad \Delta M_J = 0, \pm 1, \tag{9.26}$$

except that the transition $J = 0 \rightarrow J' = 0$ is forbidden. Under the Russell–Saunders coupling scheme where L and S are nearly good quantum numbers the selection rules are $\Delta S = 0$, $\Delta L = 0, \pm 1$. If, in addition, an electronic transition can be attributed to the excitation of an electron from one orbital to another (usually a good approximation), then for this one-electron jump there is a selection rule $\Delta l = \pm 1$.

There is one further quantity which can be used for a rigorous classification of an atomic state; this is the *parity* symbol. Because an atom has a centre of symmetry, the wave functions must either be unchanged or

† We use the notation of section 7.1.

change signs under the inversion operation. Those which are unchanged are called even states. Those which change signs are called odd states, and are distinguished by a superscript o(°) To find the parity of a state, we simply multiply the parities of each orbital occupied by an electron (odd × odd = even × even = even; odd × even = odd) or, even simpler, we just take the sum of the l values for each electron. For example, the ground state of carbon, ^3P, arises from an even configuration $2s^22p^2$. The excited configuration, $2s2p^3$, also gives a ^3P term, but this is odd, so that it is given the symbol $^3P^o$. For light absorption there is a parity selection rule that transitions are only allowed between states of different parity (the Laporte rule).

The $2J + 1$ components of an atomic level are strictly degenerate in the absence of an external field. If this were not so, it would imply that the z direction was in some way essentially different from the x and y directions. If, however, we apply an external field to the atom, then the $(2J + 1)$-fold degeneracy is removed.

An external magnetic field will split up a level into its $2J + 1$ components; each has a characteristic M_J value if we take the magnetic field to be along the z direction. If this splitting is small compared with the spin-orbit separation of the level, then we have what is called a Zeeman effect, and each level is split into $2J + 1$ equally spaced components. The separation of the components is given by $g(eh/4\pi mc)H^M$ where H^M is the magnetic field and

$$g = 1 + \frac{J(J + 1) - L(L + 1) + S(S + 1)}{2J(J + 1)}. \tag{9.27}$$

If the magnetic field gives rise to a splitting comparable with the effect of spin-orbit coupling, then we have the so-called Paschen–Back effect, and a more complicated pattern of energy levels.

An external electric field acting on an atom gives rise to what is known as the Stark effect. For a uniform field in the z direction each level is split up so that, in general, components of different $|M_J|$ have different energies. The interaction of such a field with an atom (or molecule) goes through the electric dipole moment. In the case of an atom the dipole moment is itself due to the field, the induced dipole being related to the field through the polarizability of the atom (α) (see section 18.1)

$$\mu_z = \alpha \varepsilon_z. \tag{9.28}$$

It follows that the interaction energy of the electric field with the atom is given by a term in the Hamiltonian of the type

$$\mathscr{H}^E = -\mu_z \varepsilon_z = -\alpha \varepsilon_z^2. \tag{9.29}$$

It can immediately be seen that reversing the field direction does not change the energy; reversing the field direction is equivalent to changing from a state with M_J to one with $-M_J$, so that these two states must have the same energy in the presence of the field. This simple picture is somewhat different for the hydrogen atom. In this case orbitals with the same values of n but different values of l are degenerate, and the Stark splitting of a level turns out to be linear in the field instead of quadratic. Because $\mathscr{H}^E$ does not contain any term which operates on the spin of the electron its effect on a pure Russell–Saunders term (no spin-orbit coupling) is only to remove the degeneracy of states having different $|M_L|$. The pattern of the Stark-split energy levels therefore depends on the relative strengths of spin-orbit coupling and the electric field, the same type of complication which occurs for the Zeeman–Paschen–Back effect.

The Stark effect is the simplest case of the interaction of an electric field with an atom. In chapter 13 we shall deal with the effects of the non-uniform fields which are produced by the ligands in a metal complex.

9.5 The helium atom

The helium atom is the simplest example of the atomic many-electron problem for which there is no exact solution of the Schrödinger equation. It provides a test of the techniques available for obtaining approximate solutions of the equation, and, as we shall see, it is possible in this case to obtain a solution which is at least as accurate as experiment.

We shall first look at the ground state of the atom. In the orbital approximation we say that two electrons occupy the $1s$ orbital with opposite spins. We write the wave function as (cf. 7.6)

$$\Psi = |1s\alpha\ 1s\beta|. \tag{9.30}$$

The Hamiltonian is (from 6.13)

$$\begin{aligned}\mathscr{H} &= -\frac{h^2}{8\pi^2 m}(\nabla_1{}^2+\nabla_2{}^2)-\frac{2e^2}{r_1}-\frac{2e^2}{r_2}+\frac{e^2}{r_{12}}\\ &= -\tfrac{1}{2}(\nabla_1{}^2+\nabla_2{}^2)-\frac{2}{r_1}-\frac{2}{r_2}+\frac{1}{r_{12}}\ \text{(in atomic units)},\end{aligned} \tag{9.31}$$

where the electrons are labelled 1 and 2, and the coordinates are defined by figure 9.3.

Let us now consider the potential acting on electron 1. For this purpose we use the results obtained by classical electrostatics, namely, that the potential outside a spherical charge distribution is the same as if the total

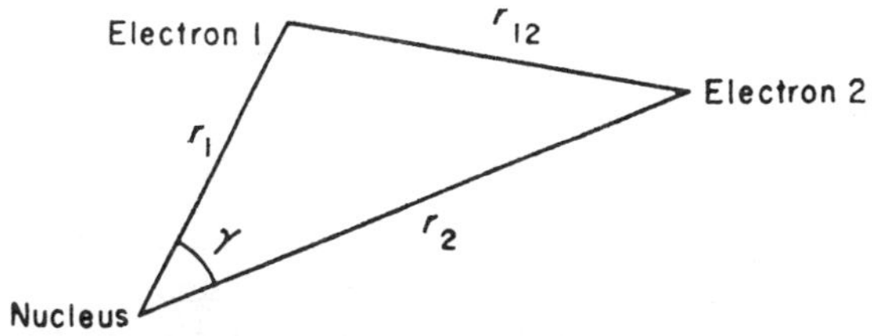

FIGURE 9.3. The coordinate system for He

charge were placed at the centre, and the potential inside a spherical charge distribution is a constant equal to the total charge.

If electron 1 is a long way from the nucleus then electron 2 is almost certainly closer to the nucleus, and so the potential acting on electron 1 is $-(2-1)/r_1$. An electron moving in this potential has a 1*s* wave function which is that of the hydrogen atom (table 3.3); in atomic units this is

$$1s = \pi^{-\frac{1}{2}}\,\mathrm{e}^{-r_1}. \tag{9.32}$$

If electron 1 is very close to the nucleus then the potential will be $1-(2/r_1)$, which is dominated by $-2/r_1$, and the 1*s* wave function for this potential is

$$1s = \left(\frac{8}{\pi}\right)^{\frac{1}{2}}\mathrm{e}^{-2r_1}. \tag{9.33}$$

Clearly the average potential acting on electron 1 will vary between these two extremes of complete screening of the nucleus by electron 2 (at large distances) and no screening (at small distances). A wave function which can reasonably be expected to be appropriate to this average potential is

$$1s = \left(\frac{\zeta^3}{\pi}\right)^{\frac{1}{2}}\mathrm{e}^{-\zeta r_1}, \tag{9.34}$$

where ζ will be somewhere between 1 and 2.†

The best value of ζ can be determined by the variation principle: by definition the best value is the one which minimizes the energy of the function. We therefore wish to evaluate‡

$$E = \iint |1s\alpha\,1s\beta|\mathscr{H}|1s\alpha 1s\beta|\,\mathrm{d}\tau_1\,\mathrm{d}\tau_2 \tag{9.35}$$

with the 1*s* orbital having the form (9.34). We now expand the Slater

† On Slater's rules (cf. Page 31) the orbital exponent ζ is given by $\dfrac{\text{Effective nuclear charge}}{\text{principal quantum number}}$, so that in the present case ζ may be equated with the effective nuclear charge.

‡ $\mathrm{d}\tau = \mathrm{d}v\,.\,ds$ is the volume element for combined space and spin coordinates.

determinant on the right-hand side of the integral; the determinant is normalized and on expansion gives (cf. 7.4)

$$\sqrt{\tfrac{1}{2}}\{1s\alpha(1)1s\beta(2) - 1s\alpha(2)1s\beta(1)\}.$$

Since the Hamiltonian is symmetrical in the two electrons, both of the terms must give the same contribution to the integral, and so we just take one term and multiply by 2. However, this 2 cancels the normalizing constants of both Slater determinants, so we are left, after expanding the left-hand side, with

$$E = \iint \{1s\alpha(1)1s\beta(2) - 1s\alpha(2)1s\beta(1)\}\,\mathscr{H}\,1s\alpha(1)1s\beta(2)\,d\tau_1\,d\tau_2. \quad (9.36)$$

This is a general procedure for simplifying integrals involving Slater determinants: only the diagonal element of the determinant on the right-hand side is taken and the normalizing constants of the determinants are neglected.

If we now integrate over the electron spin coordinates, we find, because of the orthogonality of the α and β functions, that only one of the two terms from the determinant on the left-hand side, $1s\alpha(1)1s\beta(2)$, can give a non-zero contribution. We are then left, after introducing expression (9.31), to evaluate

$$E = \iint 1s(1)1s(2)\left\{-\tfrac{1}{2}(\nabla_1{}^2 + \nabla_2{}^2) - \frac{2}{r_1} - \frac{2}{r_2} + \frac{1}{r_{12}}\right\}1s(1)1s(2)\,dv_1\,dv_2. \quad (9.37)$$

We can further simplify this by making use of the fact that the $1s$ orbital is an eigenfunction of $-\dfrac{1}{2}\nabla^2 - \dfrac{\zeta}{r}$ and has the eigenvalue $-\dfrac{\zeta^2}{2}$ (see 3.19), and thus obtain the expression

$$E = -\zeta^2 + \iint 1s(1)1s(2)\left\{-\frac{(2-\zeta)}{r_1} - \frac{(2-\zeta)}{r_2} + \frac{1}{r_{12}}\right\}1s(1)1s(2)\,dv_1\,dv_2. \quad (9.38)$$

We then split up the integral into three parts corresponding to the three operators in the central brackets. The integral involving $1/r_1$ is simplified by integrating over the coordinates of electron 2; likewise that involving $1/r_2$. Both these integrals clearly have the same value.

$$E = -\zeta^2 - 2(2-\zeta)\int 1s(1)\frac{1}{r_1}1s(1)\,dv_1 + \iint 1s(1)1s(2)\frac{1}{r_{12}}1s(1)1s(2)\,dv_1\,dv_2. \quad (9.39)$$

The mean value of $1/r$ for a hydrogen-like $1s$ orbital with nuclear charge ζ is ζ (problem 3.4). It only remains to evaluate the two-electron integral which gives the mutual repulsion energy of the two electrons. This is a bit tricky because of the variable $1/r_{12}$, but it can be done by using the cosine rule (see figure 9.3)

$$r_{12}^2 = r_1^2 + r_2^2 - 2r_1 r_2 \cos\gamma. \tag{9.40}$$

It is then possible to expand $1/r_{12}$ as the following series†

$$\frac{1}{r_{12}} = \sum_{l=0}^{\infty} \sum_{m=-l}^{l} \left(\frac{4\pi}{2l+1}\right) \frac{r_<^l}{r_>^{l+1}} Y_{lm}(\vartheta_1, \varphi_1) Y_{lm}(\vartheta_2, \varphi_2), \tag{9.41}$$

where Y_{lm} are the spherical harmonics defined by (3.6). $r_<$ means the smaller of r_1 and r_2 and $r_>$ means the larger. Although this is an infinite expansion for $1/r_{12}$, it is only necessary to go up to those values of l which appear in the electron densities involved in the integral, because the spherical harmonics are an orthogonal set of functions. In our case, $1s^2$ is a spherical density (containing Y_{00}) so that one only needs to write

$$\frac{1}{r_{12}} = \left(\frac{4\pi}{1}\right) \frac{1}{r_>} \left(\frac{1}{2\sqrt{\pi}}\right)^2 = \frac{1}{r_>}. \tag{9.42}$$

The integral over r_1 and r_2 is now separated into two parts,

$$\int_0^\infty \left\{\int_0^{r_2} dr_1\right\} dr_2 \qquad \text{where } r_> = r_2$$

and

$$\int_0^\infty \left\{\int_{r_2}^{\infty} dr_1\right\} dr_2 \qquad \text{where } r_> = r_1,$$

and by this method the two-electron integral can be shown to have the value $5\zeta/8$ (problem 9.12).

The result of the calculation is then

$$E = -\zeta^2 - 2(2-\zeta)\zeta + 5\zeta/8 = \zeta^2 - 27\zeta/8. \tag{9.43}$$

To this we apply the variation principle; differentiating gives

$$\frac{dE}{d\zeta} = 2\zeta - 27/8, \tag{9.44}$$

so that the minimum of energy arises for a value $\zeta = 27/16 = 1{\cdot}6875$. The energy E appropriate to this value is $-2{\cdot}8476$ atomic units ($-77{\cdot}483$ev).

† Eyring, Walter and Kimball, *Quantum Chemistry*, appendix V, Wiley, 1944.

The energy required to strip off both electrons from He is found spectroscopically to be $-79{\cdot}000$ ev so that our calculated energy is in error only by $1{\cdot}5$ ev or $1{\cdot}9\%$.

The improvement in energy obtained by using a $1s$ orbital with screened nuclear charge, over say, one without screening ($\zeta = 2$), is about 3 ev, which is quite considerable. However, this screened hydrogen-like function is not the best $1s$ orbital which can be obtained using the SCF procedure developed by Hartree and Fock. The best energy that can be obtained using a function of the type (9.30), without imposing any restriction on the algebraic form of the $1s$ orbital, is about $-77{\cdot}866$ ev. But this is only a further improvement of $0{\cdot}383$ ev, and there is still quite a large difference between this calculated energy and that obtained from experiment.

There are several contributions to this difference. Firstly, we have assumed that the mass of the nucleus is infinite compared with that of the electron. This means that the conversion from atomic units to electron volts should be made using a factor based on the reduced mass of the electron and nucleus $(mM)/(m+M)$ rather than the mass of the electron. However, this only produces a change of $0{\cdot}017\%$ ($0{\cdot}013$ ev). Secondly, we have used a Hamiltonian which neglects relativistic effects which manifest themselves in such things as spin-orbit coupling. For a heavy atom these can be quite large, but for He they have been estimated to give only about $0{\cdot}004$ ev. The dominant part of the error is due to the fact that the wave function (9.30), which is not an exact solution of the Schrödinger equation, does not allow for any correlation between the positions of the two electrons. This correlation energy is about $1{\cdot}1$ ev†.

The correlation between two particles is defined as follows. Suppose $P(r_1, r_2)$ is the probability of finding particle 1 at r_1 and particle 2 at r_2, $P(r_1)$ is the probability of finding particle 1 at r_1 averaged over all positions of 2, and $P(r_2)$ is the probability of finding particle 2 at r_2 averaged over all positions of 1; then a system showing no correlation is one obeying the relationship

$$P(r_1, r_2) = P(r_1).P(r_2).$$

In other words, the position of each particle depends only on the average position of the other.

The solution of the many-electron Schrödinger equation should show correlation between the electrons. There are several types of wave function which have this property, the most straightforward being one which

† Correlation energy is usually defined as the difference between the energy of the Hartree–Fock SCF function and that of an exact solution of the non-relativistic Schrödinger equation.

specifically includes the interelectron distances. This type of function was first studied by Hylleraas for the helium atom in 1929. He multiplied the uncorrelated wave function, $e^{-\zeta(r_1+r_2)}$, by a polynomial in r_1, r_2 and r_{12}. His simplest function was

$$e^{-1\cdot 849(r_1+r_2)}(1 + 0\cdot 364 r_{12}), \tag{9.45}$$

which gave an energy differing only by 0·34 ev from experiment. Using a fourteen-term polynomial the agreement with experiment was within 0·002 ev. In recent years even larger polynomials have been used and one can say that the calculated results agree completely with experiment.

The work of Hylleraas and others to obtain exact wave functions for the helium atom is important because it shows that the Schrödinger equation is the correct equation of motion for two-electron atoms. It is, therefore, probably correct also for many-electron atoms and molecules, even though we cannot yet solve it exactly in these cases. Unfortunately the type of wave function used by Hylleraas runs into extreme mathematical complications for more than two electrons, and in these cases electron correlation is usually introduced through the method of configuration interaction. This will be dealt with in chapter 12.

We now turn briefly to the excited states of helium to illustrate the concept of exchange energy. Using the orbital approach we would say that the first excited configuration is $1s\,2s$, and this would give the terms ^1S and ^3S. Written out in full, the wave functions for these terms are

$$^1S = \sqrt{\tfrac{1}{2}}\{1s(1)2s(2) + 2s(1)1s(2)\}\sqrt{\tfrac{1}{2}}\{\alpha(1)\beta(2) - \beta(1)\alpha(2)\},$$

$$^3S = \sqrt{\tfrac{1}{2}}\{1s(1)2s(2) - 2s(1)1s(2)\}\begin{cases} \alpha(1)\alpha(2) \\ \sqrt{\tfrac{1}{2}}\{\alpha(1)\beta(2) + \beta(1)\alpha(2)\}, \\ \beta(1)\beta(2) \end{cases} \tag{9.46}$$

or, using the determinantal abbreviation

$$^1S = \sqrt{\tfrac{1}{2}}\{|1s\alpha\ 2s\beta| - |1s\beta\ 2s\alpha|\},$$

$$^3S = \begin{cases} |1s\alpha\ 2s\alpha| \\ \sqrt{\tfrac{1}{2}}\{|1s\alpha\ 2s\beta| + |1s\beta\ 2s\alpha|\} \\ |1s\beta\ 2s\beta|. \end{cases} \tag{9.47}$$

There are three spin components of the triplet state ($M_s = 1, 0, -1$). The spin wave functions for the singlet and triplet were derived in section 9.4 (see (9.23) and (9.24)). The functions (9.47) all obey the antisymmetry condition (section 7.1). The singlet function has an antisymmetric spin part and hence the space part must be symmetric to electron exchange. The triplet function has symmetric spin and antisymmetric space functions. In both functions the two electrons are indistinguishable; there is as much chance of electron 1 being in the $1s$ orbital as there is of electron 2.

Let us now evaluate the energy of the ^3S term; we can take one of the components, say $|1s\alpha\ 2s\alpha|$, since all components must have the same energy in the absence of an external applied field. Following the procedure used for the ground state, we arrive at the following expression for the energy (cf. 9.36)

$$E = \iint \{1s\alpha(1)2s\alpha(2) - 1s\alpha(2)2s\alpha(1)\}\mathscr{H}1s\alpha(1)2s\alpha(2)\,d\tau_1\,d\tau_2. \tag{9.48}$$

Unlike the case of the ground state, on integrating over the electron spins one retains both terms on the left-hand side because both electrons have the same spin. The expression analogous to (9.37) is

$$E = \iint \{1s(1)2s(2) - 1s(2)2s(1)\}\left[-\tfrac{1}{2}(\nabla_1^2 + \nabla_2^2) - \frac{2}{r_1} - \frac{2}{r_2} + \frac{1}{r_{12}}\right] \times 1s(1)2s(2)\,dv_1\,dv_2. \tag{9.49}$$

We can now split up the one- and two-electron operators as described for the ground state calculation, and, making use of the fact that the $1s$ and $2s$ orbitals are orthogonal to one another, we have

$$\begin{aligned} E = &\int 1s(1)\left[-\tfrac{1}{2}\nabla_1^2 - \frac{2}{r_1}\right]1s(1)\,dv_1 + \int 2s(2)\left[-\tfrac{1}{2}\nabla_2^2 - \frac{2}{r_2}\right]2s(2)\,dv_2 \\ &+ \iint 1s(1)2s(2)\left[\frac{1}{r_{12}}\right]1s(1)2s(2)\,dv_1\,dv_2 \\ &- \iint 1s(2)2s(1)\left[\frac{1}{r_{12}}\right]1s(1)2s(2)\,dv_1\,dv_2. \end{aligned} \tag{9.50}$$

The first two terms can be taken as the energy of the $1s$ and $2s$ orbitals in the field of the helium nucleus; the third term is the repulsion between the electron densities $1s^2$ and $2s^2$. The last term is something which has no analogy in expression (9.39); it represents the mutual interaction of two electron densities $(1s2s)(1)$ and $(1s2s)(2)$. This last term is called the *exchange energy*. It is a stabilizing term and is associated with the fact that we have taken a wave function for the ^3S state which allows for the indistinguishability of the electrons. The ground state energy, on the other hand, has no exchange energy because the two electrons have opposite spins. In fact, as far as energy goes, a simple product wave function for the ground state $1s\alpha(1)1s\beta(2)$ is as good as the antisymmetrized product.

If we work out the energy of the ^1S excited state in the same way we find that the energy differs from the triplet only in the sign of the exchange integral. The singlet–triplet splitting is found from spectroscopy to be 6400 cm^{-1} (0·79 ev), and this is therefore twice the exchange integral, if we assume that the $1s$ and $2s$ orbitals are the same for both singlet and triplet states (which is a reasonable approximation). Exchange integrals are always considerably smaller than the corresponding coulomb integrals (for example, the energy of interaction between the electron densities $1s^2$ and $2s^2$ in helium is about 7 ev, which is twenty times larger than the exchange integral). This is because the electron densities involved in the exchange integral (e.g. $1s2s$) are not real densities, which are everywhere positive, but are so-called *transition densities*, which have positive and negative regions and which integrate to zero ($\int 1s2s\,dv = 0$) because of the orthogonality condition.

Suppose now we look at the correlation between the two electrons in the singlet and triplet states, which arises through the antisymmetry principle. From the definition of the physical significance of a wave function the probability density is given by Ψ^2. To find the probability density of the space coordinates of the electrons alone we must integrate Ψ^2 over the spin coordinates. The result is the function $P(r_1, r_2)$. To find the function $P(r_1)$ this must be integrated further over the coordinates of electron 2; this is the averaging procedure. Carrying out these operations one finds (see problem 9.13)

$$P(r_1, r_2) - P(r_1)P(r_2) = -\tfrac{1}{4}(1s^2 - 2s^2)(1)(1s^2 - 2s^2)(2) \pm 1s(1)2s(1)1s(2)2s(2), \qquad (9.51)$$

where the + sign is for the singlet and the − sign for the triplet states. For a simple product function $1s(1)2s(2)$, which takes no account of the antisymmetry principle, one has $P(r_1, r_2) = P(r_1)P(r_2)$.

If $P(r_1, r_2) - P(r_1)P(r_2)$ is positive, this means that there is a greater chance of finding electron 1 at r_1 and electron 2 at r_2 than one would calculate from the product $P(r_1)P(r_2)$; let us call this positive correlation. For both the singlet and triplet states the first term contributes to positive correlation if one electron is in a region where $(1s^2 - 2s^2)$ is positive and the other in a region where $(1s^2 - 2s^2)$ is negative. These regions are shown in figure 9.4. The second term in (9.51) gives a different effect for the singlet and triplet: for the singlet there is positive correlation if the two electrons are both in regions where the function $(1s2s)$ has the same sign, and for the triplet there is positive correlation if the two electrons are in regions where $(1s2s)$ has different signs.

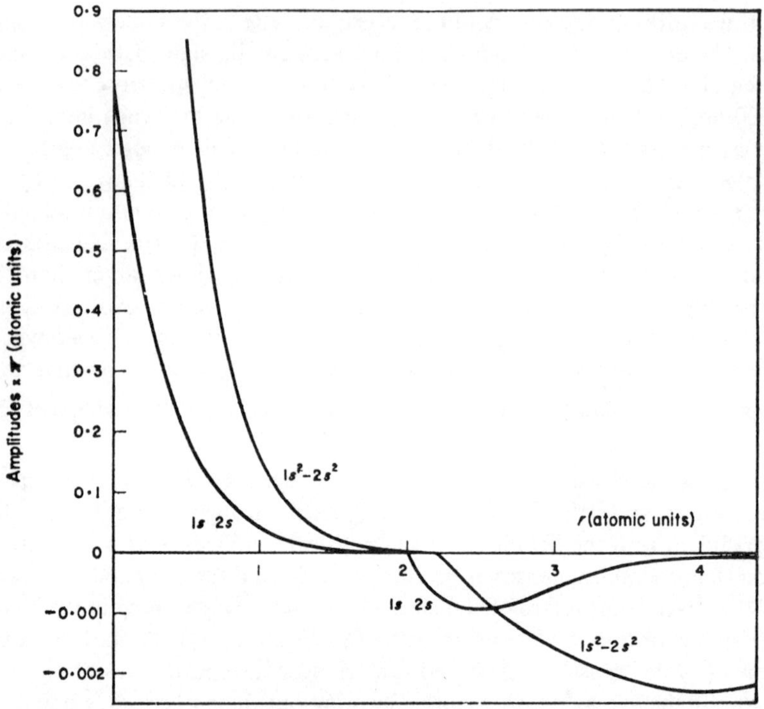

FIGURE 9.4. The electron densities determining the correlation of electrons in the singlet and triplet states

Roughly speaking, one can say that for the singlet state there is a tendency for the two electrons to get together and for the triplet to keep apart (relative to the uncorrelated wave function). This tendency manifests itself energetically through the exchange energy. For this reason Linnett proposes that the exchange energy should be called the spin-correlation energy.

9.6 The energies of Russell–Saunders terms

We saw in the last section that the energy of a state of the helium atom consists of three parts. The first represents the sum of the orbital energies, that is, the energies of the electrons in the field of bare nucleus, the second represents the coulomb repulsion between electrons in different orbitals and the third is the exchange energy. This result can be generalized to apply to many-electron atoms.

For a Russell–Saunders coupling case, all terms arising from the same configuration have the same total orbital energy but these orbital energies are now calculated using a potential made up of a nuclear part and a part due to the average potential of all closed shells of electrons. The difference in energy between terms of the same configuration only depends on the coulomb and exchange integrals between electrons in the open shells.

The coulomb and exchange integrals are conventionally given in terms of auxiliary functions F_k and G_k, which are themselves two-electron integrals but which only involve the radial coordinates of the two electrons. (The angular coordinates are integrated out following the same technique that we outlined for the ground state of the helium atom. (9.41)). These functions are called Slater–Condon parameters.

$$F_k = \frac{1}{D_k}\iint \frac{r_<^k}{r_>^{k+1}} R_{nl}^2(1)R_{n'l'}^2(2)r_1^2 r_2^2 \, dr_1 \, dr_2$$

$$G_k = \frac{1}{D_k}\iint \frac{r_<^k}{r_>^{k+1}} R_{nl}(1)R_{n'l'}(1)R_{nl}(2)R_{n'l'}(2)r_1^2 r_2^2 \, dr_1 \, dr_2. \quad (9.52)$$

$r_<$ and $r_>$ have been defined before (9.41). The functions R are the radial parts of the atomic orbitals and D_k are coefficients simply chosen to give whole number expressions for the energies. For configurations in which there is only one incomplete shell (e.g. p^2,d^3), $F_k = G_k$†.

As an example we give here the relative term energies for the p^2 configuration

$$^3P = F_0 - 5F_2$$

$$^1D = F_0 + F_2$$

$$^1S = F_0 + 10F_2.$$

The F integrals are all positive quantities so that the energies of these terms increase in the order $^3P < {}^1D < {}^1S$, in accordance with Hund's rule.

We will return to a discussion of term energies in chapter 13.

Problems 9

9.1 Prove one of the relationships (9.3).

9.2 What levels arise from the following terms ^2D, ^1G, ^6S?

† The contributions to the term energies in terms of the F_k and G_k parameters and the coefficients D_k have been worked out for all cases of chemical interest and are given in the following books.

CONDON and SHORTLEY, *The Theory of Atomic Spectra*, C.U.P., 1957.

GRIFFITHS, *The Theory of Transition Metal Ions*, C.U.P., 1961.

BALLHAUSEN, *Introduction to Ligand Field Theory*, McGraw-Hill, 1962.

9.3 What spin states can arise from three electrons in different orbitals?
9.4 Prove by considering the ways in which four electrons can be allocated to the *p* orbitals that the configuration p^4 gives the same terms as p^2.
9.5 What are the ground terms of the atoms in the first row of the periodic table (Li to Ne)?
9.6 Prove from (9.20) the Landé interval rule.
9.7 How do the ^2P terms arising from the configurations p^1 and p^5 split up through spin-orbit coupling?
9.8 What is the pattern of energy levels expected from the $3p3d$ and $3p6d$ configurations?
9.9 Obtain the wave functions of the ^1D terms from the p^2 configuration.
9.10 What are the allowed electronic transitions between the terms of the p^2 and the *pd* configurations in the case of strict Russell–Saunders coupling?
9.11 What is the effect of a weak magnetic field on a ^2P term?
9.12 Prove that the two-electron integral in (9.39) has the value $5\zeta/8$.
9.13 Prove expression (9.51) for the ^3S state.

Chapter 10

Molecular-orbital theory: its application to the electronic structure of diatomic molecules

10.1 The hydrogen molecule ion

The molecular-orbital (MO) theory of the electronic structure of molecules is the natural extension of the atomic-orbital theory of atomic structure to molecules†. It is assumed that electrons in molecules occupy orbitals which in general extend around all the nuclei in the molecule. To obtain the wave function of the ground state of the molecule we simply allocate the available electrons to the molecular orbitals of lowest energy, subject to the restriction, imposed by the Pauli exclusion principle, that not more than two electrons can occupy each orbital. This is completely analogous to the aufbau principle which is used to explain atomic structure (see chapter 4).

Most of MO theory is based on the LCAO approximation to the molecular orbitals. Near to a nucleus an electron must experience a potential due to that nucleus which swamps the potential due to any other nucleus in the molecule. It follows that near to a nucleus a molecular orbital must look like an atomic orbital of the appropriate atom. It is then assumed that a reasonable approximation to the molecular orbital in all regions of space can be obtained from a linear combination of the atomic orbitals associated with each atom in the molecule. If ϕ_ν is a general atomic orbital then the molecular orbitals are written

$$\psi = \sum_\nu c_\nu \phi_\nu, \tag{10.1}$$

where the c_ν are coefficients which have to be determined. If a large number

† This theory was first used by Hund, Mulliken and Lennard-Jones (1927–1929).

of atomic orbitals is included in the expansion, it is possible to obtain very good molecular orbitals, the coefficients being chosen according to the variation theorem. We shall see, however, that qualitatively very useful results can be obtained even if we use a very small number of atomic orbitals and work with what must be rather poor molecular orbitals.

At the end of this chapter we shall look briefly at some other molecular orbitals based on elliptic coordinates, which may be used for diatomic molecules. Also, in chapter 15 we shall examine the free-electron molecular orbitals which are useful for conjugated systems. The only molecular orbitals which have been generally used, however, are those based on the LCAO approximation.

We shall first examine the simplest molecule H_2^+. This molecule plays the same role amongst molecules as the hydrogen atom plays amongst atoms, that is, within the limits of the Born–Oppenheimer approximation we can say that the wave equation for the electron can be solved exactly. We shall not concentrate on the exact solution for the moment, however, but look at the LCAO forms of the molecular orbitals, since these will form the basis of our general description of homonuclear diatomic molecules.

The lowest energy atomic orbital of the hydrogen atom is the $1s$ orbital. We therefore expect the lowest energy molecular orbital of H_2^+ to look like a $1s$ atomic orbital near the nuclei, and in the LCAO approximation this is expressed as

$$\psi = c_a 1s_a + c_b 1s_b, \tag{10.2}$$

where the two nuclei are labelled a and b.

An electron in ψ has a probability density

$$\psi^2 = c_a^2(1s_a)^2 + c_b^2(1s_b)^2 + 2c_a c_b(1s_a 1s_b). \tag{10.3}$$

Because the two atoms in H_2^+ are indistinguishable the density around nucleus a must be equal to that around nucleus b, and this is only so if

$$c_a^2 = c_b^2. \tag{10.4}$$

There are only two orbitals which satisfy this condition, they are

$$\begin{aligned} c_a &= c_b & \psi_g &= N_g(1s_a + 1s_b), \\ c_a &= -c_b & \psi_u &= N_u(1s_a - 1s_b), \end{aligned} \tag{10.5}$$

where the N are normalizing constants. These orbitals are shown schematically in figure 10.1. In arriving at expressions (10.5) we used an argument based on the symmetry of the molecule; we could have used the more formal methods of group theory to obtain the same result (cf. p. 99). The symbols g and u are the conventional group theory symbols which

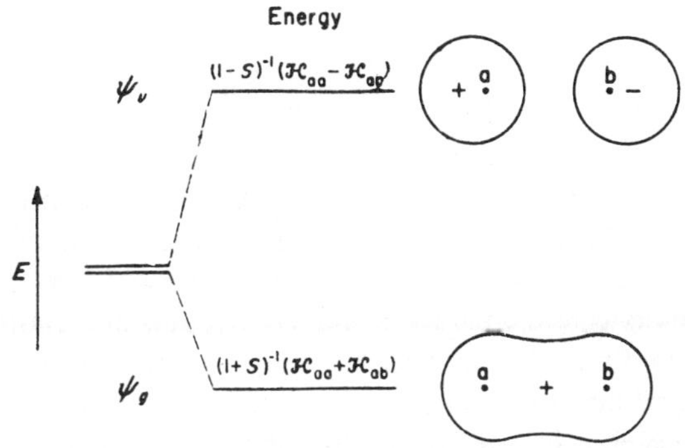

FIGURE 10.1. Representation of the molecular orbitals of H_2^+

indicate that the orbital is symmetric (g) or antisymmetric (u) to inversion through the centre of symmetry (page 98).

The probability densities for our two orbitals are

$$\psi_g^2 = N_g^2(1s_a^2 + 1s_b^2 + 2\ 1s_a 1s_b),$$

$$\psi_u^2 = N_u^2(1s_a^2 + 1s_b^2 - 2\ 1s_a 1s_b), \qquad (10.6)$$

and for the orbitals to be normalized we have

$$\int \psi_g^2 dv = N_g^2 \int (1s_a^2 + 1s_b^2 + 2\ 1s_a 1s_b) dv = 1,$$

$$\int \psi_u^2 dv = N_u^2 \int (1s_a^2 + 1s_b^2 - 2\ 1s_a 1s_b) dv = 1. \qquad (10.7)$$

Introducing the conventional symbol S for the overlap integral

$$S = \int 1s_a 1s_b\, dv, \qquad (10.8)$$

and assuming that the atomic orbitals themselves are normalized, we have

$$N_g^2(2 + 2S) = 1,$$

$$N_u^2(2 - 2S) = 1. \qquad (10.9)$$

In chapter 5 we gave a qualitative account of the importance of the overlap integral, and noted that it was a rough measure of the strength of a covalent bond. Using the explicit form of the hydrogen $1s$ atomic orbital (table 3.3)

$$1s_a = \pi^{-\frac{1}{2}} e^{-r_a}, \qquad (10.10)$$

the overlap integral is found to be

$$S = e^{-R}\left(1 + R + \frac{R^2}{3}\right), \tag{10.11}$$

where R is the internuclear distance in atomic units. The inverse exponential function of R which appears in this expression should be noted.

The product $(1s_a 1s_b)$ is called an overlap density and has its largest value in the regions of space where $1s_a$ and $1s_b$ overlap one another, which is between the two nuclei. This overlap density contributes positively to the density $\psi_g{}^2$ and negatively to $\psi_u{}^2$; as a result, the density $\psi_g{}^2$ is greater in the region between the two nuclei than the density of an electron shared equally between $1s_a$ and $1s_b$, and the density $\psi_u{}^2$ is less in this region.

Simple electrostatics would suggest that an electron in ψ_g would have a lower energy than one in ψ_u, because the density $\psi_g{}^2$ is greater than $\psi_u{}^2$ between the two nuclei where the combined field of both nuclei is strong. To calculate the energies of these orbitals exactly we must evaluate the integral

$$E = \int \psi \mathscr{H} \psi \, dv \tag{10.12}$$

for each case.

Introducing expressions (10.5) and (10.9) into (10.12), we have

$$E_g = (2 + 2S)^{-1} \int (1s_a + 1s_b) \mathscr{H} (1s_a + 1s_b) dv,$$

$$E_u = (2 - 2S)^{-1} \int (1s_a - 1s_b) \mathscr{H} (1s_a - 1s_b) dv. \tag{10.13}$$

Introducing the abbreviations

$$\mathscr{H}_{aa} = \int 1s_a \mathscr{H} 1s_a \, dv = \mathscr{H}_{bb}, \tag{10.14}$$

and

$$\mathscr{H}_{ab} = \int 1s_a \mathscr{H} 1s_b \, dv = \mathscr{H}_{ba}, \tag{10.15}$$

these expressions become

$$E_g = (1 + S)^{-1} (\mathscr{H}_{aa} + \mathscr{H}_{ab}),$$

$$E_u = (1 - S)^{-1} (\mathscr{H}_{aa} - \mathscr{H}_{ab}). \tag{10.16}$$

The electronic Hamiltonian for $H_2{}^+$ is (in atomic units)

$$\mathscr{H} = -\tfrac{1}{2}\nabla^2 - \frac{1}{r_a} - \frac{1}{r_b} + \frac{1}{R}. \tag{10.17}$$

We can say that $\mathscr{H}_{aa}$ represents the energy of a hydrogen atom (a) perturbed by a proton (b), and this is quite close to the energy of an unperturbed hydrogen atom except for very small values of R. Integrals of the type $\mathscr{H}_{ab}$ are generally called *resonance integrals*. The resonance integral represents the energy of the overlap density under the influence of the attractive field of both nuclei. $\mathscr{H}_{ab}$ is a negative quantity which stabilizes ψ_g and destabilizes ψ_u, relative to a hydrogen atom and a bare proton. ψ_g is called a bonding molecular orbital, ψ_u is an antibonding molecular orbital.

We can obtain more explicit expressions for the energy if we make use of the fact that the $1s$ orbital is a solution of the Schrödinger equation for a hydrogen atom, that is,

$$\left(-\frac{1}{2}\nabla^2 - \frac{1}{r_a}\right)1s_a = E_H 1s_a,$$

and

$$\left(-\frac{1}{2}\nabla^2 - \frac{1}{r_b}\right)1s_b = E_H 1s_b, \tag{10.18}$$

where E_H is the energy of the hydrogen $1s$ atomic orbital. Expression (10.16) then becomes

$$E_g = E_H + \frac{1}{R} + (1+S)^{-1}(\varepsilon_{aa} + \varepsilon_{ab}),$$

$$E_u = E_H + \frac{1}{R} + (1-S)^{-1}(\varepsilon_{aa} - \varepsilon_{ab}), \tag{10.19}$$

where

$$\varepsilon_{aa} = \int 1s_a\left(-\frac{1}{r_b}\right)1s_a\, dv,$$

and

$$\varepsilon_{ab} = \int 1s_a\left(-\frac{1}{r_b}\right)1s_b\, dv. \tag{10.20}$$

If these integrals are evaluated using the functional form (10.10) for the $1s$ orbital we find

$$\varepsilon_{aa} = -\frac{1}{R}\left\{1 - e^{-2R}(1+R)\right\}; \qquad \varepsilon_{ab} = -e^{-R}(1+R). \tag{10.21}$$

The dependence of these integrals on the internuclear distance R should be noted: when R is large ε_{aa} varies as $1/R$, and ε_{ab} varies inverse exponentially with R (like the overlap integral). ε_{aa} is largely cancelled out by the nuclear repulsion $1/R$; so one might say that ε_{ab} is the term which, except for small values of R, is mainly responsible for ψ_g being more stable and ψ_u being less stable than a separate hydrogen atom and a proton. The potential energy curves for these two states of $H_2{}^+$ are shown in figure 10.2.

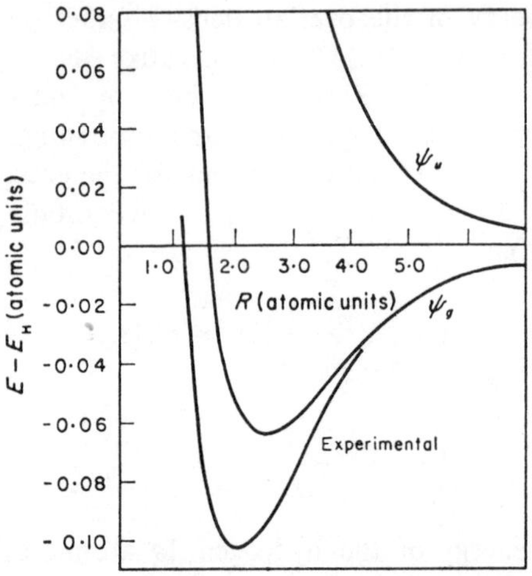

FIGURE 10.2. The potential energy curves for H_2^+

The binding energy that is obtained for $H_2{}^+$ by the calculation described above is about 60% of the correct value. At the end of the chapter we shall discuss some more sophisticated calculations which have been carried out in the molecular-orbital framework; at this point, however, we shall go on to a qualitative discussion of the electronic structure of H_2 and the homonuclear diatomic molecules, Li_2 to F_2, of the first row elements.

10.2 Molecular orbitals of homonuclear diatomic molecules

We have seen how two molecular orbitals can be formed when two $1s$ atomic orbitals overlap one another. We expect a similar situation to arise when two $2s$ orbitals overlap: there will be a bonding orbital $2s_a + 2s_b$ and an antibonding orbital $2s_a - 2s_b$, and the general form of these orbitals will be similar to that of their $1s$ counterpart except that each $2s$ atomic orbital has a radial node near to the nucleus and these will be present in both the molecular orbitals as well.

When two sets of $2p$ atomic orbitals overlap, six molecular orbitals can be obtained. We can have the $2p$ orbitals pointing towards each other along the internuclear axis (which we shall take to be the z axis), and these will give a bonding molecular orbital ($2p_{za} + 2p_{zb}$) and an antibonding orbital

$(2p_{za} - 2p_{zb})$, as shown in figure 10.3. Alternatively, the $2p$ orbitals can be perpendicular to the internuclear axis, for example, pointing in the x direction, and we again have a bonding orbital $(2p_{xa} + 2p_{xb})$ and an antibonding orbital $(2p_{xa} - 2p_{xb})$, as shown in figure 10.3. Since the x and y axes are completely equivalent in space there will be corresponding molecular orbitals $(2p_{ya} + 2p_{yb})$ and $(2p_{ya} - 2p_{yb})$ which, by symmetry, must be degenerate with their x counterparts.

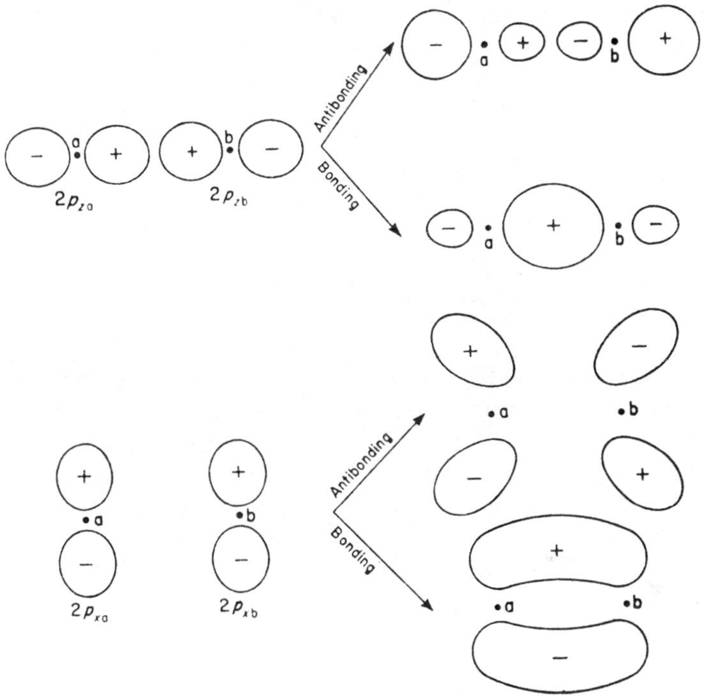

FIGURE 10.3. Molecular orbitals of a homonuclear diatomic molecule formed from $2p$ atomic orbitals

The bonding or antibonding nature of an orbital can be decided simply from the distribution of density of an electron in that orbital: if this is increased between the nuclei then the orbital is bonding, if it is depleted in this region it will be antibonding.

The molecular orbitals of a homonuclear diatomic molecule can be described in terms of symmetry symbols and angular momentum symbols (which are really also symmetry symbols of the $\mathbf{C}_{\infty v}$ and $\mathbf{D}_{\infty h}$ groups) in a similar way to that in which atomic orbitals are labelled. Atomic orbitals are labelled by two angular momentum symbols l and m. For each l there are $2l + 1$ values of m and these $2l + 1$ atomic orbitals are degenerate. However, a diatomic molecule has a lower symmetry than an atom. As an analogy we can say that the symmetry of diatomic molecule is the same as that of an atom in the presence of an axial electric field, a situation in which the Stark effect is observed (page 125). Under these circumstances the $(2l + 1)$-fold degeneracy of the atomic orbital is lost and the atomic orbitals are characterized only by their m quantum numbers, which indicate the component of the angular momentum of the electron along the direction of the electric field. There is an analogous quantum number λ for the molecular orbitals of a diatomic molecule, which also indicates the component of the orbital angular momentum along the internuclear axis. It is important to realize that λ is analogous to m and not to l.

The orbitals of a diatomic molecule can be written in the form

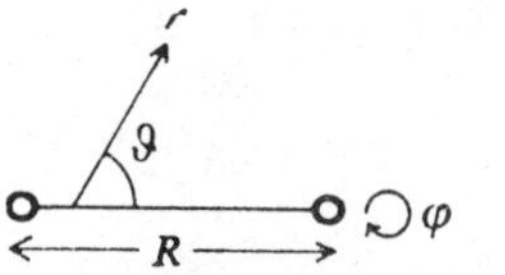

$$\psi = F(r, \vartheta)e^{i\lambda\varphi}$$

$$\lambda = 0, \pm 1, \pm 2, \ldots \qquad (10.22)$$

This expression should be compared with that for atoms (3.6) where, due to the higher symmetry, the function $F(r, \vartheta)$ is further separable. The component of angular momentum along the internuclear axis is $\pm\lambda h/2\pi$. Except when $\lambda = 0$, the orbitals are all *doubly* degenerate corresponding to a positive or negative angular momentum along the nuclear axis (cf. the Stark effect where the orbitals $+m$ and $-m$ are degenerate).

An orbital is labelled according to its value of λ in the following way

$$|\lambda| = 0, 1, 2, 3, \ldots$$

$$\text{orbital type} = \sigma, \pi, \delta, \varphi \ldots$$

The angular momentum symbols apply to both homo- and heteronuclear diatomic molecules. Homonuclear diatomic molecules have a higher symmetry ($\mathbf{D}_{\infty h}$) than heteronuclear ($\mathbf{C}_{\infty v}$) and their orbitals may be further classified by the behaviour of the orbital on inversion through the centre of symmetry. From figures 10.1 and 10.3 the orbital characteristics shown in table 10.1 can be derived.

A further label can be introduced which is analogous to the principal quantum number of atomic orbitals and indicates the order of increasing energy of the orbitals. For example, the orbital $1s_a + 1s_b$ will be $1\sigma_g$, $1s_a - 1s_b$ will be $1\sigma_u$, $2s_a + 2s_b$ will be $2\sigma_g$, $2p_{za} + 2p_{zb}$ will be $3\sigma_g$, etc. For heteronuclear diatomic molecules, where there is no g–u symmetry,

TABLE 10.1. The characteristics of the molecular orbitals formed from 1*s*, 2*s* and 2*p* atomic orbitals for a homonuclear diatomic molecule

Atomic-orbital combination	Angular momentum symbol	Inversion	Bonding properties
$1s_a + 1s_b$	σ	g	bonding
$1s_a - 1s_b$	σ	u	antibonding
$2s_a + 2s_b$	σ	g	bonding
$2s_a - 2s_b$	σ	u	antibonding
$2p_{za} + 2p_{zb}$	σ	g	bonding
$2p_{za} - 2p_{zb}$	σ	u	antibonding
$2p_{xa} + 2p_{xb}$, $2p_{ya} + 2p_{yb}$	π	u	bonding
$2p_{xa} - 2p_{xb}$, $2p_{ya} - 2p_{yb}$	π	g	antibonding

these orbitals will simply be $1\sigma, 2\sigma, 3\sigma$, etc. An alternative scheme, which we use in this book, is to label the orbitals by the atomic orbitals into which they dissociate when the nuclei are pulled apart: $\sigma_g 1s$, $\sigma_u 1s$, $\sigma_g 2s$, etc.

There are two things that determine the energy of a molecular orbital: the type of atomic orbital from which it is built up, and the overlap between these atomic orbitals. For example, $\sigma_g 1s$ and $\sigma_u 1s$ have a much lower energy than $\sigma_g 2s$ simply because the 1*s* atomic orbital is much lower in energy than the 2*s*. Likewise, $\sigma_g 2s$ is lower than $\sigma_g 2p$ because, except for the hydrogen atom, the 2*s* atomic orbitals have an appreciably lower energy than the 2*p*. Also, except at rather small internuclear distances, the overlap between two 2*s* orbitals or two $2p_z$ orbitals is much greater than the overlap between two $2p_x$ or $2p_y$ orbitals. It follows that the separation of the bonding and antibonding π orbitals is less than the separation of the bonding and antibonding σ orbitals. From these arguments we expect the following order of energies

$$\sigma_g 1s < \sigma_u 1s < \sigma_g 2s < \sigma_u 2s < \sigma_g 2p < \pi_u 2p < \pi_g 2p < \sigma_u 2p. \quad (10.23)$$

However, the 2*s* and 2*p* atomic orbitals are not so far apart in energy that they can be treated as non-interacting. Indeed when hybridization was introduced to explain the facts of directed valency (chapter 5) its introduction was justified by the fact that the 2*s* and 2*p* atomic orbitals have about the same energy.

If the 2*s* and 2*p* atomic orbitals are rather close in energy, then the σ molecular orbitals associated with the L $(n = 2)$ shell of atomic orbitals must be written as a mixture of 2*s* and $2p_z$ orbitals

$$\sigma_g = c_1(2s_a + 2s_b) + c_2(2p_{za} + 2p_{zb}),$$
$$\sigma_u = c_3(2s_a - 2s_b) + c_4(2p_{za} - 2p_{zb}). \quad (10.24)$$

We can conclude from perturbation theory (expression 6.54) that the interaction between the $2s$ and $2p_z$ orbitals (which occurs because $2s_a$ and $2p_{zb}$ orbitals overlap one another), will lower the energy of the first two σ orbitals of this group, $\sigma_g 2s$ and $\sigma_u 2s$, and raise the energy of the second two σ orbitals $\sigma_g 2p$ and $\sigma_u 2p$. The extent of this interaction decreases as the difference between the energies of the $2s$ and $2p$ atomic orbitals increases, and so it becomes less important from Li_2 to F_2. Because of $2s$–$2p$ interaction, it is found that for the lighter molecules of this series the orbital $\sigma_g 2p$ (we still use this label although it contains some $2s$) gets pushed above $\pi_u 2p$, and this gives the order

$$\sigma_g 1s < \sigma_u 1s < \sigma_g 2s < \sigma_u 2s < \pi_u 2p < \sigma_g 2p < \pi_g 2p < \sigma_u 2p. \qquad (10.25)$$

This order holds for all the first row homonuclear diatomic molecules except O_2 and F_2, but for these molecules both $\pi_u 2p$ and $\sigma_g 2p$ are completely filled by electrons so that their order is only important if one needs to describe the excited states of O_2 and F_2. The electronic spectrum of O_2 shows that the first excited electronic configuration is associated with the transition $\pi_u 2p \rightarrow \pi_g 2p$ rather than $\sigma_g 2p \rightarrow \pi_g 2p$, so that $\sigma_g 2p$ is below $\pi_u 2p$. Leaving these excited states aside, however, the order of orbital energies (10.25) can explain the symmetries of the ground states of all the first row homonuclear diatomic molecules and their positive ions in the same way that the order of atomic-orbital energies can explain the symmetries of the ground states of the elements (chapters 4 and 9). Table 10.2 shows the lowest energy electron configurations of the homonuclear diatomic molecules H_2 to F_2 and some of their positive ions.

The strength of a bond depends roughly on the net excess of electrons in bonding orbitals over those in antibonding orbitals. Thus the dissociation energy of H_2 is about twice that of H_2^+, and the latter is about the same as that of He_2^+. He_2, which would have two bonding and two antibonding electrons, is not stable in its ground state, although some excited states exist which have a positive dissociation energy relative to one excited and one ground state helium atom. If there are equal numbers of bonding and antibonding electrons then the net antibonding effect always just outweighs the net bonding effect; thus Be_2 is also not a stable molecule.

The strongest bond is that of N_2 which has six more bonding electrons than antibonding (which fits in with the chemical structure $N \equiv N$). When an electron is removed from N_2 the bond becomes weaker and longer. O_2, on the other hand, has only four net bonding electrons, and the electron most easily removed is one from the antibonding orbital $\pi_g 2p$. For this reason the bond in O_2^+ is stronger (and shorter) than that of O_2.

TABLE 10.2. The ground states of the homonuclear diatomic molecules up to F_2

Molecule	Electronic configuration	Net excess of bonding electrons	Ground state	Dissociation energy[a] (ev)	Bond length[b] (Å)
H_2^+	$\sigma_g 1s$	1	$^2\Sigma_g^+$	2·65	1·06
H_2	$(\sigma_g 1s)^2$	2	$^1\Sigma_g^+$	4·48	0·74
He_2^+	$(\sigma_g 1s)^2(\sigma_u 1s)$	1	$^2\Sigma_u^+$	(3·1)	1·08
He_2	$(\sigma_g 1s)^2(\sigma_u 1s)^2$	0	$^1\Sigma_g^+$	—	—
Li_2	$KK(\sigma_g 2s)^2$	2	$^1\Sigma_g^+$	1·1	2·67
Be_2	$KK(\sigma_g 2s)^2(\sigma_u 2s)^2$	0	$^1\Sigma_g^+$	—	—
B_2	$[Be_2](\pi_u 2p)^2$	2	$^3\Sigma_g^-$	3·0 ± 0·5	1·59
C_2	$[Be_2](\pi_u 2p)^4$	4	$^1\Sigma_g^+$	6·2	1·24
N_2^+	$[Be_2](\pi_u 2p)^4\sigma_g 2p$	5	$^2\Sigma_g^+$	8·73	1·12
N_2	$[Be_2](\pi_u 2p)^4(\sigma_g 2p)^2$	6	$^1\Sigma_g^+$	9·76	1·09
O_2^+	$[Be_2](\sigma_g 2p)^2(\pi_u 2p)^4\pi_g 2p$	5	$^2\Pi_g$	6·48	1·12
O_2	$[Be_2](\sigma_g 2p)^2(\pi_u 2p)^4(\pi_g 2p)^2$	4	$^3\Sigma_g^-$	5·08	1·21
F_2	$[Be_2](\sigma_g 2p)^2(\pi_u 2p)^4(\pi_g 2p)^4$	2	$^1\Sigma_g^+$	1·6 ± 0·35	1·44

[a] GAYDON, *Dissociation Energies*, Chapman and Hall, 1953.
[b] HERZBERG, *Spectra of Diatomic Molecules*, van Nostrand, 1950.

(BALLIK and RAMSEY (J. *Chem. Phys*., **31**, 1128 (1959)) showed that the $^1\Sigma_g^+$ state for C_2 is lower than the state $(\pi_u 2p)^3\ (\sigma_g 2p)^1\ ^3\Pi_u$, which was earlier thought to be the ground state, by about 0·1 ev.

The symbol *KK* indicates that at the equilibrium internuclear distance the 1*s* orbitals do not overlap one another to any significant extent and are best considered as being non-bonding.

The dissociation energy of He_2^+ is taken from reference *b* but its value is rather uncertain.

The overall state of an atom is described by the quantum numbers of the total spin and orbital angular momentum operators. We likewise describe the state of a diatomic molecule by the total spin angular momentum and the total orbital angular momentum along the internuclear axis. The total spin multiplicity $2S + 1$ is written as a superior prefix as for atomic states. The total orbital angular momentum is indicated by a capital Greek letter following the scheme

$$|\textstyle\sum\lambda| = 0, 1, 2, 3 \ldots$$

$$\text{Symbol} = \Sigma, \Pi, \Delta, \Phi \ldots$$

An electron configuration which has all orbitals completely filled or empty, and no degenerate sets of orbitals partly filled, is called a *closed-shell* structure and is a $^1\Sigma$ state; there must be as many electrons with α spin as with β spin, and as many with angular momentum $+\lambda h/2\pi$ as with $-\lambda h/2\pi$. Thus H_2, Li_2, C_2, N_2 and F_2 have $^1\Sigma$ ground states.

The overall symmetry of a state to inversion (which applies only to homonuclear diatomic molecules), is obtained by multiplying all the symmetries of the individual electrons according to the scheme $g \times g = u \times u = g$, $g \times u = u$. Only if there is an odd number of electrons in u orbitals can a state of u symmetry (e.g. He_2^+) exist.

There is one more symmetry operation for a diatomic molecule which we have not mentioned so far: any plane containing the internuclear axis is a plane of symmetry, and wave functions must be symmetric or antisymmetric to reflection in this plane. All σ orbitals are symmetric to this operation, so the label is not used since it does not tell us anything new. Likewise, the doubly-degenerate orbitals $\pi, \delta, \ldots$ can all be chosen in such a way that for a given symmetry plane one orbital of the pair will be symmetric and one antisymmetric to reflection; for this reason this symmetry behaviour is again not very informative. The same applies to the doubly-degenerate states $\Pi, \Delta, \Phi, \ldots$; we can always choose one component which is symmetric and one which is antisymmetric. It is only in the case of Σ states that the behaviour under reflection in a symmetry plane is informative. The symbol Σ just indicates that there are as many electrons with angular momentum $+\lambda h/2\pi$ as with $-\lambda h/2\pi$. We can have a Σ state from an open-shell configuration and this can be symmetric or antisymmetric to reflection in a symmetry plane. This is illustrated by the following example.

Suppose there are two electrons occupying a degenerate pair of π orbitals and that their orbital angular momenta cancel. There are two electron arrangements of this type, which we can describe by the wave functions (see (10.22))

$$R(1,2)\{e^{+i\varphi}(1)e^{-i\varphi}(2)\}, \qquad R(1,2)\{e^{+i\varphi}(2)e^{-i\varphi}(1)\},$$

where $e^{+i\varphi}(1)$ indicates that electron 1 is in the orbital with angular momentum $+h/2\pi$, and $R(1,2)$ is the part of the wave function independent of φ. To make the two electrons equivalent we must take wave functions

$$\Psi_\pm = R(1,2)\{e^{+i\varphi}(1)e^{-i\varphi}(2) \pm e^{+i\varphi}(2)e^{-i\varphi}(1)\}.$$

If we now reflect these functions in the plane containing $\varphi = 0$, then we just replace φ by $-\varphi$ and obtain ($\mathcal{R}$ stands for the reflection operator)

$$\mathcal{R}\,\Psi_\pm = R(1,2)\{e^{-i\varphi}(1)e^{+i\varphi}(2) \pm e^{-i\varphi}(2)e^{+i\varphi}(1)\} = \pm\,\Psi. \quad (10.26)$$

The function Ψ_+ is unchanged by this operation and is labelled Σ^+; Ψ_- changes sign and is labelled Σ^-. All closed-shell electron configurations give only Σ^+ states; an open-shell configuration e.g. $(\pi_u 2p)^2$ can give both Σ^+ and Σ^- states.

Two molecules in table 10.2 have half-filled π shells. These electron configurations can give more than one electronic state depending on how the spin and orbital angular momenta of the two electrons are arranged. We have here a similar situation to the open-shell atomic configuration and we shall examine it in the same way that we looked at the p^2 configuration (page 116). We can allocate electrons to the π orbitals, taking account of the Pauli exclusion principle, as follows (cf. table 9.3).

$\lambda = 1$	$\lambda = -1$	$\Sigma\lambda$	Σm_s	
↑ ↓		2	0	$^1\Delta$
↑	↑	0	1	$^3\Sigma^-$
↑	↓	0	0	$\{^1\Sigma^+, {}^3\Sigma^-$
↓	↑	0	0	
↓	↓	0	−1	$^3\Sigma^-$
	↑ ↓	−2	0	$^1\Delta$

The arrangement with $\sum\lambda = 2$ must be a component of a Δ state and since its total spin is zero this must be a $^1\Delta$ state. This state is doubly degenerate (not five-fold degenerate as is an atomic ^1D state), the other component having $\sum\lambda = -2$. There are four arrangements which give $\sum\lambda = 0$ and these must all belong to Σ states. Clearly three of these are the components of a triplet, since there is an arrangement with $\sum m_s = 1$, and the fourth must be $^1\Sigma$. A triplet spin function for two electrons is symmetric to exchange of two electrons (9.23), hence if the Pauli exclusion principle is to be satisfied the space part of the $^3\Sigma$ state must be antisymmetric to the exchange of electrons (since the whole function must be antisymmetric). It follows that the space part of the $^3\Sigma$ state must be of the form of Ψ_- in (10.26), so that the state is $^3\Sigma^-$. The singlet spin function is antisymmetric to exchange of electrons so it must be combined with a symmetric space function and is therefore a $^1\Sigma^+$ state.

The configuration π^2 for a homonuclear diatomic molecule therefore gives rise to the states $^1\Delta_g$, $^3\Sigma_g{}^-$, $^1\Sigma_g{}^+$. In agreement with Hund's rule the state of lowest energy is the triplet, and $^1\Delta_g$ is lower than $^1\Sigma_g{}^+$.

Both B_2 and O_2 have been shown to have $^3\Sigma_g{}^-$ ground states by spectroscopy. The triplet spin also means that these molecules are paramagnetic. The simple way in which the nature of the ground states of these molecules has been explained is one of the triumphs of molecular-orbital theory. To explain their structure by valence-bond theory is, as we shall see in the next chapter, much more difficult.

10.3 Correlation diagrams

Let us now return to H_2^+ and examine the relative energies of the molecular orbitals as a function of the internuclear distance; these are shown in figure 10.4. On the right-hand side of the figure we have the hydrogen energy levels $-1/2n^2$ and on the left the energy levels of He^+ $-2/n^2$, both relative to the energy of a free electron.

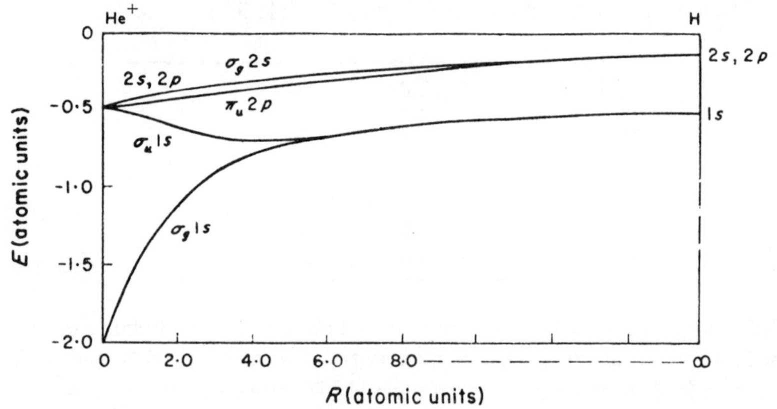

FIGURE. 10.4. Correlation diagram for orbitals of H_2^+

The orbital $\sigma_g 1s$ will pass smoothly into the helium $1s$ orbital when $R = 0$. It is clear from the diagram that this orbital is strongly bonding. The orbital $\sigma_u 1s$ does not correlate with the helium $2s$ orbital, since this has the wrong symmetry under inversion, but with one of the $2p$ orbitals; when one includes the nuclear repulsion energy this orbital is antibonding. $\sigma_g 2s$ will correlate with the helium $2s$, and its energy will be lowered, but $\sigma_u 2s$ must correlate with the next orbital of σ_u symmetry not so far accounted for. This is one of the $3p$ orbitals, and at small R it will be strongly antibonding. The bonding $\pi_u 2p$ correlates with the remaining $2p$ orbitals, but the bonding $\sigma_g 2p$ must correlate with $3s$. It can now be seen that for short internuclear distances as well as small $2s - 2p$ energy differences the bonding $\pi_u 2p$ will have a lower energy than the bonding $\sigma_g 2p$.

Figure 10.4 is called a correlation diagram. H_2^+ of course is a special case because the $2s$ and $2p$ energies are degenerate both for the separate and united atoms. In general this will not be so and we shall have a situation illustrated schematically in figure 10.5. These diagrams tell us the general pattern of molecular-orbital energies as a function of internuclear distance,

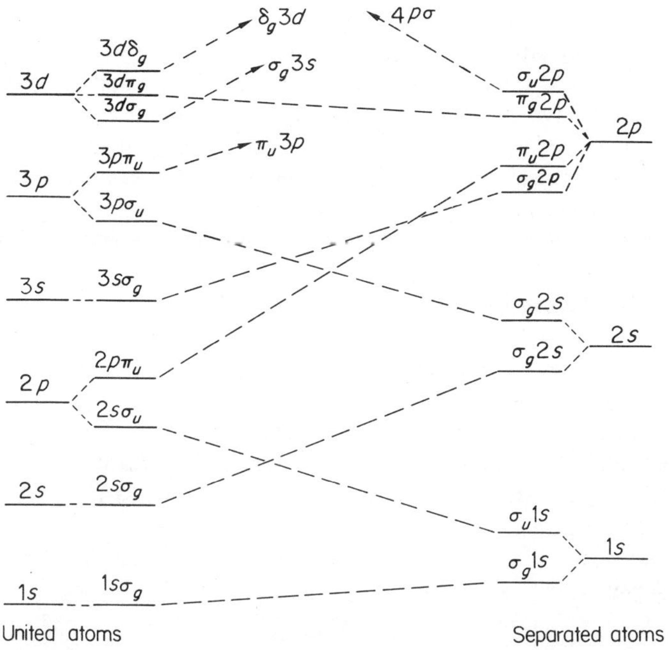

FIGURE 10.5. Correlation diagram for the orbitals of a homonuclear diatomic molecule

and knowing this we can, as we have seen, predict the symmetries of the low energy electronic states of the molecules.

Under some circumstances a molecular orbital is better represented by the atomic-orbital wave function of the united atom than by a linear combination of the atomic orbitals of separate atoms. For example, all molecules show a series of excited states called *Rydberg states* whose energies converge to the energy of the ionized molecule. The energies of these excited states follow roughly the pattern of the hydrogen-like orbitals ($-1/n^2$), indicating that they represent the energy of an electron in an orbital which is so large compared with size of the molecular positive ion that the potential acting on this electron is roughly the same as if the net charge of the ion ($+1$) were placed at the centre of charge. These Rydberg orbitals are, to a good approximation, represented by the corresponding Rydberg atomic orbitals of the united atom.

Another situation in which united atom orbitals have been used is when we have a molecule in which most of the electrons come from one atom. For example, in HF the electron density is almost entirely due to the fluorine atom. In these cases the molecular orbitals will approximate to the orbitals of the united atom except near the two nuclei; if we are calculating some property which does not depend too sensitively on the electron density near the nuclei then united atom orbitals may be very useful. Calculations have been made along these lines for the first row hydrides CH_4 to HF starting with the atomic orbitals of Ne. Platt, in particular, has given a very useful analysis of bond lengths and force constants of these molecules by such an approach†.

10.4 *Heteronuclear diatomic molecules*

For heteronuclear diatomic molecules the LCAO coefficients cannot be determined by symmetry. Instead they are determined by a method based on the variation theorem. Since the determination of molecular-orbital energies and their LCAO coefficients is extremely important, we shall derive the necessary equations in a somewhat simpler but less rigorous manner than that given in chapter 6.

Let us start by assuming that the molecular orbitals, written in their LCAO form

$$\psi = \sum_{\nu} c_\nu \phi_\nu, \tag{10.27}$$

are eigenfunctions of some Hamiltonian **H**.

$$\mathbf{H}\psi = E\psi. \tag{10.28}$$

At this point we shall not discuss the nature of **H** except to emphasize that it is only a one-electron operator (since ψ is a one-electron function) which is only unambiguously defined for a molecule having one electron (e.g. H_2^+): in this case **H** is the actual electronic Hamiltonian for the molecule. We shall leave further discussion on this point to the end of the chapter and say for the moment that **H** contains the kinetic energy operator for one electron and the average effective potential acting on that electron due to all the nuclei and the other electrons in the molecule.

If we substitute (10.27) into (10.28), we get

$$\sum_{\nu} c_\nu(\mathbf{H} - E)\phi_\nu = 0. \tag{10.29}$$

To determine the coefficients we now multiply from the left-hand side by one of the atomic orbitals ϕ_μ, say (or its complex conjugate if we are dealing

† PLATT, J. *Chem. Phys.*, **18**, 932 (1950).

with complex functions), and integrate over all the three-dimensional space occupied by the electron. This leads to

$$\sum_{\nu} c_{\nu}(\mathbf{H}_{\mu\nu} - S_{\mu\nu}E) = 0, \tag{10.30}$$

where

$$\mathbf{H}_{\mu\nu} = \int \phi_{\mu}\mathbf{H}\phi_{\nu}\, dv \tag{10.31}$$

and

$$S_{\mu\nu} = \int \phi_{\mu}\phi_{\nu}\, dv. \tag{10.32}$$

There is one equation of the type (10.30) for every atomic orbital which is taken in the function (10.27). If we take n atomic orbitals, we then obtain n simultaneous equations in the n unknown coefficients; these are called the secular equations. These equations only have a non-trivial solution for the coefficients (the trivial solution is all $c_{\nu} = 0$) if the $n \times n$ secular determinant is zero; that is (cf. 6.68)

$$|\mathbf{H}_{\mu\nu} - S_{\mu\nu}E| = 0. \tag{10.33}$$

On expanding this determinant we obtain an equation of the nth degree in E which has n solutions. These are the energies of the n molecular orbitals which can be formed by taking linear combinations of n atomic orbitals.

Suppose E_r is one solution of (10.33). To find the coefficients of the orbital ψ_r having this energy, we substitute this value of E into the secular equations (10.30), and solve in the usual way to obtain the ratios of the coefficients. The absolute values of the coefficients are determined by the normalization condition.

The secular equations (10.30) have exactly the same form as the equations (6.67) which result from applying the variation principle to a linear combination of functions with unknown coefficients. It follows that the molecular orbitals which are obtained by the above method will be the best orbitals, according to the criterion of the variation principle, even when we choose only a small number of atomic orbitals in our expansion —too small for us to be able to assert that ψ is an eigenfunction of $\mathbf{H}$.

As an example of a molecular-orbital calculation using the equations given above we take the lithium hydride molecule. We shall make only a very crude estimate of the integrals, but this will enable us to grasp the qualitative features of the molecular orbitals of heteronuclear diatomic molecules. Lithium in its ground state has the electron configuration $1s^2 2s$, hydrogen has an electron in a $1s$ orbital. We anticipate that the bond will arise from the delocalization of the lithium 2*s* electron and the hydrogen 1*s* electron over the whole molecule. In the simplest application of MO theory this delocalization would be introduced by forming molecular orbitals which are a combination of $2s_{\mathrm{Li}}$ and $1s_{\mathrm{H}}$. To make the orbitals a bit more flexible, some of the lithium $2p\sigma$ orbital† will also be included in the wave function. We shall assume that the lithium 1*s* orbital plays no part in the bond formation, since it does not overlap the hydrogen 1*s* orbital to any great extent. This means that we treat LiH essentially as a two-electron problem.

The molecular orbitals of σ symmetry are taken to have the form

$$\psi = c_h\phi_h + c_s\phi_s + c_p\phi_p, \tag{10.34}$$

where ϕ_h is the hydrogen 1*s* orbital, ϕ_s is the lithium 2*s* and ϕ_p the lithium $2p\sigma$ orbital. Karo and Olsen have evaluated all the necessary integrals for LiH with this set of atomic orbitals, and we shall use their data‡. At the equilibrium bond length of 3 atomic units (1·6 Å) the overlap integrals have the values§

$$S_{hh} = S_{ss} = S_{pp} = 1, \qquad S_{sp} = 0, \qquad S_{hs} = 0{\cdot}469, \qquad S_{hp} = 0{\cdot}506.$$

We come now to the question of the Hamiltonian integrals. Suppose first that we were looking at the molecule LiH^+. We could then say that the outer electron is moving in the potential of the +1 charge on the proton, the +3 charge of the lithium nucleus and the two lithium 1*s* electrons. If $V(1s^2)$ represents the potential due to a 1*s* electron then the Hamiltonian for LiH^+ would be

$$\mathbf{H} = -\tfrac{1}{2}\nabla^2 - \frac{1}{r_{\mathrm{H}}} - \frac{3}{r_{\mathrm{Li}}} + 2V(1s^2). \tag{10.35}$$

In LiH, however, we have to include the potential due to the other outer electron. But we do not know this potential until we have found the appropriate molecular orbital. A cyclic calculation must be carried out, first assuming some crude form for the molecular orbital and calculating the potential due to an electron in this orbital. This potential is added to (10.35), and the molecular orbitals appropriate to this Hamiltonian are found. These molecular orbitals are then used to give a better potential, and the process is repeated until a self-consistent situation is reached: a

† We use the label $2p\sigma$ for the *p* orbital pointing along the bond axis; this is a common abbreviation which avoids the necessity of specifying coordinate axes.

‡ KARO and OLSEN, *J. Chem. Phys.*, **30**, 1232 (1959).
KARO, *J. Chem. Phys.*, **30**, 1241 (1959).

§ We have changed the sign of the 2*s* wave function used by Karo and Olsen in order to have all overlap integrals positive.

potential produces orbitals identical to those from which it was derived. This type of calculation is analogous to the self-consistent-field (SCF) treatment for atoms, which was described in chapter 4.

We shall simply perform what could be the first cycle in an SCF calculation on LiH. Let us assume that the lowest energy molecular orbital (apart from $1s_{\text{Li}}$), which is going to receive the two electrons, is equally distributed between the two atoms, and we shall take the density of an electron in this orbital to be

$$\tfrac{1}{2}\phi_h^2 + \tfrac{1}{4}\phi_s^2 + \tfrac{1}{4}\phi_p^2.$$

We know this is not going to be very close to the final result, since hydrogen is more electronegative than lithium, but at least we have some point at which to start the calculation.

The Hamiltonian is then

$$\mathbf{H} = -\frac{1}{2}\nabla^2 - \frac{1}{r_{\text{H}}} - \frac{3}{r_{\text{Li}}} + 2V(1s_{\text{Li}}{}^2) + \frac{1}{2}V(\phi_h{}^2) + \frac{1}{4}V(\phi_s{}^2) + \frac{1}{4}V(\phi_p{}^2). \tag{10.36}$$

Consider, for example, the integral $\mathbf{H}_{ss}$. One part of this is

$$\int \phi_s\left(-\frac{1}{2}\nabla^2 - \frac{3}{r_{\text{Li}}} + 2V(1s^2)\right)\phi_s\,\mathrm{d}v, \tag{10.37}$$

which we can take to be the energy of the 2s orbital of the lithium atom; from spectroscopic parameters this has been found to be $-0{\cdot}198$ atomic units. The remainder will consist of integrals like

$$\int \phi_s\left(\frac{1}{r_{\text{H}}}\right)\phi_s\,\mathrm{d}v \tag{10.38}$$

and†

$$\int \phi_s V(\phi_h{}^2)\phi_s\,\mathrm{d}v \equiv \iint \phi_s(1)\phi_h(2)\left(\frac{1}{r_{12}}\right)\phi_s(1)\phi_h(2)\mathrm{d}v_1\mathrm{d}v_2, \tag{10.39}$$

which can be obtained from Karo and Olsen's papers. Proceeding in this way, one finds the following values for the Hamiltonian integrals

$$\mathbf{H}_{hh} = -0{\cdot}391, \quad \mathbf{H}_{ss} = -0{\cdot}226, \quad \mathbf{H}_{pp} = -0{\cdot}177,$$
$$\mathbf{H}_{hs} = -0{\cdot}212, \quad \mathbf{H}_{hp} = -0{\cdot}219, \quad \mathbf{H}_{sp} = -0{\cdot}059.$$

† This follows because the potential due to an electron in ϕ_h is

$$V(\phi_h{}^2)(1) = \int \phi_h(2)\left(\frac{1}{r_{12}}\right)\phi_h(2)\mathrm{d}v_2.$$

The secular equations for LiH then have the form

$$\begin{aligned}
&c_h(-0{\cdot}391-E)+c_s(-0{\cdot}212-0{\cdot}469E)+c_p(-0{\cdot}219-0{\cdot}506E) = 0\\
&c_h(-0{\cdot}212-0{\cdot}469E)+c_s(-0{\cdot}226-E)+c_p(-0{\cdot}059) = 0\\
&c_h(-0{\cdot}219-0{\cdot}506E)+c_s(-0{\cdot}059)+c_p(-0{\cdot}177-E) = 0,
\end{aligned} \tag{10.40}$$

and, after multiplying throughout by -1, this leads to the secular determinant

$$\begin{vmatrix}
0{\cdot}391+E & 0{\cdot}212+0{\cdot}469E & 0{\cdot}219+0{\cdot}506E\\
0{\cdot}212+0{\cdot}469E & 0{\cdot}226+E & 0{\cdot}059\\
0{\cdot}219+0{\cdot}506E & 0{\cdot}059 & 0{\cdot}177+E
\end{vmatrix} = 0. \tag{10.41}$$

Expanding (10.41) gives the polynomial

$$0{\cdot}524E^3+0{\cdot}305E^2+0{\cdot}0407E+0{\cdot}00096 = 0, \tag{10.42}$$

which can be solved by the usual methods to give

$$E = -0{\cdot}398, -0{\cdot}154, -0{\cdot}030$$

The lowest root of the polynomial corresponds to the energy of the lowest molecular orbital. Substituting this back into the secular equation (10.40), we find that if $c_h = 1$, then

$$c_s = 0{\cdot}196$$

$$c_p = 0{\cdot}134.$$

The molecular orbital in its un-normalized form is then

$$\psi = \phi_h+0{\cdot}196\phi_s+0{\cdot}134\phi_p. \tag{10.43}$$

To normalize this we multiply by

$$\left\{\int \psi^2\, dv\right\}^{-\frac{1}{2}} = \{1+0{\cdot}038+0{\cdot}018+0{\cdot}392S_{hs}+0{\cdot}268S_{hp}\}^{-\frac{1}{2}} = 0{\cdot}853, \tag{10.44}$$

and obtain

$$\psi = 0{\cdot}853\phi_h+0{\cdot}167\phi_s+0{\cdot}114\phi_p. \tag{10.45}$$

Our calculation ends up with about 70% of the electron density in the hydrogen 1s orbital, which is rather far from the assumption of 50% which we adopted to construct the Hamiltonian (10.36). To obtain a better orbital, (10.45) should now be used to obtain a better potential for the Hamiltonian. Karo and Olsen have found, for the self-consistent solution, the following ratios of coefficients

$$c_s/c_h = 0{\cdot}468, \qquad c_p/c_h = 0{\cdot}292,$$

which would lead to an orbital

$$\psi = 0{\cdot}700\phi_h + 0{\cdot}328\phi_s + 0{\cdot}204\phi_p. \tag{10.46}$$

The energies obtained from the solution of the secular determinant can be equated to the ionization potential of an electron in that orbital—within the approximation that on ionization the wave functions of the other electrons in the molecule do not change (see section 10.6). Our calculation gives the ionization potential of LiH to be 10·8 ev. Karo and Olsen's SCF calculation gives 8·3 ev which is in good agreement with the estimated experimental value of 8 ev.

The LCAO coefficients in MO theory give us a direct measure of the ionic character of a bond. For example, the calculation on LiH clearly shows that there is an excess of charge on the hydrogen atom, which is in agreement with the chemical properties of the molecule.

If the wave function of a molecule is represented by a product of molecular orbitals, antisymmetrized to allow for the Pauli exclusion principle (7.5), then the electron density is obtained by squaring the molecular orbitals and multiplying by the number of electrons n (which may be 0, 1 or 2) occupying that orbital

$$\rho = \sum_k n_k \psi_k^2. \tag{10.47}$$

For LiH, using the orbital (10.46), one finds

$$\begin{aligned}\rho = 2(1s_{\text{Li}})^2 &+ 0{\cdot}98\phi_h^2 + 0{\cdot}22\phi_s^2 + 0{\cdot}08\phi_p^2 \\ &+ 0{\cdot}92\phi_h\phi_s + 0{\cdot}57\phi_h\phi_p + 0{\cdot}27\phi_s\phi_p.\end{aligned} \tag{10.48}$$

This density consists of atomic-orbital densities like ϕ_h^2 and overlap densities like $\phi_h\phi_p$. To get an estimate of the relative distribution of charge between the two atoms, we divide up an overlap density $\phi_a\phi_b$ equally between the two orbitals ϕ_a and ϕ_b, multiplying by the overlap integral so as to conserve the net charge†:

$$\phi_a\phi_b = \frac{S_{ab}}{2}(\phi_a^2 + \phi_b^2).$$

† The conservation of charge follows if this expression is integrated over all space.

This procedure gives what Mulliken calls the *gross atomic population numbers*, which add up to the total number of electrons in the molecule. From the calculation on LiH we obtain the following population numbers

$$1s_{\text{Li}} = 2, \qquad 2s_{\text{Li}} = 0{\cdot}432, \qquad 2p\sigma_{\text{Li}} = 0{\cdot}228, \qquad 1s_h = 1{\cdot}340,$$

which lead to an ionic character for the bond that can be described by the formula $\text{Li}^{+0\cdot34}\text{H}^{-0\cdot34}$.

The dipole moment of a discrete set of charges q_μ is given in classical physics by

$$\boldsymbol{\mu} = \sum_\mu q_\mu \mathbf{r}_\mu. \tag{10.49}$$

If the charge distribution is continuous, with a density ρ, then

$$\boldsymbol{\mu} = \int \rho \, . \, \mathbf{r} \, dv. \tag{10.50}$$

For a molecule we have a discrete set of charges $Z_\mu e$ from the nuclei and a continuous charge density from the electrons given by the square of the wave function multiplied by the electron charge. The dipole moment is then

$$\boldsymbol{\mu} = -e\int \psi^2 \mathbf{r} \, dv + \sum_\mu Z_\mu e \mathbf{r}_\mu. \tag{10.51}$$

For a neutral molecule, the dipole moment calculated by (10.51) is independent of the origin of the vectors **r**; for a charged molecule this is not so, but it is usual to measure **r** from the centre of charge.

The calculation on LiH gives a dipole moment of about 6 D. It should be noted that for small molecules in particular it is usually a poor approximation to calculate the dipole moments just from the population numbers; on this basis only 2·6 D would be calculated for LiH. Overlap densities like $\phi_s\phi_p$ make a very important contribution to the dipole moment but no contribution to population numbers because $S_{sp} = 0$.

Figure 10.6 gives the correlation diagram for the orbitals of a heteronuclear diatomic molecule with atom A being more electronegative than atom B. The most important difference from figure 10.5 is that, because of the absence of the *g–u* labels, the second molecular orbital, which for a homonuclear diatomic molecule would be $\sigma_u 1s$, does correlate with the $2s$ orbital of the united atom. The labelling of an orbital as bonding or antibonding does not necessarily follow the pattern of the homonuclear molecular orbitals. For example, we have seen that the orbital 2σ is bonding for LiH, whereas $\sigma_u 1s$ is always antibonding.

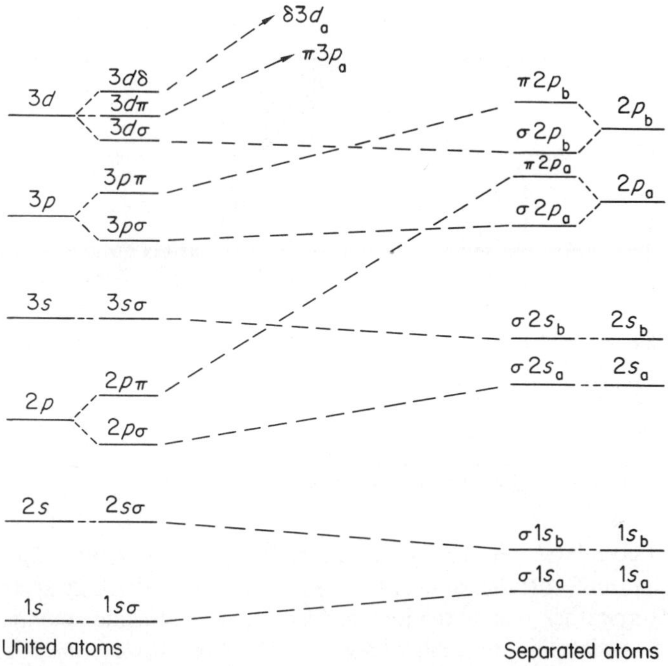

FIGURE 10.6. Correlation diagram for the orbitals of a heteronuclear diatomic molecule

In some cases the states of a heteronuclear diatomic molecule can be thought of as being a perturbation on the states of the isoelectronic homonuclear diatomic molecule. For example, CO and N_2, although chemically very different, have a rather similar set of electronic states, as can be seen from table 10.3. However, the SCF orbitals that have been evaluated for CO do show a fairly considerable distortion from the N_2 orbitals, as can be seen from the example of the bonding π orbital†

$$N_2 \qquad \psi(\pi_u 2p) = 0{\cdot}62(2p\pi_a + 2p\pi_b)$$

$$\mathrm{CO} \qquad \psi(1\pi) = 0{\cdot}47(2p\pi_C) + 0{\cdot}77(2p\pi_O).$$

10.5 *The non-crossing rule*

There is one point about the correlation diagram of figure 10.5 that we have glossed over so far. How, for example, does one know that the $\sigma_g 2s$

† RANSIL, *Rev. Mod. Phys.*, **32**, (1960).

TABLE 10.3. Comparison of some states of N_2 and CO

	N₂		CO	
	Symbol	Energy (ev)	Symbol	Energy (ev)
π bonding to π antibonding excitations	$^1\Sigma_u^+$	12·85	$^1\Sigma^+$	—
	$^1\Delta_u$	9·26	$^1\Delta$	—
	$^1\Sigma_u^-$	8·76	$^1\Sigma^-$	—
	$^3\Sigma_u^-$	8·76	$^3\Sigma^-$	8·11
	$^3\Delta_u$	7·47	$^3\Delta$	7·22
	$^3\Sigma_u^+$	6·17	$^3\Sigma^+$	6·92
σ bonding to π antibonding excitations	$^1\Pi_g$	8·55	$^1\Pi$	8·07
	$^3\Pi_g$	7·35	$^3\Pi$	6·04
Ground state	$^1\Sigma_g^+$	0	$^1\Sigma^+$	0

From MULLIKEN, *Can. J. Chem.*, **36**, 10 (1958).

orbital correlates with the united atom $2s$ orbital, and the $\sigma_g 2p$ with the united atom $3s$ orbital, and not the other way around? Our knowledge is based on a rather general theorem concerning the electronic potential energy curves of diatomic molecules known as the *non-crossing rule*. This states that if the potential energy curves for a diatomic molecule are plotted as functions of the internuclear distance, then the curves for two states having the same symmetry cannot cross one another.

The proof of this theorem is as follows. Suppose that at a given internuclear distance we know the wave function of all the electronic states of the molecule except two, Ψ_1 and Ψ_2. Further, let us have a complete orthonormal set of functions Θ, which can be used to expand the wave functions of the molecule, and let two of these Θ_1 and Θ_2 be orthogonal to all the *known* wave functions and to each other. In this case Ψ_1 and Ψ_2 must be some linear combination of the two functions Θ_1 and Θ_2:

$$\Psi_1 = c_{11}\Theta_1 + c_{12}\Theta_2$$

$$\Psi_2 = c_{21}\Theta_1 + c_{22}\Theta_2.$$

To find the energies of these states Ψ_1 and Ψ_2 we solve the secular determinant (6.68)

$$\begin{vmatrix} \mathscr{H}_{11} - E & \mathscr{H}_{12} \\ \mathscr{H}_{12} & \mathscr{H}_{22} - E \end{vmatrix} = 0$$

$$(S_{11} = S_{22} = 1, \qquad S_{12} = 0),$$

and the solutions of the resulting quadratic in E are

$$E = \frac{\mathcal{H}_{11} + \mathcal{H}_{22}}{2} \pm \frac{1}{2}\{(\mathcal{H}_{11} - \mathcal{H}_{22})^2 + 4\mathcal{H}_{12}{}^2\}^{\frac{1}{2}}. \quad (10.52)$$

It is only possible to have the two solutions of the secular determinant equal to one another if both $\mathcal{H}_{11} - \mathcal{H}_{22} = 0$ and $\mathcal{H}_{12} = 0$. But $\mathcal{H}_{11} - \mathcal{H}_{22}$ and $\mathcal{H}_{12}$ will be unrelated functions of the internuclear distance and it is extremely unlikely that there will be a value of the internuclear distance where both are simultaneously zero. In general, therefore, two electronic states of the molecule will not have the same energy for any value of the internuclear distance, so that the crossing of potential energy curves is forbidden. However, if $\mathcal{H}_{12}$ is zero for all values of the internuclear distance, as it will be if Ψ_1 and Ψ_2 have different symmetries or spins, then there can be some point where $\mathcal{H}_{11} - \mathcal{H}_{22} = 0$; it follows that the potential energy curves for states of different symmetries or spins can cross one another.

The non-crossing rule applies to states (cf. figures 5.2 and 5.6) and to orbitals, and the rule applied to orbitals allows us to construct unambiguously the correlation diagrams given in this chapter.

For polyatomic molecules the non-crossing rule is a bit more complicated. For example, if in a molecule there are two variable coordinates X and Y, then the potential energies are represented as surfaces $E(X, Y)$, and if two states have the same symmetry, their surfaces can cross, but only at a point and not along a line†.

10.6 *More sophisticated calculations*

We shall look briefly at the attempts that have been made to obtain accurate molecular-orbital wave functions for diatomic molecules.

The simplest LCAO MO function for $H_2{}^+$ gives a dissociation energy of 1·77 ev and an equilibrium distance of 1·32 Å; the best calculation gives a dissociation energy of 2·78 ev and an equilibrium distance of 1·06 Å. What are the important steps for improving the calculation?

The correlation diagram for $H_2{}^+$ shows that the $\sigma_g 1s$ orbital must get smaller as the two nuclei come together, since the $1s$ orbital of He^+ has an exponential constant which is twice as large as that of the $1s$ orbital of hydrogen. This suggests that a better wave function for the $\sigma_g 1s$ orbital

† HERZBERG and LONGUET-HIGGINS, *Discussions Faraday Soc.*, **35**, 77 (1963).

would be obtained by taking a linear combination of 1*s* orbitals with variable effective nuclear charges

$$1s = \left(\frac{\zeta^3}{\pi}\right)^{\frac{1}{2}} e^{-\zeta r}.$$

Coulson† applied the variation principle to this wave function and found that at the equilibrium distance of 1·06 Å the best value of ζ was 1·24, which gives a dissociation energy of 2·25 ev.

Another effect to be considered is that the potential field of one proton will distort the orbitals of the other. This is best allowed for by including some $2p\sigma$ atomic orbitals into the calculation. Dickinson‡ calculated a ground state molecular orbital of the form

$$(1s_a + 1s_b) + k(2p\sigma_a + 2p\sigma_b), \tag{10.53}$$

where both the 1*s* and $2p\sigma$ functions contained variable exponents. He minimized the energy with respect to both exponents and to k and at the equilibrium distance found $\zeta(1s) = 1{\cdot}247$, $\zeta(2p) = 1{\cdot}868$ and $k = 0{\cdot}145$. This calculation gave a dissociation energy of 2·73 ev which is very close to that of the exact solution of the Schrödinger equation for H_2^+.

Orbital contraction is an extremely important aspect of simple molecular wave functions. If calculations are made with atomic orbitals appropriate to the free atoms, and not with orbitals optimized for the molecular calculation, then the total energy may not be too bad. However if one looks at the two contributions to the total energy, kinetic and potential, then orbital optimization has a dramatic effect on both.

This is most clearly seen with reference to a general theorem of classical or quantum mechanics called the virial theorem (see problem 3.4). This states, that for an equilibrium system and coulomb forces, the kinetic energy is the negative of the total energy and the potential energy is twice the total energy:

$$E = -T = \tfrac{1}{2}V.$$

Because the total energy of H_2^+ is more negative than that of $H + H^+$, it follows that the kinetic energy must increase on bond formation and the potential energy become more negative. Calculations on H_2^+ using contracted atomic orbitals give energies in agreement with this theorem, but if uncontracted hydrogen 1*s* orbitals are used, one in fact finds that the kinetic energy *decreases* and the potential energy *increases* on bond

† Coulson, *Trans. Faraday Soc.*, **33**, 1479 (1937).
‡ Dickinson, *J. Chem. Phys.*, **1**, 317 (1933).

formation, and at the equilibrium position the virial theorem is not obeyed. The same is found for calculations on H_2.

It is possible to find an exact solution of the electronic wave equation for H_2^+ without resorting to the LCAO procedure. To do this the problem is expressed in terms of elliptical coordinates, and the differential equation is then separable. These elliptical coordinates are defined by

$$\mu = \frac{r_a + r_b}{R}, \qquad \nu = \frac{r_a - r_b}{R}, \qquad \varphi, \tag{10.54}$$

where r_a and r_b are the distances of the electron from the two nuclei, and φ is the angle about the internuclear axis. A solution of the wave equation can then be obtained of the form

$$\Psi = M(\mu)N(\nu)\, e^{i\lambda\varphi}, \tag{10.55}$$

where M and N are functions which are solutions of different but coupled differential equations†.

The use of elliptical coordinates for many-electron diatomic molecules has been investigated‡. The main limitation of this approach is that it cannot be extended to polyatomic molecules, unlike the cruder but more flexible LCAO approach.

If one takes a molecular-orbital wave function for H_2 which puts two electrons into the orbital $\psi_g = N_g(1s_a + 1s_b)$ with opposite spins,

$$\Psi = |\psi_g\alpha \ \psi_g\beta|,$$

then the total electronic energy is calculated to be 93% of that found experimentally. This looks a very satisfying result, but the total energy is large compared with the dissociation energy of the molecule, which is the quantity of chemical interest, and this simple wave function gives a dissociation energy of only 45% of that observed (2·65 ev calculated, compared with 4·74 ev observed). Also the calculated equilibrium bond length is rather poor (0·85 Å compared with 0·74 Å observed), so that overall the simple MO wave function is not very good.

If SCF functions are used for the orbital ψ_g, then one calculates a bond length of about 0·73 Å which is quite satisfactory, but a dissociation energy of 3·64 ev which is still rather far from the observed value§. This means that to calculate a good dissociation energy it is necessary to step outside

† A detailed account of the solutions of these equations has been given by BATES, LEDSHAM and STEWART, *Phil. Trans. Roy Soc. London*, **246**, 215 (1953).

‡ HARRIS, *J. Chem. Phys.*, **32**, 3 (1960).

§ KOLOS and ROOTHAAN, *Rev. Mod. Phys.*, **32**, (1960).

the restrictive form of the single determinant MO wave function, a step which we also found necessary for the helium atom (section 9.5). This point will be taken up again in chapter 12.

10.7 Self-consistent-field equations and molecular-orbital energies

The orbital is undoubtedly the most useful concept in the theory of valency and yet it is difficult to give it a firm mathematical foundation. Strictly speaking an orbital (atomic or molecular) is simply a useful one-electron function for constructing many-electron wave functions. It is an eigenfunction of a one-electron operator, which may be explicitly defined in terms of the positions of the electrons and nuclei in the system. But this operator is not related to the complete Hamiltonian except through some approximation which involves an averaging of the electron-interaction terms (cf. 4.2). Likewise, the energy of a molecular orbital, an eigenvalue of the one-electron operator, can only be related in an approximate manner to the observable energy levels of the system.

We shall see in chapter 12 that the lack of a strict definition of a molecular orbital has some practical advantages. At this point, however, we shall examine the mathematical basis of the self-consistent-field (SCF) orbital, since this is the most rigorously defined of all types of orbital, and is moreover the one whose energy is most closely related to experimental energies.

Suppose we write a wave function for a system of n electrons as a single product of n spin-orbitals (cf. 7.1)

$$\Psi = \psi_a(1)\psi_b(2)\ldots\psi_k(n). \tag{10.56}$$

We now evaluate the energy of this wave function by

$$E = \int \Psi\mathscr{H}\Psi\,\mathrm{d}\tau, \tag{10.57}$$

where $\mathscr{H}$ is the complete electronic Hamiltonian (the $\mathscr{H}_e$ of 7.10). We shall write $\mathscr{H}$ in the form

$$\mathscr{H} = \sum_i \mathrm{H}^c(i) + \sum_{i>j}\left(\frac{1}{r_{ij}}\right) + V_{nn}, \tag{10.58}$$

where $\mathrm{H}^c(i)$ is the so-called *core Hamiltonian* which consists of the kinetic energy operator and the electron nuclear attraction terms for electron i, and where V_{nn} is the nuclear repulsion energy.

Substituting (10.56) and (10.58) into (10.57) it is easily seen that

$$E = \sum_{r=a}^{k} \mathbf{H}^c_{rr} + \sum_{\substack{\text{pairs}\\ rs}} J_{rs} + V_{\text{nn}}, \tag{10.59}$$

where

$$\mathbf{H}^c_{rr} = \int \psi_r \mathbf{H}^c \psi_r \, \mathrm{d}\tau \tag{10.60}$$

and

$$J_{rs} = \iint \psi_r(i)\psi_s(j)\left(\frac{1}{r_{ij}}\right)\psi_r(i)\psi_s(j)\mathrm{d}\tau_i\,\mathrm{d}\tau_j. \tag{10.61}$$

Expression (10.59) consists of two parts: the first, involving $\mathbf{H}^c$, is the sum of the energies that each electron would have if all other electrons were absent, the second is the total of all electron repulsion energy.

The wave function (10.56) does not satisfy the antisymmetry requirement, but this is easily corrected by converting the single product into a Slater determinant (7.6)

$$\Psi = |\psi_a\psi_b \cdots \psi_k|. \tag{10.62}$$

Providing that the spin-orbitals are mutually orthogonal, one finds, on substituting (10.62) into (10.57) (problem 10.7),

$$E = \sum_{r=a}^{k} \mathbf{H}^c_{rr} + \sum_{\substack{\text{pairs}\\ rs}} (J_{rs} - K_{rs}) + V_{\text{nn}}. \tag{10.63}$$

This differs from (10.59) by the appearance of the new integral

$$K_{rs} = \iint \psi_r(i)\psi_s(j)\left(\frac{1}{r_{ij}}\right)\psi_s(i)\psi_r(j)d\tau_i\, d\tau_j. \tag{10.64}$$

We met integrals like J_{rs} and K_{rs} before, when we evaluated the energy of the ^3S excited state of helium, except that we are now dealing with spin-orbitals rather than just space orbitals. J_{rs} is called the coulomb integral, K_{rs} the exchange integral. If ψ_r and ψ_s have different associated spin functions (one α and the other β) then it follows from the integration over the spin coordinates that K_{rs} is zero.

We can now apply the variation principle to the wave functions (10.56) or (10.62) and ask for the conditions that the respective energies (10.58 and 10.63) shall be minimized. Applying the condition is sufficient to define the orbitals ψ, and orbitals evaluated in this way are called SCF orbitals. The orbitals defined with respect to the single product function (10.56) are called Hartree orbitals, those defined with respect to the antisymmetrized product (10.62) are called Hartree–Fock orbitals, after the originators of the theory for atomic orbitals.

For an open-shell configuration a single antisymmetrized product is not necessarily a good wave function. For example, two electrons in different orbitals can give a singlet or a triplet spin state and the singlet has a wave function which involves two determinants (cf. 9.47). In principle one can define SCF orbitals for any chosen type of wave function, but the mathematics is only simple for the Hartree and Hartree–Fock orbitals.

We shall now examine the conditions which define the SCF Hartree–Fock orbitals as these are the ones most frequently encountered in modern work. Any later references to SCF orbitals in this book will imply Hartree–Fock orbitals unless stated otherwise.

Suppose that the function (10.62) does not give the lowest energy of the state. Then there is some other function, say

$$\Psi' = |\psi_a'\psi_b \dots \psi_k|, \tag{10.65}$$

which has a lower energy. Let us further assume that ψ_a' differs only slightly from ψ_a, and can be written

$$\psi_a' = \psi_a + c_t\psi_t \tag{10.66}$$

where ψ_t is a spin-orbital which is orthogonal to the set $\psi_a \dots \psi_k$. Providing c_t is small ψ_a' will still be normalized, since re-normalization only involves a term in c_t^2. Using (10.66) we can clearly write (10.65) in the form

$$\begin{aligned}\Psi' &= |\psi_a\psi_b \dots \psi_k| + c_t|\psi_t\psi_b \dots \psi_k| \\ &= \Psi + c_t\Psi_a^t \text{ (say).}\end{aligned} \tag{10.67}$$

In other words, Ψ' is formed by adding to Ψ a small amount of the state Ψ_a^t which arises from the excitation of an electron from ψ_a to ψ_t.

In order that Ψ shall be the best wave function of its type it is necessary that c_t be zero, and this further requires (from 6.42) that the Hamiltonian integral between Ψ and Ψ_a^t, to which we give the symbol $\mathbf{F}_{at}$, shall be zero.

$$\mathbf{F}_{at} \equiv \int \Psi\mathscr{H}\Psi_a^t \, d\tau = 0. \tag{10.68}$$

If this integral is expressed in terms of the spin-orbitals we find (problem 10.8)

$$\begin{aligned}\mathbf{F}_{at} = \mathbf{H}_{at}^c + \sum_{s=a}^{k}\Bigg\{&\iint\psi_a(i)\psi_s(j)\left(\frac{1}{r_{ij}}\right)\psi_t(i)\psi_s(j)d\tau_i\,d\tau_j \\ &-\iint\psi_a(i)\psi_s(j)\left(\frac{1}{r_{ij}}\right)\psi_s(i)\psi_t(j)d\tau_i\,d\tau_j\Bigg\}.\end{aligned} \tag{10.69}$$

For this to be zero for any spin-orbital, not just ψ_a, it is necessary that the ψ be eigenfunctions of the operator $\mathbf{F}$ (problem 6.1). This operator is rather strange in that it depends on its own eigenfunctions. We can write

$$\mathbf{F} = \mathbf{H}^c + \sum_{s=a}^{k} (\mathbf{J}_s - \mathbf{K}_s), \tag{10.70}$$

where $\mathbf{J}_s$ and $\mathbf{K}_s$ are coulomb and exchange operators which are defined by their integrals

$$(\mathbf{J}_s)_{at} \equiv \iint \psi_a(i)\psi_s(j)\left(\frac{1}{r_{ij}}\right)\psi_t(i)\psi_s(j)\mathrm{d}\tau_i\,\mathrm{d}\tau_j$$

$$(\mathbf{K}_s)_{at} \equiv \iint \psi_a(i)\psi_s(j)\left(\frac{1}{r_{ij}}\right)\psi_s(i)\psi_t(j)\mathrm{d}\tau_i\,\mathrm{d}\tau_j. \tag{10.71}$$

It should be noted that $(\mathbf{J}_s)_{rr}$ and $(\mathbf{K}_s)_{rr}$ are the integrals J_{rs} and K_{rs} given by (10.61) and (10.64), and that $(\mathbf{J}_a)_{at} = (\mathbf{K}_a)_{at}$.

We note in passing that the operator determining the Hartree SCF orbitals does not contain the exchange operator $\mathbf{K}_s$ and that the sum over s should not include the term $s = \mathrm{a}$. Also, that the equations define only the space part of the spin-orbital, the spin part (α or β) being defined by the original choice of wave function.

The potential governing the SCF orbitals therefore consists of the core potential, the coulomb potential of all the electrons and an exchange potential for each electron. Since the coulomb and exchange potentials depend on the orbitals themselves an iteration method has to be adopted to calculate the SCF orbitals (as has been described earlier), and the condition of self-consistency is reached when the orbitals are consistent with the potential from which they were determined.

The eigenvalues of $\mathbf{F}$ may be called the orbital energies. Thus from expression (10.69).

$$E_r \equiv \mathbf{F}_{rr} = \mathbf{H}^c_{rr} + \sum_{s=a}^{k} (J_{rs} - K_{rs}). \tag{10.72}$$

The sum of the energies of all occupied spin-orbitals is

$$\sum_{r=a}^{k} E_r = \sum_{r=a}^{k} \mathbf{H}^c_{rr} + \sum_{r=a}^{k}\sum_{s=a}^{k} (J_{rs} - K_{rs}). \tag{10.73}$$

Comparing this with (10.63), and noting since $J_{rr} = K_{rr}$ that

$$\sum_{r=a}^{k}\sum_{s=a}^{k} (J_{rs} - K_{rs}) = 2 \sum_{\substack{\text{pairs}\\ rs}} (J_{rs} - K_{rs}), \tag{10.74}$$

we have

$$E = \sum_{r=a}^{k} E_r - \sum_{\substack{\text{pairs} \\ rs}} (J_{rs} - K_{rs}) + V_{nn}. \tag{10.75}$$

This emphasizes the fact that even for SCF orbitals the total electronic energy is not just the sum of the orbital energies plus the nuclear repulsion energy.

Suppose now we remove an electron from spin-orbital ψ_a but leave the wave functions of the other electrons unchanged. Then the energy of the positive ion having the wave function

$$|\psi_b \cdots \psi_k| \tag{10.76}$$

is, from (10.63),

$$E^+ = \sum_{r=b}^{k} \mathbf{H}^c_{rr} + \sum_{\substack{\text{pairs } rs \\ r \neq a,\, s \neq a}} (J_{rs} - K_{rs}) + V_{nn}, \tag{10.77}$$

and clearly from (10.72)

$$E^+ = E - \mathbf{H}^c_{aa} - \sum_{s=a}^{k} (J_{as} - K_{as}) = E - E_a. \tag{10.78}$$

It follows that $-E_a$ can be equated to the ionization potential $(E^+ - E)$, which is the energy required to ionize the molecule, *providing that the ionization process is adequately represented by the removal of an electron from an orbital without change in the wave functions of the other electrons.* This is known as Koopmans' theorem.

An important special case of the SCF orbitals is a closed-shell configuration, each orbital being occupied by two electrons with α and β spins. Expressions (10.63), (10.69) and (10.72) are, in this case,

$$E = 2\sum_r \mathbf{H}^c_{rr} + 2\sum_{\substack{\text{pairs} \\ rs}} (2J_{rs} - K_{rs}) + \sum_r J_{rr} + V_{nn}, \tag{10.79}$$

$$\mathbf{F}_{at} = \mathbf{H}^c_{at} + \sum_{s=a}^{k} \left\{ 2\iint \psi_a(i)\psi_s(j)\left(\frac{1}{r_{ij}}\right)\psi_t(i)\psi_s(j)dv_i\, dv_j - \iint \psi_a(i)\psi_s(j)\left(\frac{1}{r_{ij}}\right)\psi_s(i)\psi_t(j)dv_i\, dv_j \right\}, \tag{10.80}$$

$$E_r = \mathbf{H}^c_{rr} + \sum_{s=a}^{k} (2J_{rs} - K_{rs}) = \mathbf{F}_{rr}. \tag{10.81}$$

where the summations are over all occupied *orbitals*, not spin-orbitals.

If the SCF orbitals are to be represented by the LCAO approximation

$$\psi = \sum_v c_v \phi_v, \tag{10.82}$$

then by substituting (10.82) into (10.80) and picking out the terms in $c_{a\mu}c_{t\nu}$ one finds

$$\mathbf{F}_{\mu\nu} = \mathbf{H}^c_{\mu\nu} + \sum_{s=a}^{k}\sum_{\rho}\sum_{\sigma} c_{\rho s}c_{\sigma s}\left\{2\iint \phi_\mu(i)\phi_\rho(j)\left(\frac{1}{r_{ij}}\right)\phi_\nu(i)\phi_\sigma(j)\mathrm{d}v_i\,\mathrm{d}v_j \right.$$
$$\left. - \iint \phi_\mu(i)\phi_\rho(j)\left(\frac{1}{r_{ij}}\right)\phi_\sigma(i)\phi_\nu(j)\mathrm{d}v_i\,\mathrm{d}v_j\right\}. \quad (10.83)$$

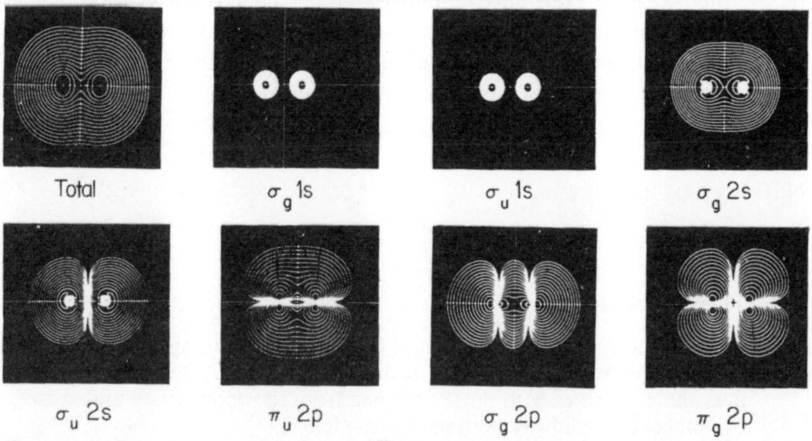

FIGURE 10.7(a). Contour diagrams of the Hartree–Fock molecular orbitals of the fluorine molecule (A. C. Wahl *et al.*, *Int. J. Quant. Chem.*, **1s**, 123 (1967).)

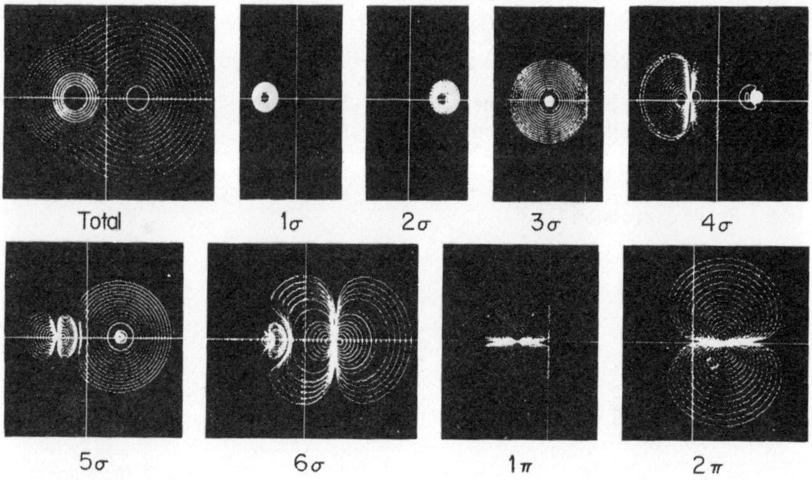

FIGURE 10.7(b). Contour diagrams of the Hartree–Fock molecular orbitals of the sodium fluoride molecule (A. C. Wahl *et al.*, *Int. J. Quant. Chem.*, **1s**, 123 (1967).)

The SCF orbitals for a closed-shell system in this form are then determined by solving the secular equations

$$\sum_{\nu} c_{\nu}(\mathbf{F}_{\mu\nu} - ES_{\mu\nu}) = 0, \tag{10.84}$$

through the determinant

$$|\mathbf{F}_{\mu\nu} - ES_{\mu\nu}| = 0. \tag{10.85}$$

These are usually known as Roothaan's equations although they should be credited also to Hall and Lennard-Jones who derived them independently. Again an iterative procedure is necessary for their solution because they are simultaneous cubic equations in the coefficients ($F_{\mu\nu}$ itself depends on the squares of the coefficients through (10.83)).

Figure 10.7 shows contour diagrams of the Hartree–Fock orbitals of F_2 and NaF. The F_2 orbitals are labelled as in table 10.2; the NaF orbitals are labelled as σ or π in order of increasing energy.

Problems 10

10.1 Prove expression (10.19).

10.2 Plot the value of the functions ψ_g^2 and ψ_u^2 (10.6) along the line passing through the nuclei for an internuclear distance of 2 atomic units. Compare these densities with those obtained from the appropriate united atom orbital centred midway between the nuclei.

10.3 What is the ground state of C_2^+?

10.4 Write down wave functions for all the states arising from the configuration $(\pi_g 2p)^2$.

10.5 Draw a correlation diagram for the orbitals of $(HeH)^{2+}$.

10.6 Evaluate the energies of the $\sigma_g 1s$ and $\sigma_u 1s$ orbitals of H_2^+ using the secular equation method described in section 10.4.

10.7 Prove expression (10.63).

10.8 Prove expression (10.69).

Chapter 11

The valence-bond theory: its application to diatomic molecules

11.1 The hydrogen molecule

The first successful theory of the chemical bond was published in 1927 by Heitler and London, and their theory forms the basis of what is now called valence-bond (VB) theory. The essential feature of this theory is the assumption that the electronic structure of the atoms is largely preserved in a molecule, and that the energy of the covalent bond is associated with the exchange of electrons between atoms. We shall start first with the Heitler–London theory of the hydrogen molecule.

A hydrogen atom in its ground state has an electron in a 1*s* orbital. Let us label the nucleus a, and the electron 1, and write the wave function $1s_a(1)$. A second atom (nucleus b, electron 2) has the wave function $1s_b(2)$. If these two atoms are far apart then the wave function of the whole system is

$$\Psi = 1s_a(1)1s_b(2). \tag{11.1}$$

The Hamiltonian for the system is

$$\mathscr{H} = \mathbf{H}_a + \mathbf{H}_b + \mathscr{H}', \tag{11.2}$$

where $\mathbf{H}_a$ and $\mathbf{H}_b$ are the Hamiltonians for the two atoms, and

$$\mathscr{H}' = -\frac{1}{r_{a2}} - \frac{1}{r_{b1}} + \frac{1}{r_{12}} + \frac{1}{R} \tag{11.3}$$

can be considered as the terms in the Hamiltonian which express the interaction between the two atoms. The energy of the function (11.1) is then

$$E = \int \Psi\mathscr{H}\Psi \, dv = \iint 1s_a(1)1s_b(2)(\mathbf{H}_a + \mathbf{H}_b + \mathscr{H}')1s_a(1)1s_b(2)\, dv_1\, dv_2. \tag{11.4}$$

Making use of the fact that $1s_a$ is an eigenfunction of $\mathbf{H}_a$ (10.18), and $1s_b$ is an eigenfunction of $\mathbf{H}_b$, (11.4) becomes

$$E = 2E_H + \iint 1s_a(1)1s_b(2)\mathscr{H}'1s_a(1)1s_b(2)\,dv_1\,dv_2. \tag{11.5}$$

This integral, which is given the symbol Q, can be expanded using (11.3); after integration over the coordinates not involved in the operators, we have

$$Q = -\int 1s_a(1)\left(\frac{1}{r_{b1}}\right)1s_a(1)\,dv_1 - \int 1s_b(2)\left(\frac{1}{r_{a2}}\right)1s_b(2)\,dv_2 + \iint 1s_a(1)1s_b(2)\left(\frac{1}{r_{12}}\right)1s_a(1)1s_b(2)\,dv_1\,dv_2 + \frac{1}{R}. \tag{11.6}$$

The first term represents the attraction of the electron density $1s_a{}^2$ to the nucleus b; the second the attraction of $1s_b{}^2$ to a; the third is the repulsion energy of the two electron densities $1s_a{}^2$ and $1s_b{}^2$; the last is the nuclear repulsion energy. Q is therefore simply the classical electrostatic energy between the two hydrogen atoms and it is called the coulomb integral. When this is evaluated it is found to give a negligible binding energy: only 5% of that observed. The wave function (11.1) is too restrictive for the hydrogen molecule because it represents a situation in which an electron is localized around one nucleus. An equally good function would be $1s_a(2)1s_b(1)$. The true wave function of H_2 must give an equal chance of each electron being around the two nuclei, and this condition is satisfied by either of the functions

$$\Psi_{\pm} = (2 \pm 2S^2)^{-\frac{1}{2}}\{1s_a(1)1s_b(2) \pm 1s_a(2)1s_b(1)\}, \tag{11.7}$$

where the factor in front is a normalizing constant, with

$$S = \int 1s_a 1s_b\,dv. \tag{11.8}$$

The energies of these functions can be evaluated in the same way as before. The result is

$$E_+ = 2E_H + \frac{Q+A}{1+S^2};\quad E_- = 2E_H + \frac{Q-A}{1-S^2}, \tag{11.9}$$

where Q is defined as before, and A is the so-called exchange integral†

$$A = \iint 1s_a(1)1s_b(2)\mathscr{H}'1s_a(2)1s_b(1)\,dv_1\,dv_2. \tag{11.10}$$

† It is unfortunate that the terms coulomb integral and exchange integral are used for different things in VB and MO theories (cf. 10.64).

Expanding $\mathscr{H}'$ we find

$$A = -S\int 1s_a(1)\left(\frac{1}{r_{b1}}\right) 1s_b(1)dv_1 - S\int 1s_b(2)\left(\frac{1}{r_{a2}}\right)1s_a(2)\,dv_2$$
$$+\iint 1s_a(1)1s_b(2)\left(\frac{1}{r_{12}}\right)1s_a(2)1s_b(1)\,dv_1\,dv_2 + \frac{S^2}{R}. \quad (11.11)$$

The first two integrals were encountered in the MO theory of H_2^+ (cf. 10.20). The third one is a two-electron exchange integral. Calculations show that the sum of the first two terms is larger than the sum of the second two, hence from (11.9) the function Ψ_+ will have a lower energy than Ψ_-. From it we calculate a binding energy about 60% of that observed for the hydrogen molecule. Ψ_+ is generally known as the Heitler–London function.

We have not yet included spin in our wave function: because Ψ_+ is symmetric to electron exchange, then, by the Pauli principle, it must be combined with an antisymmetric spin function, and this will give a singlet spin state (9.24)

$$^1\Psi_+ = (2+2S^2)^{-\frac{1}{2}}\{1s_a(1)1s_b(2) + 1s_a(2)1s_b(1)\}\ \sqrt{\tfrac{1}{2}}\{\alpha(1)\beta(2) - \alpha(2)\beta(1)\}$$
$$\equiv (2+2S^2)^{-\frac{1}{2}}\{|1s_a\alpha\ 1s_b\beta| - |1s_a\beta\ 1s_b\alpha|\}. \quad (11.12)$$

The function Ψ_-, which gives a state of repulsion of the two atoms, must be combined with a symmetric spin state, which is a triplet (9.23)

$$^3\Psi_- = (2-2S^2)^{-\frac{1}{2}}\{1s_a(1)1s_b(2) - 1s_a(2)1s_b(1)\}\begin{cases}\alpha(1)\alpha(2)\\ \sqrt{\tfrac{1}{2}}\{\alpha(1)\beta(2)+\alpha(2)\beta(1)\}.\\ \beta(1)\beta(2)\end{cases} \quad (11.13)$$

It should be noted, from (11.9), that when the two nuclei come close together the electronic energy of $^3\Psi_-$ becomes very large. This large repulsion might be associated with the tendency to violate the exclusion principle for $^3\Psi_-$ as $1s_a$ and $1s_b$ become identical.

The VB description of the covalent bond is based on the Heitler–London wave function (11.12). If two atomic orbitals ϕ_a and ϕ_b overlap one another, and there is one electron in each orbital, and if the spins of these electrons are paired to give a singlet spin state, then a wave function of the type (11.12),

$$\Psi = (2+2S^2)^{-\frac{1}{2}}\{|\phi_a\alpha\ \phi_b\beta| - |\phi_a\beta\ \phi_b\alpha|\}, \quad (11.14)$$

describes this situation, which is bonding. If one or both of the atomic orbitals contain two electrons one cannot write such a function without violating the exclusion principle. Further, if ϕ_a and ϕ_b have a zero overlap integral, say by virtue of their symmetries, then this is also not a bonding situation because the first two terms in (11.11), which are the stabilizing terms, will be zero. In this case the triplet state has a lower energy than the singlet.

11.2 Homonuclear diatomic molecules

Let us now consider the nitrogen molecule. The ground configuration of a nitrogen atom is $1s^2 2s^2 2p^3$. The ground state is ^4S, one electron being in each of the three p orbitals. It follows that each of these three electrons is available to form a bond with the corresponding electron of the other atom: the electrons in the p_z orbitals give a σ bond, the electrons in the p_x orbitals one π bond and those in the p_y orbitals another π bond. This is as good a description as that given by MO theory. However, the wave function describing these bonds turns out to be rather complicated.

The wave function for the σ bond is (taking the z axis to be the internuclear axis)†

$$|p_{\mathrm{za}}\overline{p_{\mathrm{zb}}}| - |\overline{p_{\mathrm{za}}}p_{\mathrm{zb}}|;$$

that for the π_x bond is

$$|p_{\mathrm{xa}}\overline{p_{\mathrm{xb}}}| - |\overline{p_{\mathrm{xa}}}p_{\mathrm{xb}}|;$$

and for the π_y bond

$$|p_{\mathrm{ya}}\overline{p_{\mathrm{yb}}}| - |\overline{p_{\mathrm{ya}}}p_{\mathrm{yb}}|. \tag{11.15}$$

The wave function for the whole molecule (ignoring the $1s$ and $2s$ electrons which are assumed to be non-bonding) could then be

$$(|p_{\mathrm{z_1}}\overline{p_{\mathrm{zb}}}| - |\overline{p_{\mathrm{z_1}}}p_{\mathrm{zb}}|)(|p_{\mathrm{xa}}\overline{p_{\mathrm{xb}}}| - |\overline{p_{\mathrm{xa}}}p_{\mathrm{xb}}|)(|p_{\mathrm{ya}}\overline{p_{\mathrm{yb}}}| - |\overline{p_{\mathrm{ya}}}p_{\mathrm{yb}}|). \tag{11.16}$$

But this function does not allow for the exchange of electrons between, say, the p_z and p_x atomic orbitals. To introduce this degree of flexibility it is necessary to write the wave function in terms of 6×6 determinants as follows

$$\begin{aligned}
\{&|p_{\mathrm{za}}\overline{p_{\mathrm{zb}}}p_{\mathrm{xa}}\overline{p_{\mathrm{xb}}}p_{\mathrm{ya}}\overline{p_{\mathrm{yb}}}| - |\overline{p_{\mathrm{za}}}p_{\mathrm{zb}}\overline{p_{\mathrm{xa}}}p_{\mathrm{xb}}p_{\mathrm{ya}}\overline{p_{\mathrm{yb}}}| \\
&-|p_{\mathrm{za}}\overline{p_{\mathrm{zb}}}\overline{p_{\mathrm{xa}}}p_{\mathrm{xb}}p_{\mathrm{ya}}\overline{p_{\mathrm{yb}}}| - |p_{\mathrm{za}}\overline{p_{\mathrm{zb}}}p_{\mathrm{xa}}\overline{p_{\mathrm{xb}}}\overline{p_{\mathrm{ya}}}p_{\mathrm{yb}}| \\
&+|\overline{p_{\mathrm{za}}}p_{\mathrm{zb}}\overline{p_{\mathrm{xa}}}p_{\mathrm{xb}}p_{\mathrm{ya}}\overline{p_{\mathrm{yb}}}| + |\overline{p_{\mathrm{za}}}p_{\mathrm{zb}}p_{\mathrm{xa}}\overline{p_{\mathrm{xb}}}\overline{p_{\mathrm{ya}}}p_{\mathrm{yb}}| \\
&+|p_{\mathrm{za}}\overline{p_{\mathrm{zb}}}\overline{p_{\mathrm{xa}}}p_{\mathrm{xb}}\overline{p_{\mathrm{ya}}}p_{\mathrm{yb}}| - |\overline{p_{\mathrm{za}}}p_{\mathrm{zb}}\overline{p_{\mathrm{xa}}}p_{\mathrm{xb}}\overline{p_{\mathrm{ya}}}p_{\mathrm{yb}}|\}.
\end{aligned} \tag{11.17}$$

† We now use the convention described on page 83 to distinguish between α and β spin-orbitals.

To construct this function one just multiplies the factors in (11.16) together, neglecting the fact that they are determinants, and then each product is made into a determinant. The wave function (11.17) changes sign if the spins associated with any bonding pair of orbitals are exchanged: this is the general method for obtaining a Heitler–London function. If there are n electron-pair bonds, then 2^n determinants are needed to describe this function if it is written out in full.

Let us now compare (11.17) with the corresponding MO wave function for N_2 (table 10.2)

$$|\pi_u 2p_x \overline{\pi_u 2p_x} \pi_u 2p_y \overline{\pi_u 2p_y} \sigma_g 2p_z \overline{\sigma_g 2p_z}|. \tag{11.18}$$

It is certainly easier to write down and is generally easier to manipulate. However, this does not mean that it is a better wave function, defining the better function as the one which gives the lower energy.

It is evidently convenient to have a shorthand notation for valence-bond functions. We use the symbol $\overset{\frown}{\phi_a \phi_b}$ to indicate that the wave function describes a bond between orbitals ϕ_a and ϕ_b. The wave function (11.17) is then written

$$|\overset{\frown}{p_{za}p_{zb}} \overset{\frown}{p_{xa}p_{xb}} \overset{\frown}{p_{ya}p_{yb}}|. \tag{11.19}$$

Let us now look at the VB interpretation of O_2. The oxygen atom has four electrons in the $2p$ shell, which means that two of these can be unpaired and are capable of forming bonds. Since σ bonds give a lower energy than π bonds, we expect for O_2 that there should be one σ and one π bond. We could then describe this by the wave function

$$|p_{xa}\overline{p_{xa}}\, p_{xb}\overline{p_{xb}}\, \overset{\frown}{p_{za}p_{zb}}\, \overset{\frown}{p_{ya}p_{yb}}|, \tag{11.20}$$

where $\overset{\frown}{p_{za}p_{zb}}$ is the σ bond and $\overset{\frown}{p_{ya}p_{yb}}$ the π bond. The other orbitals of π symmetry p_{xa} and p_{xb} contain non-bonding pairs of electrons.

There are two things wrong with the function (11.20). Firstly, it cannot represent the ground state of O_2 since this is a triplet spin state and a Heitler–London function is always a singlet state. Secondly, it is neither a Σ state nor a Δ state but a mixture of the two.

We have previously seen that in VB theory a triplet function is antibonding if the two orbitals involved overlap one another, but bonding if they are orthogonal to one another. This suggests that the π bond in O_2 is to be considered as being due to two electrons, which are in orthogonal orbitals, whose spins combine to give a triplet state.

If one starts with the MO wave function for the π bonds of the ${}^3\Sigma_g^-$ state of O_2 (say the component with $M_s = 1$), then this is, in terms of the complex form for the π orbitals π^+ and π^-†,

$$|\ldots(\pi_u{}^+2p)(\overline{\pi_u{}^+2p})(\pi_u{}^-2p)(\overline{\pi_u{}^-2p})(\pi_g{}^+2p)(\pi_g{}^+2p)|. \qquad (11.21)$$

If the molecular orbitals are then expanded into their atomic orbitals

$$\begin{aligned} \pi_u{}^+2p &= \pi_a{}^+ + \pi_b{}^+, \qquad & \pi_u{}^-2p &= \pi_a{}^- + \pi_b{}^-, \\ \pi_g{}^+2p &= \pi_a{}^+ - \pi_b{}^+, \qquad & \pi_g{}^-2p &= \pi_a{}^- - \pi_b{}^-, \end{aligned} \qquad (11.22)$$

the terms which have as many electrons on atom a as on atom b are as follows

$$|\ldots\pi_a{}^+\overline{\pi_a{}^+}\pi_b{}^-\overline{\pi_b{}^-}\pi_a{}^-\pi_b{}^+| - |\ldots\pi_a{}^-\overline{\pi_a{}^-}\pi_b{}^+\overline{\pi_b{}^+}\pi_a{}^+\pi_b{}^-|. \qquad (11.23)$$

Each determinant represents an electron pair on atom a, an electron pair on atom b, and a triplet-bonding situation between a π^+ orbital on one atom and a π^- on the other. In addition there will be other terms which represent ionic structures for O_2.

There seems to be no really satisfactory way of representing the situation in O_2 by a chemical bond structure. If we want to describe a normal Heitler–London electron-pair bond between atoms A and B we write A—B. If we have a triplet-state antibonding situation it is conventional to write A· ·B. The triplet-state bonding situation occurs very infrequently, but for the sake of completeness we suggest A~B would be appropriate for this case.

We have seen, in the case of LiH, that the σ molecular orbitals have an appreciable mixture of $2s$ and $2p\sigma$ atomic orbitals. This is true also for the homonuclear diatomic molecules although the extent of $2s$–$2p$ interaction will decrease as we move along the series from Li_2 to F_2 (page 34). In VB theory this $2s$–$2p$ interaction is introduced through the concept of hybridization which has already been discussed at length in chapter 5. Let us consider Li_2 to illustrate this point.

The Heitler–London wave function for the single σ bond in Li_2 in its simplest form is (11.14)

$$(2+2S^2)^{-\frac{1}{2}}\{|2s_a\overline{2s_b}| - |\overline{2s_a}2s_b|\}. \qquad (11.24)$$

† $\pi^+ = \sqrt{\tfrac{1}{2}}(\pi_x + i\pi_y)$, $\pi^- = \sqrt{\tfrac{1}{2}}(\pi_x - i\pi_y)$ (cf. the real and complex forms of p orbitals given in chapter 3).

But we can always improve this wave function by introducing more flexibility and applying the variation principle. In particular, the energy of the function

$$(2 + 2S^2)^{-\frac{1}{2}}\{|\phi_a\overline{\phi_b}| - |\overline{\phi_a}\phi_b|\}, \quad (11.25)$$

where

$$\phi_a = (1 + \lambda^2)^{-\frac{1}{2}}(2s_a + \lambda 2p\sigma_a)$$

and

$$\phi_b = (1 + \lambda^2)^{-\frac{1}{2}}(2s_b + \lambda 2p\sigma_b) \quad (11.26)$$

can be minimized with respect to the parameter λ.

As λ is increased from zero there will be an increase in the overlap of the two atomic orbitals ϕ_a and ϕ_b. This is illustrated in figure 11.1. We have taken Slater functions for the $2s$ and $2p\sigma$ orbitals of Li (chapter 4)

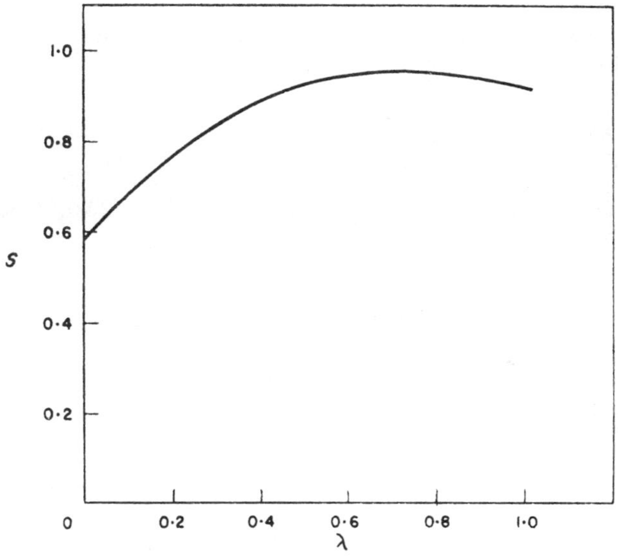

FIGURE 11.1. The effect of *sp* hybridization on the overlap integral between the two Li σ orbitals in Li_2

and calculated the overlap integrals for the equilibrium bond length of 2·67 Å. It is seen that the maximum is at $\lambda = 0{\cdot}7$ and at this value the overlap integral has the surprisingly large value of 0·95 (although if we took a $2s$ orbital with a radial node in it instead of a simple Slater function this would be reduced slightly).

The energy of the function (11.25) is lowered by an increase in the overlap integral between ϕ_a and ϕ_b but raised by increasing the $2p\sigma$ part of the hybrids (since the energy of the $2p$ orbital is about 1·85 ev greater than that of the $2s$). The best wave function will therefore be some compromise between these two effects, with λ probably being in the range 0·2 to 0·4.

11.3 Heteronuclear diatomic molecules

The Heitler–London wave function is usually taken to represent the pure covalent bond, but for a heteronuclear diatomic molecule this function does not give an equal charge density on each atom. For example, squaring the wave function (11.14) and integrating over the spin coordinates one obtains the two-electron probability density (cf. 9.51)

$$P(1, 2) = (2 + 2S^2)^{-1}\{\phi_a^2(1)\phi_b^2(2) + \phi_b^2(1)\phi_a^2(2) + 2\phi_a(1)\phi_b(1)\phi_a(2)\phi_b(2\ \}. \quad (11.27)$$

The density of just one electron is obtained by integrating over the coordinates of the other electron

$$P(1) = \int P(1, 2)\mathrm{d}v_2 = (2 + 2S^2)^{-1}\{\phi_a^2(1) + \phi_b^2(1) + 2S\phi_a(1)\phi_b(1)\}$$

$$P(2) = \int P(1, 2)\mathrm{d}v_1 = (2 + 2S^2)^{-1}\{\phi_a^2(2) + \phi_b^2(2) + 2S\phi_a(2)\phi_b(2)\}, \quad (11.28)$$

and the total electron density is

$$\rho = P(1) + P(2) = (1 + S^2)^{-1}\{\phi_a^2 + \phi_b^2 + 2S\phi_a\phi_b\}. \quad (11.29)$$

We note in passing that the electron density is built up between the two nuclei in the same way as for a bonding molecular orbital.

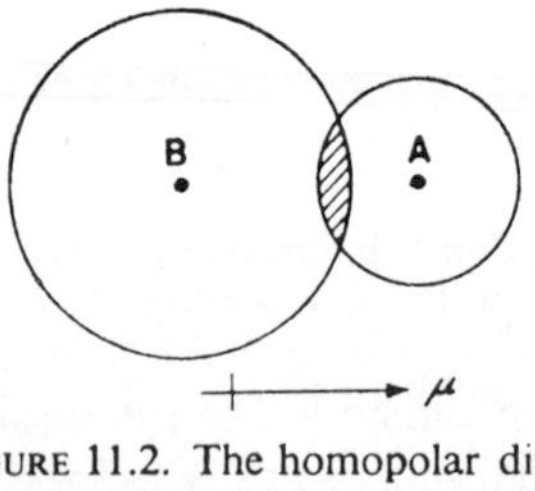

FIGURE 11.2. The homopolar dipole

The overlap density $\phi_a\phi_b$ is not equally distributed over both nuclei. If the effective nuclear charge of ϕ_a is larger than that of ϕ_b (roughly speaking ϕ_a is a smaller orbital than ϕ_b), then the overlap density is found more on A than on B. This is illustrated in figure 11.2. It follows that for a heteronuclear diatomic molecule the Heitler–London function does give rise to a dipole moment—the so-called homopolar dipole. For HCl it has been estimated that this dipole is about 1D†.

Although the Heitler–London function does give some polarity to the bond, polarity is mainly introduced in VB theory by including in the total wave function some contributions which represent an ionic structure of the molecule. If we write for A—B the function (11.14)

$$\Psi(\text{A—B}) = (2 + 2S^2)^{-\frac{1}{2}}\{|\phi_a\bar{\phi}_b| - |\bar{\phi}_a\phi_b|\} \tag{11.30}$$

then for A^+B^- we have an electron pair on atom B

$$\Psi(A^+B^-) = |\phi_b\bar{\phi}_b| \tag{11.31}$$

and for A^-B^+

$$\Psi(A^-B^+) = |\phi_a\bar{\phi}_a|. \tag{11.32}$$

The total wave function is then written as a linear combination of these three functions

$$\Psi = \lambda\Psi(\text{A—B}) + \mu\Psi(A^+B^-) + \nu\Psi(A^-B^+), \tag{11.33}$$

and the variation principle is used to determine the values of the coefficients λ, μ and ν. A similar wave function can be written for a homonuclear diatomic molecule but in this case the two ionic structures must have equal weighting in the wave function.

For most polar molecules one of the ionic functions can generally be neglected on energetic grounds. For example, the energy of H^+F^-, relative to two separate neutral atoms is

$$E(H^+F^-) = I_H - A_F + Q(H^+F^-), \tag{11.34}$$

where I_H is the ionization potential of a hydrogen atom, A_F the electron affinity of a fluorine atom and $Q(H^+F^-)$ the coulomb energy between H^+ and F^-, which is negative. Likewise, if we neglect the exchange energy between the orbitals of H^- and F^+

$$E(H^-F^+) = I_F - A_H + Q(H^-F^+). \tag{11.35}$$

Now $I_H - A_F = 10$ ev and $I_F - A_H = 17$ ev so that, providing the two coulomb terms are comparable in magnitude, the state corresponding to H^-F^+ can be assumed to play no great part in the ground state wave function.

The coulomb energy for H^+F^-, calculated using the Slater orbitals for F^-, is $-14{\cdot}1$ ev, so that we predict the state H^+F^- to be about 4 ev more stable than separate hydrogen and fluorine atoms. However, the observed

† COULSON, *Valence*, O.U.P., 1952.

dissociation energy of HF is about 6 ev so that a completely ionic wave function will not be adequate to describe the bond. The further 2 ev of stability will be accounted for by the interaction of the covalent and ionic wave functions. That is, if we evaluate the secular determinant arising in the variation method (6.68), then it is the off-diagonal element involving both the covalent and the ionic functions which will be important. This interaction leads to the concept of *resonance* in VB theory.

Valence-bond theory is built around two rather dangerous concepts: 'structure' and 'resonance'. They are dangerous because there is a tendency to lift the terms from the level of a concept to the level of reality. A structure is a certain way of pairing electrons in a molecule, and a wave function can be written which describes the pairing scheme. Resonance means that stability arises through the off-diagonal terms of the secular determinant which is built up from these structures. It is important to note that a structure does not represent a stationary state of the molecule, and resonance does not mean that the molecule is jumping back and forth between two resonating structures (which is the way the term is used in classical mechanics). It was because the VB terms structure and resonance do not have any physical reality that VB theory was stated to be incompatible with Materialism by the Soviet Academy of Sciences in the 1940's. But if such a rigid line is to be adopted then MO theory must go as well, which would not leave very much of this book.

Pauling built up an electronegativity scale of the elements through the concept of resonance energy, which in fact predated the Mulliken scale already described in chapter 4. Pauling first defined a quantity Δ_{AB} which was the difference (in ev) between the actual dissociation energy of a molecule AB and that calculated on the basis of a pure covalent structure A—B. He then suggested, on purely empirical grounds, that $\sqrt{\Delta_{AB}}$ was a measure of the difference between the electronegativities x of atoms A and B.

$$\sqrt{\Delta_{AB}} = |x_A - x_B|. \qquad (11.36)$$

The difficulty with the Pauling scale of electronegativities lies in evaluating the energy of the covalent structure. The procedure that has been adopted is to take this as the geometric mean of the dissociation energies of A_2 and B_2, although there is little theoretical foundation for this method.

The Pauling and Mulliken scales of electronegativity are approximately linearly related:

$$\chi\,(\text{Mulliken}) = 2{\cdot}78\,x\,(\text{Pauling}). \qquad (11.37)$$

Since the Mulliken scale is based on a simpler and physically more reasonable approach, we think it is to be preferred.

In the next chapter, when we come to compare MO and VB theories, we shall describe some of the more sophisticated VB calculations that have been carried out.

Problems 11

11.1 Prove that the functions (11.7) are normalized.

11.2 Draw the potential energy curves for the $^1\Sigma_g^+$ and $^3\Sigma_g^+$ states of H_2 using the VB wave functions and the following table.

R (atomic units)	0	1	2	3	4
S	1·00	0·86	0·59	0·35	0·19
$\int 1s_a\left(\frac{1}{r_b}\right)1s_a dv$	1·00	0·73	0·47	0·33	0·25
$\int 1s_a\left(\frac{1}{r_b}\right)1s_b dv$	1·00	0·74	0·41	0·20	0·09
$\iint 1s_a(1)1s_b(2)\left(\frac{1}{r_{12}}\right)1s_a(1)1s_b(2)dv_1dv_2$	0·63	0·55	0·43	0·32	0·25
$\iint 1s_a(1)1s_b(2)\left(\frac{1}{r_{12}}\right)1s_b(1)1s_a(2)dv_1dv_2$	0·63	0·44	0·18	0·06	0·02
$\frac{1}{R}$	∞	1·00	0·50	0·33	0·25

11.3 Write down a VB wave function for the ground state of C_2.

11.4 Given the VB wave function of a hydrogen halide HX in the form

$$\Psi = N\{\Psi(\text{H—X}) + \lambda\Psi(\text{H}^+\text{—X}^-)\}.$$

Calculate the percentage ionic character defined by $P = \dfrac{100\lambda^2}{(1+\lambda^2)}$ from the following data.

	HF	HCl	HBr	HI
Dipole moment (D)	1·74	1·03	0·78	0·38
R (Å)	0·92	1·27	1·41	1·61

Assume the following:

(a) $\Psi(\text{H—X})$ and $\Psi(\text{H}^+\text{X}^-)$ are orthogonal and separately normalized.
(b) $\Psi(\text{H—X})$ has no dipole moment.
(c) $\Psi(\text{H}^+\text{—X}^-)$ has a dipole $\mu = eR$.
(d) The transition dipole moment $e\int\Psi(\text{H—X})\sum \mathbf{r}_i \Psi(\text{H}^+\text{X}^-)d\tau$ is zero.

To what extent are these assumptions valid?

11.5 Evaluate a VB wave function for the ground state of LiH of the form

$$\Psi = c_1\Psi_1 + c_2\Psi_2,$$

where Ψ_1 represents a Heitler–London function between $2s_{\mathrm{Li}}$ and $1s_{\mathrm{H}}$, and Ψ_2 a Heitler–London function between $2p\sigma_{\mathrm{Li}}$ and $1s_{\mathrm{H}}$ (see 12.32). The following data are sufficient:

$$\mathscr{H}_{11} = -9{\cdot}48, \quad \mathscr{H}_{22} = -10{\cdot}19, \quad \mathscr{H}_{12} = -2{\cdot}12 \text{ (atomic units)}$$
$$S_{11} = 1{\cdot}19, \quad S_{22} = 1{\cdot}29, \quad S_{12} = 0{\cdot}26.$$

It should be noted that Ψ_1 and Ψ_2 are not normalized functions.

Chapter 12

The bonding in polyatomic molecules

12.1 A comparison of MO and VB theories for diatomic molecules

In this chapter we shall first compare the simple MO and VB theories of the hydrogen molecule in order to show their limitations and to see what improvements must be made to obtain more accurate wave functions.

The MO wave function for H_2 is obtained by allocating two electrons with opposite spins into the orbital $\sigma_g 1s = N(1s_a + 1s_b)$.

$$\Psi(\text{MO}) = N^2 |1s_a + 1s_b\ \overline{1s_a + 1s_b}|. \tag{12.1}$$

This determinant can now be expanded to give

$$\Psi(\text{MO}) = N^2\{|1s_a\overline{1s_b}| + |1s_b\overline{1s_a}| + |1s_a\overline{1s_a}| + |1s_b\overline{1s_b}|\}. \tag{12.2}$$

The first two terms make up the Heitler–London wave function for H_2 (11.12), the second two represent the ionic structures H^-H^+ and H^+H^-. We can say that the MO wave function is an equal mixture of the Heitler–London covalent and the ionic wave functions

$$\Psi_{\text{cov}} = (2 + 2S^2)^{-\frac{1}{2}}\{|1s_a\overline{1s_b}| - |\overline{1s_a}1s_b|\}$$

$$\Psi_{\text{ion}} = (2+2S^2)^{-\frac{1}{2}}\{|1s_a 1s_a| + |1s_b 1s_b|\}. \tag{12.3}$$

On the other hand, the simple VB theory of H_2 is based entirely on the covalent function. Which situation is closer to the truth?

Clearly the MO function is poor at large internuclear separations. It correlates with dissociation products which are an equal mixture of atoms and ions, whereas the lowest energy dissociation products of H_2 are atoms. On the other hand, for very short internuclear distances one expects that there will be as much chance of finding both electrons in the same atomic orbital as in different orbitals—which is the situation described by the MO function.

The Heitler–London wave function for H_2 gives a dissociation energy of 3·14 ev and an internuclear distance of 0·87 Å. The simple MO wave function gives a much poorer energy (2·65 ev) but a slightly better equilibrium distance 0·85 Å. However, both functions are really so far from the observed result (4·75 ev, 0·74 Å) that there is little to choose between them. A better wave function than either is one which allows the ratio of ionic to covalent structures to vary

$$\Psi = \Psi_{\text{cov}} + \lambda \Psi_{\text{ion}}. \tag{12.4}$$

A calculation using the variation theorem gives $\lambda = 0{\cdot}16$ if the covalent and ionic wave functions are built up with hydrogen orbitals, but this function is hardly any improvement on the Heitler–London function (3·23 ev, 0·88 Å). However, if the orbital exponent of the $1s$ function is allowed to vary, then with $\zeta_{\text{cov}} = \zeta_{\text{ion}} = 1{\cdot}193$ and $\lambda = 0{\cdot}25$ quite a striking improvement is obtained (4·02 ev, 0·75 Å) over either the covalent function alone or the best MO wave function. These results are summarized in table 12.1 (page 189).

The role of ionic–covalent resonance in VB theory is filled by *configurational interaction* in MO theory. This term means that we take a wave function which is a linear combination of functions arising from different electronic configurations. For example, the doubly-excited configuration, which has two electrons in the $\sigma_u 1s$ antibonding orbital of H_2, also gives a function of ${}^1\Sigma_g{}^+$ symmetry†. A better wave function than the simple MO is then obtained by mixing the two ${}^1\Sigma_g{}^+$ functions, one obtained from the ground configuration $(\sigma_g 1s)^2$ and one from the doubly-excited configuration $(\sigma_u 1s)^2$

$$\Psi = |(\sigma_g 1s)\overline{(\sigma_g 1s)}| + \mu|(\sigma_u 1s)\overline{(\sigma_u 1s)}|. \tag{12.5}$$

This wave function can be shown to be identical to (12.4) (apart from a normalizing factor) if

$$\mu = \frac{(\lambda - 1)(1 - S)}{(\lambda + 1)(1 + S)}. \tag{12.6}$$

Because λ and μ are both chosen by the variation theorem to give the best wave function, these functions must be identical.

The MO wave function for H_2 has no correlation between the position of the electrons (page 130)—from it one deduces that if one electron is in orbital $1s_a$ the other is as likely to be in $1s_a$ as in $1s_b$. The VB function on the other hand over-emphasizes electron correlation (one can check from (11.27) and (11.28) that the VB function does not satisfy $P(1, 2) = P(1)\,P(2)$): if

† The configuration $(\sigma_g 1s)(\sigma_u 1s)$ will give states which cannot contribute to the ground state wave function because they have a different symmetry.

one electron is in orbital $1s_a$ then the other must be in $1s_b$. The correct amount of correlation is somewhere between these two extremes, but rather closer to that of the Heitler–London function. The correlation energy for H_2, which is the difference between the SCF MO wave function and the exact solution of the Schrödinger equation (cf. page 130), is about 1·1 ev.

In VB theory the one-electron functions are atomic orbitals; in MO theory they are molecular orbitals, which, for a homonuclear diatomic molecule, are equally distributed over both nuclei. Wave functions can also be constructed from one-electron functions which are between these two extremes. For example, if we have two identical atomic orbitals on different atoms ϕ_a and ϕ_b, then the orbitals

$$\psi_1 = (1 + 2\lambda S + \lambda^2)^{-\frac{1}{2}}(\phi_a + \lambda\phi_b)$$
$$\psi_2 = (1 + 2\lambda S + \lambda^2)^{-\frac{1}{2}}(\phi_b + \lambda\phi_a) \qquad (12.7)$$

are equivalent delocalized orbitals. If we then construct a Heitler–London function from these

$$\Psi = N\{|\psi_1\bar{\psi}_2| - |\bar{\psi}_1\psi_2|\}, \qquad (12.8)$$

then we have in the limit $\lambda = 0$ the simple VB function, and in the limit $\lambda = 1$ the simple MO function†. If the energy of this function is minimized with respect to λ then we again have a function which is identical to the VB function with ionic–covalent resonance (12.4) or the MO function with configuration interaction (12.5).

It is possible to choose the parameter λ in (12.7) such that the orbitals ψ_1 and ψ_2 are orthogonal, a property that leads to a considerable simplification of the mathematics when one is dealing with a many-electron problem. The condition for this is that

$$\int \psi_1\psi_2 \, dv = (1 + 2\lambda S + \lambda^2)^{-1}(2\lambda + S(1 + \lambda^2)) = 0, \qquad (12.9)$$

which has the solution

$$\lambda = [-1 \pm (1 - S^2)^{\frac{1}{2}}]/S. \qquad (12.10)$$

This orthogonalization of atomic orbitals is illustrated in figure 12.1.

Orthogonalized atomic orbitals can be used as a basis for constructing molecular orbitals (by linear combination of orthogonalized orbitals) and this is a useful simplification for polyatomic molecules. However, it is important to realize that orthogonalization completely changes the character of the Heitler–London wave function. If we evaluate the exchange

† COULSON and FISCHER, *Phil. Mag.*, **40**, 386 (1949).

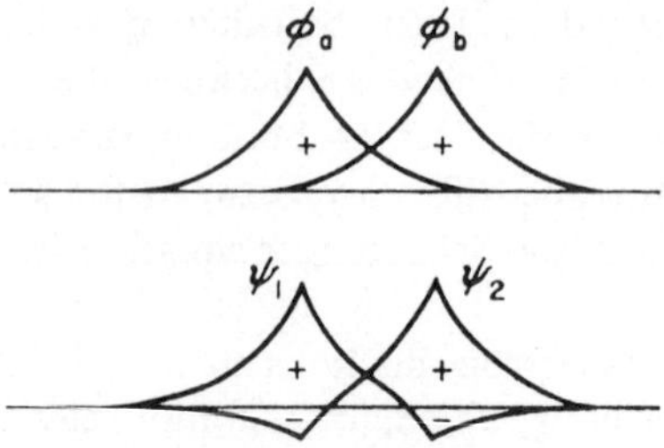

FIGURE 12.1. The orthogonalization of the two 1*s* atomic orbitals; the wave functions are plotted along the internuclear axis

integral for these orbitals (11.11) then the only term which is non-zero is the two-electron integral, and this makes A positive. It follows that the Heitler–London function now represents a state of repulsion, and if a VB formulation is to be followed through using these orthogonal orbitals it is the 'ionic' functions $|\psi_1 \bar{\psi}_1|$ and $|\psi_2 \bar{\psi}_2|$ which dominate the ground state wave function.

For the hydrogen molecule it is possible to construct only two independent wave functions which have the symmetry of a $^1\Sigma_g{}^+$ state from two 1*s* atomic orbitals: these are the functions (12.3) in VB theory, and the ground and doubly-excited functions in MO theory. It follows, as we have seen, that the two theories will give the same answer if both functions are included in the total wave function. On the other hand, let us consider the lowest state of He_2. In MO theory this is described by the function

$$\Psi(\text{MO}) = |(\sigma_g 1s)(\overline{\sigma_g 1s})(\sigma_u 1s)(\overline{\sigma_u 1s})|, \tag{12.11}$$

and in VB theory by

$$\Psi(\text{VB}) = |1s_a \overline{1s_a} 1s_b \overline{1s_b}|. \tag{12.12}$$

But these are the only functions in each theory that can be obtained by using only the 1*s* atomic orbitals and so they must be equivalent. This can be confirmed by adding and subtracting columns in the determinant (12.11) when one can arrive at (12.12). *For completely filled atomic shells the simple wave functions used in the two theories are therefore equivalent.*

At the present time most work on the theory of the chemical bond is based on molecular orbitals. This is because for polyatomic molecules MO theory is mathematically much simpler than VB theory, and it is also easier to programme for an electronic computer. The simplification lies in the fact that molecular orbitals, if they all arise from one secular determinant, are orthogonal. Because of this the Hamiltonian integrals between wave functions based on these orbitals can all be reduced to a sum of one- and two-electron integrals.

To illustrate this point we shall consider the following integral between two three-electron functions

$$\mathscr{I} = \iiint \psi_a(1)\psi_b(2)\psi_c(3)\mathscr{H}\psi_d(1)\psi_e(2)\psi_f(3)\, dv_1\, dv_2\, dv_3. \quad (12.13)$$

The Hamiltonian consists of one-electron and two-electron operators and the nuclear repulsion terms V_{nn} (10.58)

$$\mathscr{H} = \mathbf{H}_1^c + \mathbf{H}_2^c + \mathbf{H}_3^c + \frac{1}{r_{12}} + \frac{1}{r_{23}} + \frac{1}{r_{31}} + V_{nn}. \quad (12.14)$$

Consider the contribution from $\mathbf{H}_1$

$$\iiint \psi_a(1)\psi_b(2)\psi_c(3)\mathbf{H}_1^c\psi_d(1)\psi_e(2)\psi_f(3)\, dv_1\, dv_2\, dv_3. \quad (12.15)$$

We can integrate over the coordinates of electrons 2 and 3 and obtain

$$S_{be}S_{cf}\int \psi_a(1)\mathbf{H}_1^c\ \psi_d(1)\, dv_1, \quad (12.16)$$

where S is an overlap integral defined in the usual way. If we build up our wave functions from an orthonormal set of molecular orbitals then S_{be} is either one (b = e) or zero (b ≠ e). It is clear that the integrals of one-electron operators like $\mathbf{H}_1^c$ between two many-electron functions of this type will be zero if more than one orbital is different in the two functions. Similarly the contributions from the two-electron operators like $1/r_{12}$ will be zero if more than two orbitals are different in the two functions, and the nuclear repulsion term only comes in when the two functions are identical. If we do not have an orthonormal set of one-electron functions, as is the case in VB theory, then these simplifications do not arise.

The mathematics of VB theory is therefore rather complicated, but the results of a simple VB theory may be better than those of a simple MO theory, particularly with regard to dissociation energies. In fact, we shall see at the end of this chapter, that the best method at present available for calculating good dissociation energies is a modification of the VB method called the *method of atoms in molecules.*

12.2 *Accurate wave functions for the hydrogen molecule*

Before proceeding with a general discussion of the bonding in polyatomic molecules we shall look briefly at some of the accurate wave functions for the hydrogen molecule.

There have been several extended configuration interaction treatments of H_2 which give a dissociation energy within 0·2 ev of the observed energy.

For example, using a wave function which contained terms from the configurations $(\sigma_g 1s)(\sigma_g 1s')$, $(\sigma_u 1s)(\sigma_u 1s')$, $(\sigma_g 2s)(\sigma_g 2p)$, $(\pi_u 2p)^2$ and $(\pi_g 2p)^2$, where $\sigma_g 1s$ and $\sigma_g 1s'$ are based on $1s$ orbitals that have different screening constants ($(\sigma_g 1s)(\sigma_g 1s')$ is a so-called open configuration), the calculated energy was 4·543 ev†.

James and Coolidge in 1933 constructed a wave function for H_2 based on elliptical coordinates (10.54), which explicitly included the inter-electronic distance. In this they were following the procedure used by Hylleraas for the helium atom, which we described in chapter 9. Using a thirteen-term function they obtained an energy only 0·03 ev greater than that observed. 27 years later Kolos and Roothaan extended their calculation to a fifty-term function and obtained an energy which is more accurate than that obtained from experiment.

Modern electronic computers have brought us to the point where the Schrödinger equation has, in effect, been solved exactly for small molecules. The wave functions are not very elegant, but why should they be? A more important thing to ask is what new insight have these calculations given into the nature of the chemical bond. Probably the advance from James–Coolidge to Kolos–Roothaan has given very little in this direction. These very accurate calculations have, however, shown two things of importance. Firstly, that the basic equations of quantum mechanics do apply to chemical systems, otherwise we would not get agreement with experiment. Secondly, that even with the computers of the foreseeable future we are not going to be able to calculate the energy levels of a large molecule like benzene to the level where it is better to do a calculation than an experiment.

Table 12.1 summarizes the calculations on H_2 that have been described in chapters 10, 11 and 12.

12.3 *The role of hybridization in* VB *and* MO *theories*

In chapter 5 we showed that in a theory which emphasized the electron-pair bond it was necessary to introduce the concept of hybridization to explain the shapes of molecules like methane. The VB wave function appropriate to the structure

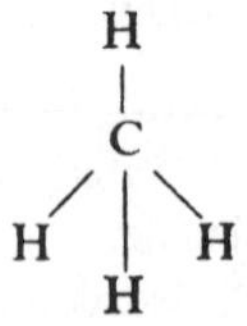

† McLean, Weiss and Yoshimine, *Rev. Mod. Phys.*, **32**, 211 (1960).

TABLE 12.1. Calculations on H_2

Type of wave function	Dissociation energy (ev)	Equilibrium bond length (Å)
Heitler–London ($\zeta = 1$)	3·14	0·87
Simple molecular-orbital ($\zeta = 1$)	2·65	0·85
$\Psi_{cov} + \lambda\Psi_{ion}$ ($\zeta = 1$)	3·23	0·88
Heitler–London (Wang) with screening ($\zeta = 1{\cdot}166$)	3·78	0·743
Molecular-orbital with screening ($\zeta = 1{\cdot}197$)	3·49	0·732
$\Psi_{cov} + \lambda\Psi_{ion}$ ($\zeta = 1{\cdot}193$) (Weinbaum)	4·02	0·748
SCF molecular-orbital	3·62 at 0·74 Å	
Best configuration interaction without r_{12} terms	4·71 at 0·74 Å	
James–Coolidge (13-term function)	4·72	0·740
Kolos–Roothaan (50-term function)	4·7467	0·741
Experiment	4·7466 ± 0·0007	0·741

A complete bibliography and references are given by MCLEAN, WEISS and YOSHIMINE, *Rev. Mod. Phys.*, 32, 211 (1960).

in which all four C—H bonds are equivalent, must involve electron-pair bonds between four hydrogen orbitals ($1s_a\ 1s_b\ 1s_c\ 1s_d$) and four sp^3 hybrid carbon orbitals (σ_a, σ_b, σ_c, σ_d) as follows:

$$|\widehat{1s_a\sigma_a}\ \widehat{1s_b\sigma_b}\ \widehat{1s_c\sigma_c}\ \widehat{1s_d\sigma_d}|. \tag{12.17}$$

(We use the abbreviation introduced by 11.19.) It would be possible to write wave functions in which the $1s$ orbitals were forming electron-pair bonds with carbon $2s$ and $2p$ atomic orbitals, but it would then be necessary to invoke resonance between a large number of structures to get a wave function which would be consistent with four equivalent C—H bonds.

In the MO treatment of CH_4, on the other hand, molecular orbitals are formed by taking a linear combination of the hydrogen $1s$ orbitals and the carbon $2s$ and $2p$ orbitals, and then eight electrons are allocated to the four molecular orbitals of lowest energy. Hybridization does not appear in this theory. Since hybrid orbitals are linear combinations of atomic orbitals of the same atom, whether molecular orbitals are formed by taking linear combinations of hybrid orbitals or linear combinations of atomic orbitals is immaterial, the final molecular orbitals will be the same. *Hybridization is a necessary first step in* VB *theory but no more than a convenience in* MO *theory.*

MO theory emphasizes the delocalization of electrons over the whole molecule, VB theory emphasizes the electron-pair bond. Let us look at a molecule like n-hexane from these two standpoints. There are fourteen C—H bonds in this molecule and chemical evidence indicates that it requires almost the same amount of energy to break any one of them (and the same as a C—H bond in any other paraffin). This constancy of bond properties, which is found to be extremely useful in many branches of chemistry, comes naturally from VB theory. A paraffinic C—H bond is formed between a carbon sp^3 hybrid and a hydrogen $1s$ orbital and that is all that has to be said. On the other hand, the molecular orbitals of n-hexane are not localized in one C—H bond, and they are not like the orbitals of methane, so how does the constancy of bond properties come out of MO theory?

If we remove an electron from n-hexane, this electron will not just be removed from one bond but to some extent from all bonds in the molecule, that is, each of the fourteen C—H bonds will lose a small amount of electron density. This is clearly explained by MO theory, because the orbital from which the electron is removed is delocalized over the whole molecule. In VB theory, on the other hand, one has again to introduce the concept of resonance in order to fit the facts: fourteen structures each having a different one-electron C—H bond are needed even to make the C—H bonds equivalent.

It would appear from these two examples that the VB and MO theories are largely complementary, but this is going too far. VB theory with resonance *can* explain why the positive charge is delocalized over the whole molecule in the n-hexane cation. We must now see how the additivity of bond properties arises from MO theory; to do this we look at the so-called equivalent orbitals, which were first introduced by Lennard-Jones, Hall and Pople.

We first look at some SCF molecular orbitals that have been obtained for the water molecule†. The atomic orbitals are defined with respect to the coordinate axes shown in figure 12.2.

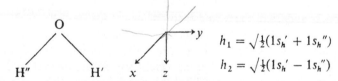

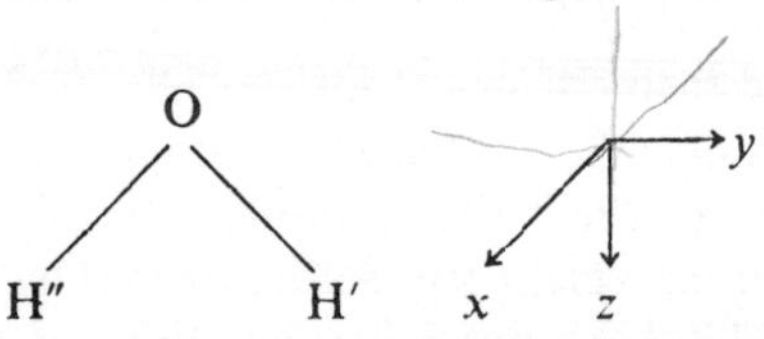

$$h_1 = \sqrt{\tfrac{1}{2}}(1s_h' + 1s_h'')$$

$$h_2 = \sqrt{\tfrac{1}{2}}(1s_h' - 1s_h'')$$

FIGURE 12.2. The coordinate axes for H_2O and the symmetry combinations of the hydrogen orbitals

The molecular orbitals must belong to some symmetry species of the symmetry group of water which is $\mathbf{C}_{2v}$. The oxygen atomic orbitals ($1s$, $2s$, $2p_x$, $2p_y$ and $2p_z$) already have this property; if the hydrogen $1s$ orbitals are taken

† AUNG, PITZER and CHAN, *J. Chem. Phys.*, **49**, 2071 (1968).

in the combination h_1 and h_2 (see page 104 where a similar operation was carried out for formaldehyde) then we can classify all the atomic orbitals according to their transformation properties in the $\mathbf{C}_{2v}$ group (table 8.5).

Symmetry species	A_1	B_2	B_1
	$1s$	$2p_y$	$2p_x$
	$2s$	h_2	
	$2p_z$		
	h_1		

Species A_1 is totally symmetric to all operations of the group; B_2 is antisymmetric under $C_2(z)$ and $\sigma_v(xz)$; B_1 is antisymmetric under $C_2(z)$ and $\sigma_v'(yz)$. There are no orbitals of the A_2 species (antisymmetric under $\sigma_v(xz)$ and $\sigma_v'(yz)$).

Only atomic orbitals belonging to the same symmetry species can be combined to form molecular orbitals. This means, for example, that $2p_x$ is a non-bonding orbital. The five SCF orbitals of lowest energy obtained by Aung, Pitzer and Chan for a bond angle of 104° are given in table 12.2.

TABLE 12.2. The SCF orbitals for water

	Symmetry species	Energy (ev)	LCAO coefficients						
			$1s$	$2s$	$2p_z$	$2p_y$	$1s_h'$	$1s_h''$	$2p_x$
ψ_5	B_1	$-11{\cdot}0$	—	—	—	—	—	—	1
ψ_4	A_1	$-12{\cdot}7$	0·093	−0·516	0·787	—	0·264	0·264	—
ψ_3	B_2	$-17{\cdot}0$	—	—	—	0·624	0·423	−0·423	—
ψ_2	A_1	$-35{\cdot}0$	−0·222	0·843	0·132	—	0·152	0·152	—
ψ_1	A_1	$-559{\cdot}1$	0·997	0·015	0·003	—	−0·004	−0·004	—

The lowest orbital is almost entirely $1s$. ψ_2 is mainly $2s$, so that to a large extent this orbital also can be considered to be non-bonding. ψ_3 and ψ_4 are the main bonding orbitals. In the ground state the ten electrons occupy the five lowest energy orbitals so that the MO wave function is

$$\Psi = |\psi_1\bar{\psi}_1\psi_2\bar{\psi}_2\psi_3\bar{\psi}_3\psi_4\bar{\psi}_4\psi_5\bar{\psi}_5|. \tag{12.18}$$

The value of a determinant is not changed by adding or subtracting columns (or rows) from one another. For example, an identical wave function to (12.18) is

$$\Psi = |(\psi_1 + \psi_2)\overline{(\psi_1 + \psi_2)}\psi_2\bar{\psi}_2\psi_3\bar{\psi}_3\psi_4\bar{\psi}_4\psi_5\bar{\psi}_5|. \tag{12.19}$$

which is derived from (12.18) by adding column 3 to 1 and column 4 to 2. We could now say that the electrons occupy orbitals $\psi_1 + \psi_2$, ψ_2, ψ_3, ψ_4 and ψ_5, and clearly we can get many other sets of five orbitals which may or may not be orthonormal. There is one set which brings out the idea of a molecular orbital localized in a bond, and this set is called the equivalent orbitals.

The molecular orbitals belong to symmetry species of the group (transform like *I.R.*'s of the group). This means for the C_{2v} group, where there are no degenerate *I.R.*'s, that any symmetry operation will simply leave a molecular orbital unchanged or change its sign. However, suppose we want orbitals to be localized in the O—H bonds of water. There must be two such orbitals, which do not belong to any symmetry species of the group, but *can be changed into one another by the symmetry operations of*

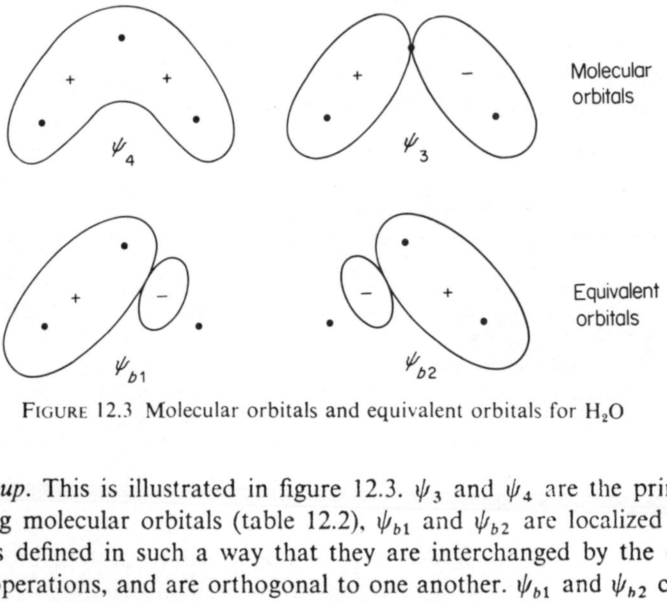

FIGURE 12.3 Molecular orbitals and equivalent orbitals for H_2O

the group. This is illustrated in figure 12.3. ψ_3 and ψ_4 are the principal bonding molecular orbitals (table 12.2), ψ_{b1} and ψ_{b2} are localized bond orbitals defined in such a way that they are interchanged by the C_2 or $\sigma(xz)$ operations, and are orthogonal to one another. ψ_{b1} and ψ_{b2} can be written as orthogonal linear combinations of ψ_3 and ψ_4

$$\psi_{b1} = \lambda\psi_4 + \mu\psi_3, \qquad \psi_{b2} = \mu\psi_4 - \lambda\psi_3. \tag{12.20}$$

It is also possible to transform the non-bonding orbitals ψ_2 and ψ_5 into two non-bonding equivalent orbitals ψ_{l1} and ψ_{l2} which are directed out of the plane of the water molecule—roughly in such a way that together with ψ_{b1} and ψ_{b2} they point to the corners of a tetrahedron. Because ψ_2 and ψ_5 are not strictly non-bonding, the equivalent orbitals are actually a mixture of all the molecular orbitals $\psi_2 \ldots \psi_5$; if we impose the restriction that the lone pair orbitals have no contribution from the hydrogen 1s orbitals, then we can obtain

$$\psi_{b1} = 0{\cdot}513\ 1s_h' - 0{\cdot}085\ 1s_h'' - 0{\cdot}021\ 1s - 0{\cdot}019\ 2s + 0{\cdot}528\ 2p_z + 0{\cdot}441\, 2p_y$$

$$\psi_{b2} = 0{\cdot}513\ 1s_h'' - 0{\cdot}085\ 1s_h' - 0{\cdot}021\ 1s - 0{\cdot}019\ 2s + 0{\cdot}528\ 2p_z - 0{\cdot}441\, 2p_y$$

$$\psi_{l1} = -0{\cdot}168\ 1s + 0{\cdot}697\ 2s - 0{\cdot}195\ 2p_z + 0{\cdot}707\ 2p_x$$

$$\psi_{l2} = -0{\cdot}168\ 1s + 0{\cdot}697\ 2s - 0{\cdot}195\ 2p_z - 0{\cdot}707\ 2p_x \qquad (12.21)$$

The wave function

$$\Psi = |\psi_1\bar{\psi}_1\psi_{b1}\bar{\psi}_{b1}\psi_{b2}\bar{\psi}_{b2}\psi_{l1}\bar{\psi}_{l1}\psi_{l2}\bar{\psi}_{l2}| \qquad (12.22)$$

is identical with (12.18) although it appears different.

We can always compound a mixture of $2p_y$ and $2p_z$ into a $2p$ orbital pointing at some direction in the yz plane. For example,

$$0{\cdot}528\ 2p_z + 0{\cdot}441\ 2p_y = 0{\cdot}687\ 2p_\varphi, \qquad (12.23)$$

where $2p_\varphi$ is a p orbital making an angle of $\varphi = 34°$ with the z axis. With this result one can see from (12.21) that the equivalent bonding orbitals are constructed mainly from a combination of a hydrogen 1s orbital and the orbital $2p_\varphi$ which is pointing roughly in its direction (although not exactly along the bond).

Hybrid atomic orbitals can be said to be equivalent orbitals of an atom. For example, suppose we consider a carbon atom in the $2s2p^3$ configuration with all its electrons having the same spin. This is one component of a ^5S state and has the wave function

$$|2s\ 2p_x\ 2p_y\ 2p_z|. \qquad (12.24)$$

An identical function is

$$|\psi_1\ \psi_2\ \psi_3\ \psi_4|, \qquad (12.25)$$

where these ψ are the four sp^3 hybrids defined by expressions (5.13,14). These tetrahedral hybrids are the equivalent orbitals of the carbon atom in a potential field of tetrahedral symmetry (such as that produced by the four hydrogen atoms in methane). If hybridization is introduced as a first step in MO theory and molecular orbitals are then formed as linear combinations of orbitals (atomic or hybrid) which are strongly overlapping one another, localized bond molecular orbitals are obtained which are usually a good approximation to the equivalent orbitals of the molecule. Taking methane as an example, if the tetrahedral hybrid σ_1 is combined with the hydrogen orbital $1s_1$ at which it is pointing, to form a bond molecular orbital

$$\psi_1 = a\sigma_1 + b1s_1, \tag{12.26}$$

then ψ_1 is a good approximation to an equivalent orbital of methane: the exact equivalent orbital differs from ψ_1 only in having a small amount of $1s_2$, $1s_3$ and $1s_4$.

Let us summarize the points we have just made. Molecular orbitals transform like one of the symmetry species of the group. A wave function written as a Slater determinant of these molecular orbitals emphasizes the delocalized nature of the electrons, and is the most appropriate form for describing, say, the ionization of the molecule. Because the wave function is written as a determinant (because of the Pauli exclusion principle one could say) the molecular orbitals can be transformed into other sets of orbitals which give an equally good wave function. One of these sets is called the equivalent orbitals. These equivalent orbitals are transformed into one another by the symmetry operations of the group. They are functions which are fairly well localized in bonds or they represent orbitals for lone pairs of electrons. They provide the best basis within MO theory for appreciating the additivity of bond properties. Whenever VB theory can adequately describe a molecule in terms of localized electron-pair bonds, then equivalent orbitals can be found to do the same thing. For molecules like benzene, where the π bonds are not localized in VB theory, then similarly it is not possible to obtain localized equivalent orbitals. The two theories are therefore in agreement in this respect.

12.4 The MO *theory of polyatomic molecules*

The application of MO theory to polyatomic molecules is basically the same as its application to diatomic molecules. The operations of group theory are valuable for reducing the mathematical labour involved in obtaining a set of molecular orbitals, and a very simple example of this was illustrated by the water molecule discussed in the last section.

In section 10.7 we pointed out that a molecular orbital is not always a well defined entity. But to some extent this increases its usefulness, as is brought out in the following discussion.

Suppose we have a set of molecular orbitals each with an associated spin function. We can use this set of spin-orbitals ψ_a, ψ_b, ... to build up wave functions for the molecule. Suppose the molecule has n electrons, then as a first step we might take a wave function in which these electrons are allocated to the n spin-orbitals of lowest energy; taking the anti-symmetry principle into account, this would give a function

$$|\psi_a \psi_b \cdots \psi_n|. \qquad (12.27)$$

Now if we really want to limit ourselves to a wave function of this type—a single Slater determinant of spin-orbitals—and yet we want this to be a good wave function, then we have to be very careful about our choice of molecular orbitals. They have really to be derived by the SCF procedure described in section 10.5.

The best wave function that can be obtained of this type using the minimum energy criterion is known as the Hartree-Fock wave function, and the molecular orbitals as Hartree-Fock orbitals. The difference between the Hartree-Fock energy and the true ground state energy associated with the non-relativistic Hamiltonian of the molecule is called the correlation energy (see page 130). This seems to be about 1ev. or 0·05 a.u. per pair of electrons in the molecule. As chemically interesting energies, such as dissociation energies, are usually only a few electron volts, the failure to calculate this correlation energy can be a serious deficiency in a calculation. Nevertheless a lot of effort has gone into obtaining wave functions of the single-determinant type, particularly for closed-shell systems; some of their properties are expected to be close to those associated with the exact wave function for the following reason.

The Hartree-Fock wave function cannot be improved by adding to it any determinant in which just one Hartree-Fock orbital has been replaced by some other orbital. This follows because we can write

$$|\psi_a \psi_b \ldots \psi_n| + \lambda|\psi_p \psi_b \ldots \psi_n| = |(\psi_a + \lambda\psi_p)\, \psi_b \ldots \psi_n| \qquad (12.28)$$

so that if $|\psi_a \psi_b \ldots \psi_n|$ is really the best single determinant wave function then λ must be zero. It was by using this property that we derived the Hartree-Fock equations in section 10.7. Thus the Hartree-Fock wave function can only be improved by adding to it determinants which differ by at least two orbitals thus:

$$|\psi_a \psi_b \ldots \psi_n| + \lambda|\psi_p \psi_q \ldots \psi_n|,\ p, q \neq a, b \qquad (12.29)$$

which we cannot contract to a single determinant.

If we now evaluate the expectation value of a one electron operator $\mathbf{M}(i)$ for this wave function we get

$$\int |\psi_a \psi_b \ldots \psi_n|\mathbf{M}(i)|\psi_a \psi_b \ldots \psi_n| \mathrm{d}\tau + \lambda^2 \int |\psi_p \psi_q \ldots \psi_n|\mathbf{M}(i)|\psi_p \psi_q \ldots \psi_n| \mathrm{d}\tau \qquad (12.30)$$

the cross term being zero. Therefore if λ is a small quantity the expectation value of $\mathbf{M}$ for this function will be approximately equal to that for the Hartree-Fock wave function. On this argument the expectation value of any one-electron operator for the *exact* wave function will approximate to that for the Hartree-Fock wave function. In particular the electron density in a molecule (and properties dependent on this, like the dipole moment) will be expected to be close to the electron density associated with the Hartree-Fock wave function.

In addition to these one electron properties, there is good reason to think that the energies associated with the Hartree-Fock orbitals (corresponding to the eigenvalues of the SCF operator) are, by Koopman's theorem (p. 168), approximately equal to the negative of the ionization potentials. Also equilibrium geometries calculated from Hartree-Fock wave functions, or at least reasonably good SCF approximations to these wave functions, seem to be in good agreement with experiment. Thus although the Hartree-Fock wave function gives in absolute terms a poor approximation to the total energy, it is certainly a useful approximation from which to calculate some molecular properties.

It is not necessary to restrict ourselves to a function like (12.27). Using the method of configuration interaction we can take a linear combination of such Slater determinants with variable coefficients, as follows:

$$\Psi = \sum_i c_i |\psi|_i, \qquad (12.31)$$

where $|\psi|_i$ indicates some determinant of n different spin-orbitals. Now if we are minimizing the energy of this function (according to the variation theorem) with respect to the coefficients c_i, we need be much less particular about our choice of orbitals. If we take enough terms in the summation it does not really matter which molecular orbitals we take (provided that they are a reasonably complete set); any deficiency that the orbitals may have is taken up through our variation of the coefficients c_i. But although we might have saved ourselves a lot of work in getting a set of orbitals, the work has now to be expended in determining the best coefficients c_i.

It is an unfortunate fact of quantum-mechanical life that to get accurate wave functions some lengthy calculation has usually to be done somewhere. In MO theory, however, one does have the choice of doing it at the orbital stage, or at the configuration interaction stage. For one problem it may be better to concentrate on the orbitals and for another to concentrate on the configuration interaction. Usually the best approach is to select orbitals that are a reasonable approximation to SCF orbitals and to carry out a limited amount of configuration interaction.

Let us look at the configuration interaction required to obtain a good ground state wave function for the water molecule. We can form seven independent molecular orbitals by taking linear combinations of the oxygen $1s$, $2s$ and $2p$ atomic orbitals, and the hydrogen $1s$ orbital. If we restrict our attention to configurations in which the lowest molecular orbital (roughly the oxygen $1s$ atomic orbital) always has two electrons, but allow the remaining eight electrons to be allocated in all possible ways amongst the other six molecular orbitals, then there are 95 different electron configurations altogether. These configurations give rise to 105 singlet spin states, and 37 of these have the symmetry of the ground state, A_1.

A configuration interaction problem with 37 terms is still rather lengthy, but some of these configurations can probably be neglected. For example, if the ground state is reasonably well represented by a wave function of the type (12.18), then, using the results of perturbation theory, we need consider only those other functions which have a non-zero Hamiltonian integral with this ground state function. These can only arise from the singly- and doubly-excited configurations, that is, those which do not have more than two electrons in orbitals different from the ground configuration (see page 187). There are only 17 singlet A_1 functions of this type for water so the problem starts to become manageable.

If one starts with the SCF wave function for the ground state, which is already a good wave function, this further simplifies the configuration interaction calculation because the Hamiltonian integral is now also zero between this wave function and any function arising from a singly-excited configuration (from 10.68). For the water molecule this brings the configuration interaction calculation down to 14 terms.

The water molecule is not what chemists would call a large molecule, and yet it is clear that a configuration interaction calculation is rather a lengthy procedure even when we take only the most important configurations. For a large molecule it is clearly not yet possible to do calculations with this level of sophistication.

Calculations in quantum mechanics generally follow the maxim of

Occam's razor†, which is that: 'Entities are not to be multiplied without necessity'. Broadly speaking, this means that a calculation is taken just to that level of sophistication that enables one to answer the question that is asked, and no further. For example, if one asks why the water molecule is bent, then this can be answered by qualitative argument or a very simple calculation. If one asks why the bond angle is 105° and not 90° or 109° then this requires a rather extensive calculation as we have seen in chapter 5. In general, the more complicated the molecule the more gross the questions that are asked and the more qualitative the methods used for answering them. This does not mean, however, that the questions are without interest.

The empirical aspects of MO theory will be illustrated in detail in the remaining chapters of this book. At this point, however, we shall give an example of the use of correlation diagrams, which have proved to be very useful in discussing the shapes of simple polyatomic molecules.

The molecules H—A—A—H can exist in four different geometrical forms, in which the two hydrogen atoms are equivalent, as shown in figure 12.4.

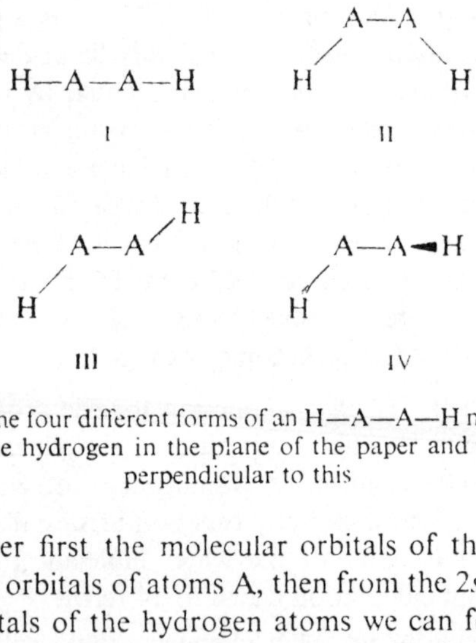

FIGURE 12.4. The four different forms of an H—A—A—H molecule. Structure IV has one hydrogen in the plane of the paper and one in a plane perpendicular to this

We shall consider first the molecular orbitals of the linear molecule. If we ignore the 1*s* orbitals of atoms A, then from the 2*s* and 2*p* orbitals of A and the 1*s* orbitals of the hydrogen atoms we can form ten molecular orbitals. Two of these will be A—H bonding orbitals (one of σ_g and one of σ_u symmetry), one will be the A—A bonding orbital (σ_g), two will be A—H antibonding (σ_g and σ_u) and one A—A antibonding (σ_u). There will

† BERTRAND RUSSELL, *History of Western Philosophy*, p. 462, George Allen and Unwin, 1946.

be a pair of A—A π bonding orbitals (π_u) and a pair of A—A π antibonding (π_g). The order of increasing energy will probably be

$$\sigma_g(\text{A—H}) < \sigma_u(\text{A—H}) < \sigma_g(\text{A—A}) < \pi_u(\text{A—A}) < \pi_g(\text{A}\overset{*}{—}\text{A}) < \sigma_u(\text{A}\overset{*}{—}\text{A}) < \sigma_g(\text{A}\overset{*}{—}\text{H}) < \sigma_u(\text{A}\overset{*}{—}\text{H}), \qquad (12.32)$$

where the star indicates an antibonding orbital.

We shall now consider the structure of the ground and excited states of C_2H_2, N_2H_2 and H_2O_2. Acetylene has ten electrons (apart from the carbon 1*s*) and these will just fill all the bonding orbitals. N_2H_2 would have two electrons in the π_g antibonding orbital and H_2O_2 four in this orbital. We must now see how these orbitals change when we bend the molecule.

Let us see what happens when one of the hydrogen atoms is bent through an angle of 60° in the *xy* plane (figure 12.5). This will involve a rehybridization of the orbitals of atom A_1 from *sp* in the linear configuration to sp^2 in the bent. This will increase slightly the energy of the A—H and A—A σ bonding orbitals because more *p* character is introduced into the bonds, and *p* orbitals have a higher energy than *s*. But this is likely to be rather small. For example, the C—H stretching vibration gives an absorption band in the infrared with a frequency of about 3300 cm^{-1} for acetylenes and about 3000 cm^{-1} for olefins. This would correspond to only a very small difference in bond energy between these two C—H bonds.

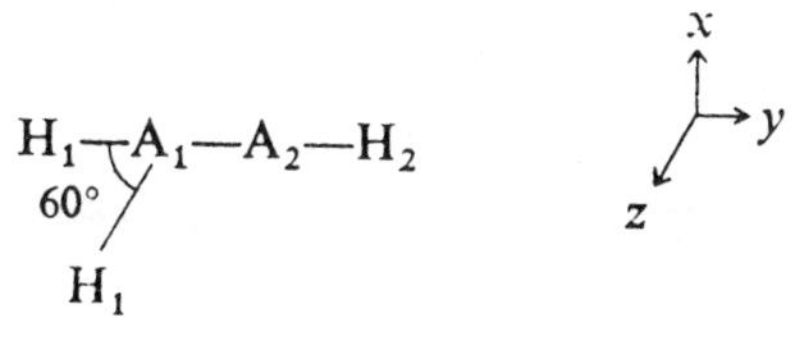

FIGURE 12.5. The coordinate system for a distorted HAAH molecule

The most important effect of the distortion is that the degeneracy of the π orbitals is removed. One component of the π orbitals, the one which has *xy* as a nodal plane, is unaffected by the distortion. The other component (which has *yz* as a nodal plane) will be strongly affected: the bonding component π_{uy} will become less bonding and the antibonding component π_{gy} will become less antibonding, because the orbital of A_1 which was p_x in the linear molecule becomes an sp^2 hybrid in the bent molecule, and this results in a much smaller overlap with the p_x orbital of A_2. However, the bending does add some s character to these components of the π orbitals, and this is a stabilizing effect. The overall result is that the antibonding π_{gy} is lowered on bending much more than the bonding π_{uy} is raised.

If we also displace H_2 in the xy plane, either to give the *cis* or *trans* configuration then we expect to have about twice the effect of displacing just H_1. However if we displace H_2 in the xz plane then we shall have the same effect on the σ orbitals, but now both components of the π orbitals will be affected. The effect of these displacements on the energy levels is shown qualitatively in figure 12.6; the effect on the antibonding σ orbitals is not important for our purposes.

If we make the reasonable assumption that the effect of distortion on the π orbitals dominates the energy changes, then we would expect that the 12-electron molecule would have a planar *cis* or *trans* structure, with *trans* being favoured since the non-bonding H—H repulsion would be minimized. Although N_2H_2 is not a known molecule, N_2F_2 would be expected to behave similarly and a planar *trans* structure has been found for this. H_2O_2 does have a non-planar structure, supporting the idea that if the

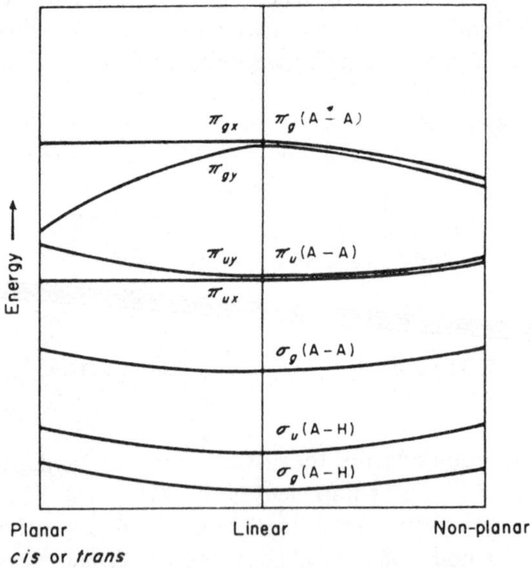

FIGURE 12.6. The effect of distortion on the energies of the molecular orbitals of H—A—A—H

antibonding π_g orbital has to contain four electrons then a distortion is favoured which reduces the antibonding energy of both components.

Lastly we consider the effect of an electron excitation on the shape of acetylene. The first excited states arise from the promotion of an electron from π_u to π_g. This would then be expected to lead to a planar distortion in order to relieve the antibonding energy of the excited electron. Acetylene has an absorption band in the region 2500–2100 Å whose vibrational fine structure shows that the excited state does have the planar *trans* configuration.

One can get useful qualitative ideas about molecular shape from such correlation diagrams, as was shown by Walsh in a long series of papers†. However, the interpretation of these diagrams requires care. It is not always clear, in cases where some orbitals are decreased in energy by a distortion and other increased in energy, which orbitals are going to exert the dominant effect. Also, if several excited states can arise from one configuration, as in the case of the excited configuration $\pi_u{}^3\pi_g$ of acetylene, then not all these states will necessarily suffer the same type of distortion. There will, for example, be states from this configuration which correlate with a configuration $\pi_{uy}{}^2\pi_{ux}\pi_{gx}$ of the planar bent molecule, and this is clearly unstable in a bent configuration.

12.5 *Valence-bond theory and the method of atoms in molecules*

The term valence-bond theory is often used to cover a wide range of type of calculation. On the one hand there are strictly non-empirical calculations, and when these use functions based on both covalent and ionic structures they are comparable to a non-empirical MO calculation with configuration interaction. On the other hand there is an empirical approach, based on the concept of resonance, which hardly reaches the level of being called a calculation. Between these two extremes there are semi-empirical calculations of which the most important is called 'the method of atoms in molecules'‡

The essential characteristic of VB theory is that the wave functions are written as a linear combination of functions which describe chemical bond structures. In the empirical theories chemical experience is used to decide which of these structures are likely to make an important contribution to the state of the molecule under consideration. The following principles are a guide.

1. A covalent (electron-pair) bond is stabilizing.
2. A short bond gives a lower energy than a long bond.

† WALSH, *J. Chem. Soc.*, 2260–2331 (1953).

‡ Empirical means based on, or guided by, the results of experiment.

3. Ionic structures in which the negative charges appear on the most electronegative atoms and the positive charges on the least electronegative atoms, will be the most stable.
4. Ionic structures, in which the positive and negative charges are close together will have a lower energy (through their coulombic energy) than those in which they are far apart.
5. The covalent homonuclear structure A—A will almost invariably have a lower energy than A^+A^-.

We can illustrate these points with the water molecule. Figure 12.7 shows the type of structure that would be considered if we assume that the oxygen 2*s* orbital is not involved in the bond formation.

Structure I will clearly dominate the ground state wave function; it represents electron-pair bonds between the $2p_x$ and $2p_y$ oxygen orbitals and the hydrogen 1*s* orbitals. Structure II will have a much higher energy because the two hydrogen atoms are rather far apart. It is unlikely to have an energy much below that of the dissociated atoms. The ionic structures III will have a lower energy than IV because oxygen is more electronegative than hydrogen; they will probably play an important role in the ground state wave function. Structures V will have rather high energy; it requires about 4 ev to produce H^+ and H^- from hydrogen atoms at the distance to which they are separated in water.

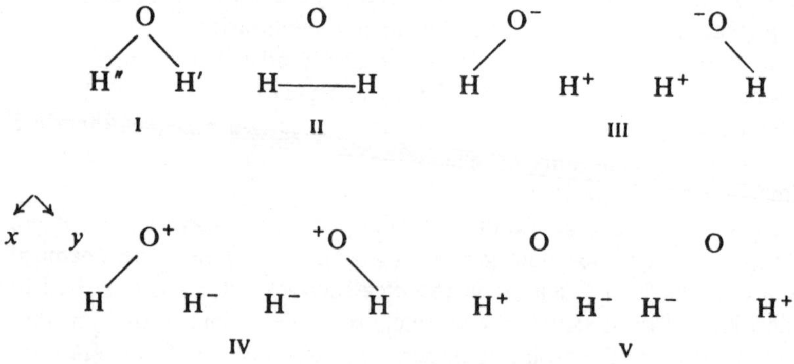

FIGURE 12.7. Valence-bond structures for H_2O

We have given only two structures for water of the Heitler–London type, I and II, and yet there are three ways of grouping four orbitals into pairs, as follows.

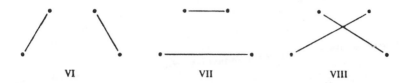

VI

VII

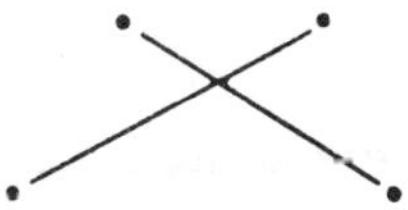

VIII

Diagrams VI and VII correspond to the structures I and II of figure 12.7, but VIII would correspond to an electron-pair bond between $2p_x$ and the $1s$ orbital of H′ and another between $2p_y$ and the $1s$ orbital of H″. However, it can be shown that VI, VII, and VIII are not independent structures. This means that if we write down the wave functions for structures VI and VII, then the wave function for VIII is a linear combination of the functions for VI and VII. In other words, if we have included VI and VII in our wave function we gain nothing by also including VIII.

There is a simple rule, discovered by Rumer, for arriving at the independent structures associated with $2n$ orbitals all involved in electron-pair bonds. We arrange the symbols for these orbitals on a circle (which need have no connection with the shape of the molecule) and pair off the orbitals by lines. We then take all structures in which no lines cross one another and these are the set of independent structures. The number of structures is given by the formula

$$\frac{(2n)!}{n!(n+1)!}. \tag{12.33}$$

For example, for six orbitals we have

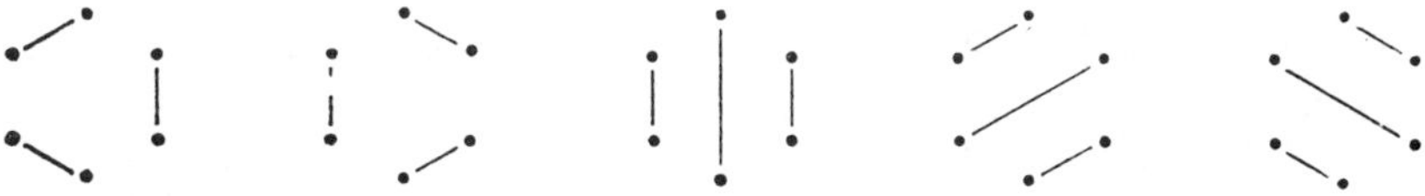

which are the Kekulé and Dewar structures for the π bonds of benzene (see chapter 15).

We turn now to an examination of the secular equations which are derived from a set of VB structures. Firstly, to derive the Hamiltonian and overlap integrals exactly, in the general case, is a complicated business and there is little point in going into it in this book. The off-diagonal elements $\mathscr{H}_{ij}$ and S_{ij} are frequently of the same order of magnitude as the diagonal elements. Particularly, it is not unusual to have S_{ij} greater than $\frac{1}{2}$. This is one of the most unsatisfactory aspects of VB theory. It means that for large molecules a very large number of structures have to be considered (particularly if ionic structures are included) but many of these are trying to do the same job. For example, if we have ten structures, and one of these, an expansion in terms of the other nine, is covered to 90%, then one can say that only 10% of this structure is really helping to improve the wave function.

We have seen in chapter 11 that the Heitler–London wave function for H_2 has an energy which consists of two terms, Q the coulomb energy and A the exchange energy. Similar terms arise in the general Hamiltonian integral of VB theory. There is one empirical approach to VB theory which expresses the total energy of the molecule in terms of Q and A in a relatively simple way; this is known as the perfect-pairing approximation. It is assumed that the molecule can be represented by a single structure in which all atomic orbitals are linked together in pairs. In addition, it is assumed that the overlap integral between different atomic orbitals is zero, and that only exchange integrals such as (11.10) are retained, integrals corresponding to multiple exchange of electrons being taken as zero. In this case the energy of the structure is given by

$$E = Q + \sum_{\substack{\text{pairs} \\ a \to b}} A_{a \to b} - \frac{1}{2} \sum_{\substack{\text{pairs} \\ a,b}} A_{a,b}, \tag{12.34}$$

where a → b indicates a pair of orbitals forming a bond and (a, b) is a pair of orbitals not bonded together.

Although the approximations leading to (12.34) are difficult to justify, they do at least lead to a formula on which a qualitative picture of the chemical bond can be built. The terms $A_{a \to b}$, which are the most important, give rise to the picture of the additivity of bond energies, whereas the terms $A_{a,b}$ are mainly responsible for the deviations from additivity. Examination of these terms has in the past led to some useful discussions of stereochemistry. However, there is little work done in this field today. The main criticism of the perfect-pairing formula is that the whole character of the Heitler–London function changes if we build it up from orthogonal orbitals (page 186), so that, although formula (12.34) is derived on the basis of zero overlap, $A_{a \to b}$ cannot be evaluated using that assumption, otherwise we do not get a bond.

There is one other development of VB theory which leads to relatively simple formulae for the matrix elements, and this is called the extended VB theory†. In this the wave functions are built up from hybrid atomic orbitals in such a way that although $S_{a \to b}$ is not zero, S_{ab} is. That is, we only have overlap between orbitals which are paired together. However, although these conditions do simplify the problem, the formulae are still too complicated to be usefully given here.

The VB method has one considerable advantage over the MO method, which we have not yet discussed, and this is that the type of wave function that it employs makes it much easier to relate the total energy of the molecule to the energies of the separated atoms. We have seen already that although quantum-mechanical calculations are quite accurate when measured by the total energy of the molecule, they are not accurate when measured by the scale of dissociation energies. For example, the best SCF calculation on N_2 gives 99·2% of the total energy, yet the calculated dissociation energy is only 2·6 ev, which is to be compared with the experimental value of 9·9 ev.

In 1951 Moffitt suggested that the reason for the large errors in calculated dissociation energies lay in the fact that the wave functions used also give relatively poor energies for the states of the individual atoms‡. He took the O_2 molecule to illustrate his point. A MO calculation of the energy difference between two excited states, ${}^3\Sigma_u{}^+$ and ${}^3\Sigma_u{}^-$, gives 10 ev, which is poor compared with the observed 2 ev. But in VB terms ${}^3\Sigma_u{}^+$ is mainly a covalent structure dissociating into neutral atoms, whereas ${}^3\Sigma_u{}^-$ is ionic $O^+—O^-$. If the calculation is carried out on the separate atoms, and the energy of $O^+ + O^-$ compared with that of two neutral atoms, then the error is again found to be about 8 ev.

Moffitt assumed then that the error in a molecular calculation is largely an error in the calculated atomic energies. What is wanted is a device for separating from the molecular energy that part which is the energy of the atoms (which may be 99% of the total), and then instead of calculating this one can use the experimental value. In other words one only attempts to calculate that part of the molecular energy which is really molecular. This semi-empirical type of calculation is called the *method of atoms in molecules*. Moffitt's original ideas have been extended by Hurley and Arai. We shall illustrate the method for LiH.

† HURLEY, LENNARD-JONES and POPLE, *Proc. Roy. Soc.*, **A220**, 446 (1953).
‡ MOFFITT, *Proc. Roy. Soc.*, **A210**, 224, 245 (1951).

We shall construct a VB wave function for LiH using three functions. These represent a covalent bond between the lithium 2s orbital and the hydrogen 1s, a covalent bond between the Li 2$p\sigma$ orbital and the hydrogen 1s and the ionic function Li^+H^-. We use the same symbols for these orbitals as in (10.34).

$$\Psi_1 = \sqrt{\tfrac{1}{2}}\{|\phi_s\bar{\phi}_h| - |\bar{\phi}_s\phi_h|\};$$
$$\Psi_2 = \sqrt{\tfrac{1}{2}}\{|\phi_p\bar{\phi}_h| - |\bar{\phi}_p\phi_h|\};$$
$$\Psi_3 = |\phi_h\bar{\phi}_h|. \tag{12.35}$$

In all these functions it is assumed that two electrons occupy the lithium 1s orbital.

The dissociation products associated with the three functions are as follows.

	Dissociation products
Ψ_1	Li (^2S) + H (^2S)
Ψ_2	Li (^2P) + H (^2S)
Ψ_3	$Li^+(^1S)$ + $H^-(^1S)$

We now go ahead and calculate the energies of these dissociation products ($\tilde{W}$), and we compare these energies with those obtained from spectroscopy (W) (in the case of H^- from an exact calculation). The following values are obtained (in atomic units)†.

	W	$\tilde{W}$	$W - \tilde{W}$
Ψ_1	−7·9779	−7·9179	−0·0600
Ψ_2	−7·9100	−7·8504	−0·0596
Ψ_3	−7·8074	−7·6953	−0·1121

The deviations $W - \tilde{W}$ are due to the fact that, even if we take the best form for the atomic orbitals, the type of wave function we have chosen cannot be an exact wave function for the atom, since electron correlation is not fully allowed for.

† HURLEY, *J. Chem. Phys.*, **28**, 532 (1958).

Suppose we have now set up the secular determinant for the three functions by a non-empirical calculation, $|\mathscr{H}_{rs} - ES_{rs}|$. The Hamiltonian integrals on the diagonal of the secular determinant are now corrected by an amount $W - \tilde{W}$. That is, instead of $\mathscr{H}_{11}$ we insert $\mathscr{H}_{11} + (W_1 - \tilde{W}_1)$ etc. However, VB calculations are characterized, as we have seen, by large off-diagonal overlap integrals. Now if S_{12} is not zero this means that Ψ_2 contains some contribution from Ψ_1, and hence $\mathscr{H}_{12}$ contains some part of $\mathscr{H}_{11}$. It follows that if we correct $\mathscr{H}_{11}$ as described above, we should also correct $\mathscr{H}_{12}$ by an amount proportional to S_{12}. But to keep the secular determinant symmetrical we need to correct $\mathscr{H}_{12}$ both for $W_1 - \tilde{W}_1$ and for $W_2 - \tilde{W}_2$. The method for doing this leads to the following determinant

$$|\mathscr{H}_{rs} + \frac{1}{2} S_{rs}[(W_r - \tilde{W}_r) + (W_s - \tilde{W}_s)] - ES_{rs}| = 0. \qquad (12.36)$$

This formula summarizes what is called the *intra-atomic correlation correction* (ICC).

The non-empirical VB calculation gives the following wave function for the ground state of LiH

$$\Psi_{VB} = 0{\cdot}573\ \Psi_1 + 0{\cdot}297\ \Psi_2 + 0{\cdot}281\ \Psi_3,$$

and gives a dissociation energy of 1·60 ev. After the ICC the wave function

$$\Psi_{ICC} = 0{\cdot}460\ \Psi_1 + 0{\cdot}230\ \Psi_2 + 0{\cdot}452\ \Psi_3$$

and the dissociation energy is 2·06 ev. The experimental dissociation energy is 2·52 ev.

The improvement in the calculation of LiH due to the ICC is in fact less spectacular than for heavier molecules, as can be seen from table 12.3. Some of the SCF calculations quoted by Hurley have now been slightly improved.

The ICC calculation increases the contribution from the ionic structure in the LiH ground state wave function because this has the largest correlation error. If all the atomic states had the same correlation error then no correction would be needed, providing dissociation energies were evaluated with respect to *calculated* atomic energies. The reason that the ICC is so important is that different atomic states have different correlation energies and the relative amounts of these states in a molecular wave function varies with the interatomic distances. The correction we should apply to a molecular energy value is the correlation errors of the so-called *valence states* of the atoms.

TABLE 12.3. Calculated dissociation energies (ev) compared with experiment

	SCF	VB	ICC	Experimental
LiH	1·34	1·67	2·09	2·5
BH	1·74	2·10	2·72	3·15
CH	1·2	1·60	2·94	3·65
NH	0·5	1·01	3·21	3·9
OH	0·8	1·14	4·00	4·58
HF	1·1	1·73	5·59	6·08
CO	5·38	7·45	11·0	11·24

The data are from the following papers:
HURLEY, *Proc. Roy. Soc.*, **A248**, 119 (1958); *Rev. Mod. Phys.*, **32**, 400 (1960); *J. Chem. Phys.*, **28**, 532 (1958); *Proc. Phys. Soc.*, **A69**, 767 (1956).

The valence state of an atom in a molecule is defined by the part of the molecular wave function that is associated with a given atom if all the atoms are considered to be separated to infinity but the form of the molecular wave function is kept constant. We shall illustrate this by considering the valence state of the oxygen in a Heitler–London wave function for the water molecule (structure I of figure 12.7).

The molecular wave function is as follows (cf. 11.17)

$$|\overset{\frown}{2p_x 1s''}\overset{\frown}{2p_y 1s'}| = N\{|2p_x\overline{1s''}2p_y\overline{1s'}| - |\overline{2p_x}1s''2p_y\overline{1s'}| - |2p_x\overline{1s''}\,\overline{2p_y}1s'| + |\overline{2p_x}1s''\overline{2p_y}1s'|\}. \quad (12.37)$$

We have omitted the contributions from the $1s$, $2s$ and $2p_z$ orbitals of the oxygen, each of which contains two electrons. To get the wave function of the valence state of the oxygen we just omit that part of the wave function which is associated with the two hydrogen atoms. This leaves

$$\Omega = N\{|2p_x 2p_y| - |\overline{2p_x}2p_y| - |2p_x\overline{2p_y}| + |\overline{2p_x}\,\overline{2p_y}|\}. \quad (12.38)$$

The ground configuration of the oxygen atom is $1s^2 2s^2 2p^4$ and this gives rise to the terms ^3P, ^1D and ^1S. The first function in (12.38) has a spin of 1 so this must belong to the ^3P term. Similarly, the last function has a spin of -1 and must be ^3P. It can be shown by examining the atomic wave functions that the second and third are both an equal mixture of ^1D and ^3P components. Adding these contributions, that is, summing the squares of the coefficients of each atomic state from each determinant, we arrive at the result that the valence state Ω is made up of $\frac{3}{4}$ ^3P and $\frac{1}{4}$ ^1D; which emphasizes that it is not a real spectroscopic state of the oxygen atom. This means that the Heitler–London wave function for water should be corrected for atomic correlation errors by $\frac{3}{4}$ the error in the ^3P term and $\frac{1}{4}$ the error in the ^1D term†.

12.6 Molecular integrals

From the very first calculation of a molecular wave function by Heitler and London, it was clear that the integrals that occur in these calculations were going to be difficult to evaluate. Heitler and London were unable to calculate accurately the two-electron exchange integral that occurs in (11.11), and this was done shortly after by Sugiura. Non-empirical calculations on polyatomic molecules, in which there were no approximations in the integrals, were not feasible until the middle 1950's when the high speed digital computers became available.

If the molecular wave functions are based on atomic orbitals the most difficult integral to evaluate is the four-centre two-electron integral

$$\iint \phi_a(1)\,\phi_b(1)\,r_{12}^{-1}\,\phi_c(2)\,\phi_d(2)\,d\tau_1\,d\tau_2. \tag{12.39}$$

If a basis of n atomic orbitals is used in the calculation then there are n^4 of these integrals, and calculating them takes the major time of any computer programme: matrix diagonalization even in a self-consistent-field programme is usually fast by comparison.

The many-centre integrals are evaluated either by direct numerical methods or by expanding the two-centre densities $\phi_a(1)\,\phi_b(1)$, in terms of functions which lead to manageable integration. The most popular expansion functions in use at the present time are gaussians of the form $\exp(-\gamma r^2)$. This is because the product of two gaussians on different centres, is a gaussian about a third centre. For example, suppose we have one gaussian at the origin, $\exp(-\gamma \mathbf{r}^2)$, and another whose centre is displaced from this by a vector $\mathbf{s}$, $\exp(-\mu(\mathbf{r}-\mathbf{s})^2)$ then the product can be written

$$e^{-\gamma \mathbf{r}^2}e^{-\mu(\mathbf{r}-\mathbf{s})^2} = Ae^{-(\gamma+\mu)\,(\mathbf{r}-\lambda\mathbf{s})^2} \tag{12.40}$$

where $\lambda = \mu/(\gamma+\mu)$ and $A = \exp-(\gamma\mu s^2/(\gamma+\mu))$. This is a gaussian orbital whose centre is displaced by $\lambda\mathbf{s}$ from the origin. It follows that in a gaussian expansion the electron repulsion integrals all reduce to *two-centre* integrals which are relatively easy to evaluate.

† A review on valence states has been given by Moffitt, *Rept. Progr. Phys.*, **17**, 173 (1954).

This property of gaussian orbitals was first noted by Boys, and has been widely used, not only for the evaluation of integrals but also for a molecular orbital approach in which the atomic orbital basis functions are taken as gaussian functions rather than the more usual Slater-type functions. Gaussian functions are not in themselves good approximations to atomic orbitals, but if in an LCAO calculation one takes two or three times as many Gaussian orbitals in the basis as one would Slater orbitals, the resulting wave function is of about the same accuracy as for the Slater basis.

If in a calculation the integrals over atomic orbitals are first calculated, then these may be stored in the computer and pulled out for later use: in the successive iterations of an SCF calculation for example.. As we have pointed out earlier, the number of these integrals for a basis of n atomic orbitals is *ca* n^4, and it is this number that effectively limits the size of molecule for which an *a priori* calculation can be made. However, if the integrals are to be evaluated by numerical integration, then there is no need to reduce all integrals to atomic orbital form: direct evaluation of the molecular orbital integrals can be performed. As Boys has pointed out, this procedure avoids the n^4 explosion, but it does have the disadvantage that one cannot store a basic programme of integrals, so that it is necessary to evaluate each integral afresh for each SCF iteration.

It is beyond the scope of this book to discuss the detailed evaluation of the molecular integrals but a brief outline will be given of the method of evaluating the overlap integrals: these are the simplest of all to solve and were the first to be extensively tabulated, in a classical paper by Mulliken and co-workers. The evaluation of overlap integrals is not only required for any *a priori* calculation, but they also form the basis of any qualitative or empirical theory of the chemical bond.

The overlap integral between two atomic orbitals is defined as follows

$$S(\phi_a, \phi_b, R) = \int \phi_a \phi_b \, \mathrm{d}v \qquad (12.41)$$

and is a function of the form of the orbitals ϕ_a and ϕ_b as well as the internuclear separation R. Taking normalized Slater orbitals in the form of (4.4) this becomes

$$S = \left[\frac{(2\zeta_a)^{2n_a+1}(2\zeta_b)^{2n_b+1}}{(2n_a)!\,(2n_b)!}\right]^{\frac{1}{2}}$$

$$\int r_a^{\,n_a-1} r_b^{\,n_b-1} \mathrm{e}^{-(\zeta_a r_a+\zeta_b r_b)}\, Y_{l_a m_a}(\vartheta_a, \varphi_a)\, Y_{l_b m_b}(\vartheta_b, \varphi_b)\, \mathrm{d}v \qquad (12.42)$$

Integrals of this type are most conveniently evaluated by converting first to elliptical coordinates (μ, ν, φ) defined in (10.54). The spherical-polar coordinates are then replaced according to the relationship

$$r_a = (\mu+\nu)R/2,\ \cos\vartheta_a = (1+\mu\nu)/(\mu+\nu),\ \sin\vartheta_a = [(\mu^2-1)(1-\nu^2)]^{\frac{1}{2}}/(\mu+\nu),$$

$$r_b = (\mu-\nu)R/2,\ \cos\vartheta_b = (1-\mu\nu)/(\mu-\nu),\ \sin\vartheta_b = [(\mu^2-1)(1-\nu^2)]^{\frac{1}{2}}/(\mu-\nu),$$

$$\varphi_a = \varphi_b = \varphi. \tag{12.43}$$

The limits of integration of these coordinates are μ from 1 to ∞, ν from -1 to $+1$, φ from 0 to 2π, with the unit volume element being $(R^3/8)\ (\mu^2-\nu^2)\ \mathrm{d}\mu\mathrm{d}\nu\mathrm{d}\varphi$. After the trivial integration over φ the overlap integral will be found to be proportional to

$$\int_1^\infty\int_{-1}^1 P(\mu,\nu)\ \mathrm{e}^{-p(\mu+t\nu)}\mathrm{d}\mu\mathrm{d}\nu \tag{12.44}$$

where $p = \frac{1}{2}(\zeta_a+\zeta_b)\ R$, and $t = (\zeta_a+\zeta_b)/(\zeta_a-\zeta_b)$ and P is a polynomial in μ and ν. The integral may now be evaluated using the auxiliary functions

$$A_k(p) = \int_1^\infty \mu^k \mathrm{e}^{-p\mu}\mathrm{d}\mu = \mathrm{e}^{-p}\sum_{\lambda=1}^{k+1}[k!/p^\lambda(k-\lambda+1)!]$$

$$B_k(pt) = \int_{-1}^1 \nu^k \mathrm{e}^{-pt\nu}\ \mathrm{d}\nu = -\mathrm{e}^{-pt}\sum_{\lambda=1}^{k+1}[k!/(pt)^\lambda(k-\lambda+1)!]$$

$$-\mathrm{e}^{pt}\sum_{\lambda=1}^{k+1}[(-1)^{k-\lambda}k!/(pt)^\lambda(k-\lambda+1)!] \tag{12.45}$$

Special formulae are needed for the cases $p=0$ or $t=0$, and as one approaches these limits the above formulae are not very useful for computation

Overlap integrals may therefore be identified first by the principal quantum numbers of the two orbitals n_a and n_b and can then be tabulated as a function of the parameters p and t.

Master formulae for the overlap integrals between s and p atomic orbitals (up to $n_a = n_b = 5$) have been given by Mullikan and his co-workers† and they also produced the first extensive tabulation of these integrals. Other overlap integrals and tabulations have been subsequently produced and now their evaluation by computer is a relatively straightforward matter. Table 12.4 gives a typical set of values which are particularly important for calculations on π electron molecules of the type discussed in chapters 15-17.

†MULLIKAN, RIECKE, ORLOFF and ORLOFF, *J. Chem. Phys.*, **17**, 1248 (1949).

TABLE 12.4. $2p\pi$–$2p\pi$ overlap integrals

p	$t=0·0$	$t=0·1$	$t=0·2$	$t=0·3$	$t=0·4$	$t=0·5$	$t=0·6$
0·0	1·000	0·975	0·903	0·790	0·647	0·487	0·328
0·5	0·976	0·951	0·882	0·772	0·633	0·478	0·323
1·0	0·907	0·887	0·823	0·723	0·596	0·453	0·308
1·5	0·809	0·790	0·737	0·652	0·542	0·416	0·287
2·0	0·695	0·680	0·638	0·568	0·4[illegible]	0·372	0·261
2·5	0·578	0·567	0·535	0·481	0·410	0·325	0·233
3·0	0·468	0·460	0·437	0·398	0·345	0·279	0·205
3·2	0·427	0·420	0·401	0·367	0·320	0·262	0·194
3·4	0·389	0·383	0·366	0·337	0·297	0·245	0·183
3·6	0·352	0·348	0·334	0·309	0·274	0·228	0·173
3·8	0·318	0·315	0·303	0·283	0·253	0·213	0·163
4·0	0·287	0·284	0·275	0·258	0·232	0·198	0·153
4·2	0·258	0·255	0·248	0·234	0·213	0·183	0·144
4·4	0·231	0·229	0·224	0·213	0·195	0·170	0·135
4·6	0·207	0·205	0·201	0·193	0·179	0·157	0·127
4·8	0·184	0·183	0·181	0·174	0·163	0·145	0·119
5·0	0·164	0·163	0·162	0·157	0·149	0·134	0·111

Problems 12

12.1 The ground state of the methyl radical is planar (group $\mathbf{D_{3h}}$). Give a qualitative account of the molecular orbitals of this molecule and identify the orbital which contains the unpaired electron. How many singly-excited configurations have the same symmetry as the ground state configuration? How do the energies of these orbitals change if the molecule is distorted to a $\mathbf{C_{3v}}$ configuration?

$\mathbf{D_{3h}}$	I	σ_h	$2C_3$	$2S_3$	$3C_2'$	$3\sigma_v$
A_1'	1	1	1	1	1	1
A_2'	1	1	1	1	−1	−1
A_1''	1	−1	1	−1	1	−1
A_2''	1	−1	1	−1	−1	1
E'	2	2	−1	−1	0	0
E''	2	−2	−1	1	0	0

12.2 The following are the N—H bonding molecular orbitals for NH_3†. What are the equivalent orbitals?

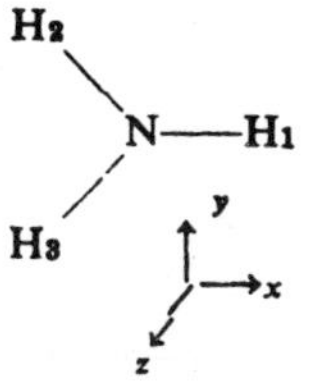

$$\psi_1 = 0{\cdot}76\ 2s + 0{\cdot}16\ 2p_z - 0{\cdot}27\ \mathrm{H}_o$$
$$\psi_2 = 0{\cdot}62\ 2p_x + 0{\cdot}49\ \mathrm{H}_x$$
$$\psi_3 = 0{\cdot}62\ 2p_y + 0{\cdot}49\ \mathrm{H}_y$$

where
$$\mathrm{H}_o = \sqrt{\tfrac{1}{3}}(h_1 + h_2 + h_3)$$
$$\mathrm{H}_x = \sqrt{\tfrac{1}{6}}(2h_1 - h_2 - h_3)$$
$$\mathrm{H}_y = \sqrt{\tfrac{1}{2}}(h_2 - h_3)$$

12.3 What is the expression for the energy of H_2O in the perfect-pairing VB scheme?

12.4 Write down a Heitler–London wave function for linear BeH_2 and deduce the valence state of the Be.

† KAPLAN, *J. Chem. Phys.*, **26**, 1704 (1957).

Chapter 13

Ligand-field theory

13.1 Simple crystal-field theory

The interesting chemistry of transition metal ions is largely that of their complexes. In the majority of these complexes a transition metal cation is surrounded by six molecules or anions (collectively called *ligands*) arranged at the corners of a more or less regular octahedron. This six-fold coordination occurs in many structures where at first sight it appears unlikely. For example, crystals of $CuSO_4.5H_2O$ and $FeCl_3$ both contain octahedrally-coordinated cations as shown in figure 13.1. In the first

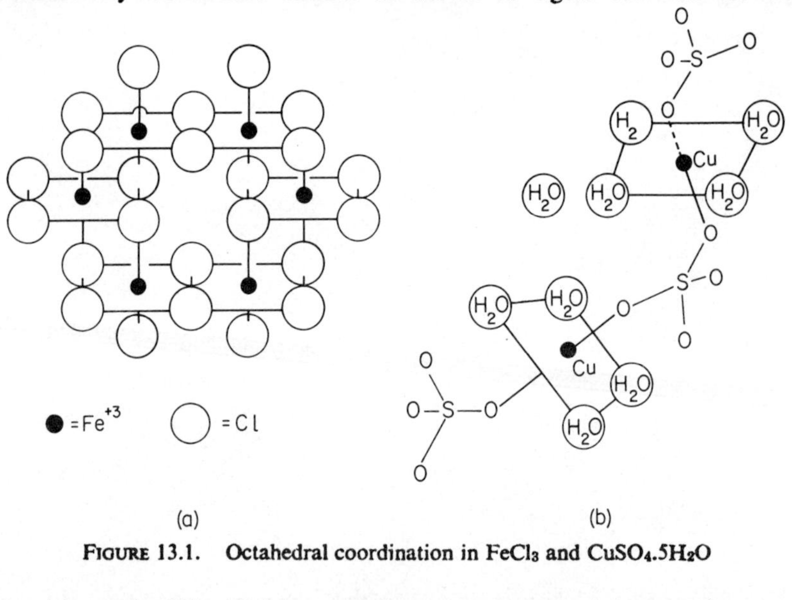

FIGURE 13.1. Octahedral coordination in $FeCl_3$ and $CuSO_4.5H_2O$

transition series (Ti to Cu) ions with from four to seven *d* electrons form two sets of complexes, one with more unpaired electrons present than the other. This is clearly shown by the electronic spectra of these complexes. For example, divalent iron ($3d^6$) forms both green $[Fe(H_2O)_6]^{2+}$ and yellow $[Fe(CN)_6]^{4-}$ complexes. The former has four unpaired electrons for each iron atom and is paramagnetic; the latter has no unpaired electrons and is diamagnetic.

Three theories have been advanced to explain these differences; in all of them the occupation of the *d* orbitals of the cation is of paramount importance. The first theory to gain general acceptance was the VB approach due to Pauling. He suggested that octahedrally-coordinated transition metal ions are either d^2sp^3 (i.e. $3d^2\,4s\,4p^3$) or sp^3d^2 (i.e. $4s\,4p^3\,4d^2$) hybridized and these hybrid orbitals are used to form bonds with the ligands. In the first case, three of the 3*d* orbitals are not used to form hybrids. In the second case, all five 3*d* orbitals are unused. This means that in the first case there is a maximum of three unpaired *d* electrons, and in the second five, which accounts for two forms of complexes. This theory has largely fallen into disuse although its nomenclature is sometimes encountered—viz. 'inner' (using 3*d* orbitals for the hybrids) and 'outer' orbital complexes (using 4*d* orbitals). Its most serious drawback is that it makes no mention of antibonding orbitals and in consequence cannot explain electronic spectra and it cannot account for the existence of fairly stable 'inner orbital' complexes with more than six *d* electrons. Despite its considerable contribution to the understanding of the chemistry of metal complexes, we shall not discuss it further here†.

A more satisfactory approach is provided by either crystal- or ligand-field theory. The former is an entirely electrostatic theory first introduced by van Vleck, the latter a mixture of electrostatic and MO theories. Crystal- and ligand-field calculations have much in common, so we shall discuss the former in some detail before explicitly introducing molecular orbitals.

Crystal-field theory regards the electrostatic attraction between the ionic or highly polar ligand (e.g. Cl^-, H_2O) and the positively charged cation as responsible for the stability of a complex. The forces involved are similar to those which hold an ionic crystal together—hence the name of the theory. The essential step made by the theory is the recognition that, although in an isolated cation the degeneracy of all five *d* orbitals is maintained, in an octahedral crystal field the degeneracy is removed.

It is usual to choose the cartesian axes of an octahedral complex to be the four-fold axes of the octahedron. It follows that *p* orbitals are still degenerate in the complex, for all have exactly equivalent positions with respect to the ligands. This is not true for the *d* orbitals. Two, $d_{x^2-y^2}$ and d_{z^2}, point along the axes whilst the other three, d_{xy}, d_{yz} and d_{zx}, are directed

† The interested reader is referred to the account given in PAULING, *Nature of the Chemical Bond*, 2nd. ed., Cornell, 1962.

between the axes. These sets will have different energies†. If the field of the ligands is approximately that of point negative charges, then clearly an electron in d_{xy}, d_{yz} or d_{xz} will have a lower energy than one in d_{z^2} or $d_{x^2-y^2}$. This is illustrated in figure 13.2. The part of the potential which affects both sets equally is not important and its effect is omitted from the figure.

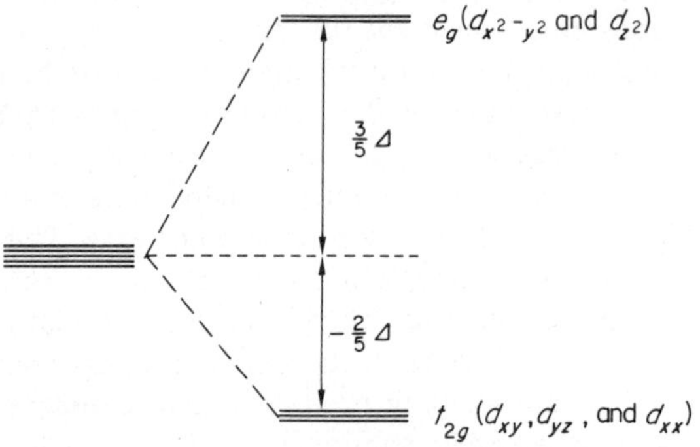

FIGURE 13.2. The splitting of d orbitals in an octahedral crystal field

In figure 13.2 the two sets of d orbitals are labelled by the *I.R.*'s of the octahedral group $\mathbf{O}_h$ under which they transform. (It is common to use lower case group theory symbols to label orbitals and upper case for states.) Alternative labelling which may be encountered replaces e_g by either d_γ or γ_3; t_{2g} by d_ε or γ_5.

The separation between the t_{2g} and e_g orbitals is denoted by Δ (or 10 Dq)‡. Δ varies from complex to complex, but for any ion it is possible to

† On page 26 we pointed out that d_{z^2} and $d_{x^2-y^2}$ were in fact equivalent although they looked different.

‡ If the potential is expanded in terms of spherical harmonics (3.6) then the first terms which can remove the degeneracy of the d orbitals are those having $l = 4$. Expressing these in cartesian coordinates we get $V_{oct} = D(x^4 + y^4 + z^4 - \frac{3}{5}r^4)$ where D is $\frac{35e^2Z}{4a^5}$, assuming that the ligands act as point charges Ze at a distance a from the cation. The separation between the two sets of d orbitals is then $\frac{5e^2Z\overline{r^4}}{3a^5}$ where $\bar{r}$ is the mean distance of a d electron from the cation nucleus. It is usual to write $\frac{5e^2Z\overline{r^4}}{3a^5}$ as $10Dq$, where $q = \frac{Z\overline{r^4}}{105}$. $10Dq$ is, then, an entirely theoretical quantity. We prefer to use Δ, and regard it entirely as an empirical quantity.

list ligands in order of the magnitude of Δ which they produce. It is found that this list is almost identical no matter which ion is considered. Since Δ is usually determined spectroscopically, the list is called the *spectrochemical series*. An abbreviated series is

$$I^- < Br^- < Cl^- < SCN^- < F^- < OH^- < H_2O < NH_3 < NO_2^- < CN^-.$$

Δ is about 12,000 cm^{-1} for divalent and 20,000 cm^{-1} for trivalent ions of the first transition series. Ions of the second and third series have rather larger values.

A consideration of figure 13.2 provides a simple explanation of the magnetic anomalies mentioned at the beginning of this chapter without involving the rather ad hoc change in bond type of valence-bond theory. The electron configuration of the di- and tri-positive cations of the elements titanium to copper are shown in table 13.1. We have to feed from one to nine electrons into the e_g and t_{2g} orbitals. There is a competition

TABLE 13.1 The number of 3*d* electrons in the di- and tri-positive ions of the elements titanium to copper. None of these ions have any 4*s* or 4*p* electrons

	Ti	V	Cr	Mn	Fe	Co	Ni	Cu
M^{2+}	2	3	4	5	6	7	8	9
M^{3+}	1	2	3	4	5	6	7	8

between the energy to be gained by having electrons in different orbitals (thus maximizing exchange energy and minimizing coulomb repulsion) and the energy to be gained by having all electrons in the lowest set of orbitals. Two limiting cases can be distinguished: Δ can be small (the 'high-spin' or 'weak-field' case) or Δ can be large (the 'low-spin' or 'strong-field' case). In the weak-field case, electrons are added successively as shown in figure 13.3, the energy gap Δ being small enough to allow Hund's rule to be obeyed. The spin multiplicity is therefore the same as that for the ground state of the free gaseous ion, reaching a maximum at the d^5 configuration.

When Δ is large the crystal-field energy dominates and the electron configurations shown in figure 13.4 are obtained.

In crystal-field theory it is therefore the balance between electron-interaction energies and the crystal-field energy which determines the magnetic properties of a complex. Another factor will become apparent when we consider ligand-field theory—the extent of σ and π bonding. The nomenclature of the two classes of complexes should by now be evident.

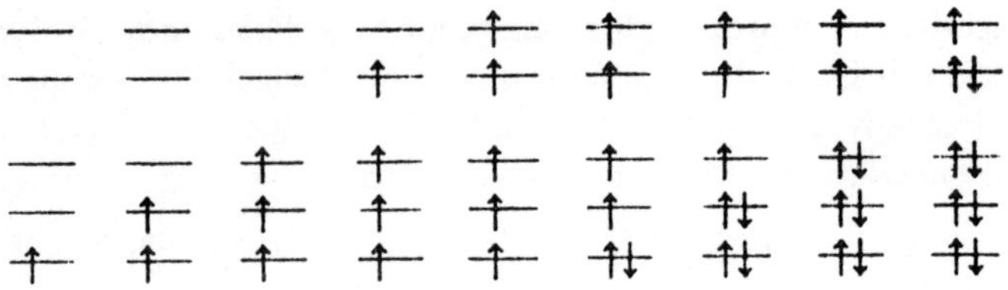

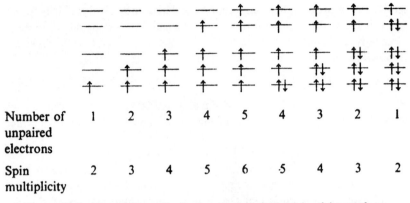

Number of unpaired electrons	1	2	3	4	5	4	3	2	1
Spin multiplicity	2	3	4	5	6	5	4	3	2

FIGURE 13.3. The electron distribution in weak-field (high-spin) complexes (but see page 228)

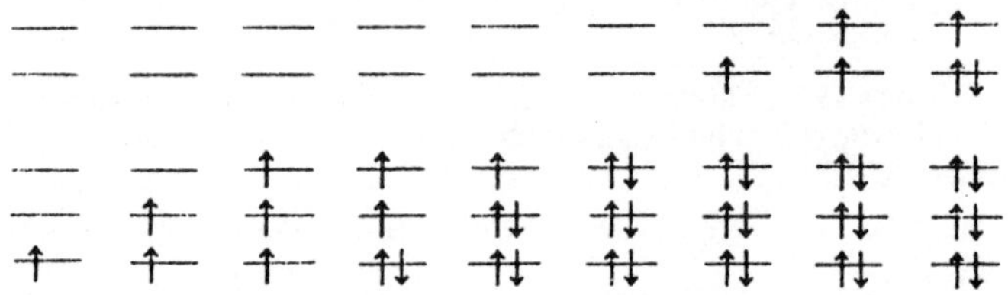

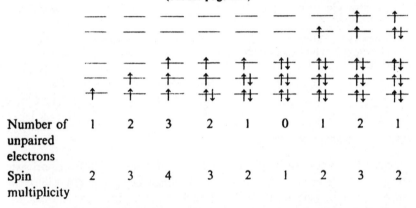

Number of unpaired electrons	1	2	3	2	1	0	1	2	1
Spin multiplicity	2	3	4	3	2	1	2	3	2

FIGURE 13.4. The electron distribution in strong-field (low-spin) complexes

'High spin' and 'low spin' are phenomenological terms, 'weak field' and 'strong field' describe the magnitude of the crystal field.

Crystal-field theory also provides a simple explanation for the regularities found in the heats of formation of series of complexes when the central cation is varied. Since an electron in a t_{2g} orbital is stabilized by $\frac{2}{5}\Delta$ with respect to the energy zero (which is taken to be the average crystal-field energy of all the d orbitals), and an electron in an e_g orbital is destabilized by $\frac{3}{5}\Delta_g$ we can calculate the total stabilization for any electron configuration. The results of such a calculation are given in table 13.2 for weak-field complexes. An analogous table can be compiled for strong-field complexes.

As an example we consider the heats of hydration of divalent ions of the first transition series. The experimental results lie on two curves, intersecting

TABLE 13.2. Crystal-field stabilization energies for weak-field octahedral complexes

Number of d electrons	Electron configuration	Crystal-field stabilization energy
0	$t_{2g}{}^{0}e_g{}^{0}$	0
1	$t_{2g}{}^{1}e_g{}^{0}$	$-\frac{2}{5}\Delta$
2	$t_{2g}{}^{2}e_g{}^{0}$	$-\frac{4}{5}\Delta$
3	$t_{2g}{}^{3}e_g{}^{0}$	$-\frac{6}{5}\Delta$
4	$t_{2g}{}^{3}e_g{}^{1}$	$-\frac{3}{5}\Delta$
5	$t_{2g}{}^{0}e_g{}^{3}$	0
6	$t_{2g}{}^{4}e_g{}^{2}$	$-\frac{2}{5}\Delta$
7	$t_{2g}{}^{5}e_g{}^{2}$	$-\frac{4}{5}\Delta$
8	$t_{2g}{}^{6}e_g{}^{2}$	$-\frac{6}{5}\Delta$
9	$t_{2g}{}^{6}e_g{}^{3}$	$-\frac{3}{5}\Delta$
10	$t_{2g}{}^{6}e_g{}^{4}$	0

in a minimum at Mn^{2+}. If the calculated crystal-field stabilization energy appropriate to each ion is subtracted (Δ being obtained from the electronic spectrum of the hydrate), an almost straight line results. This is shown in figure 13.5†.

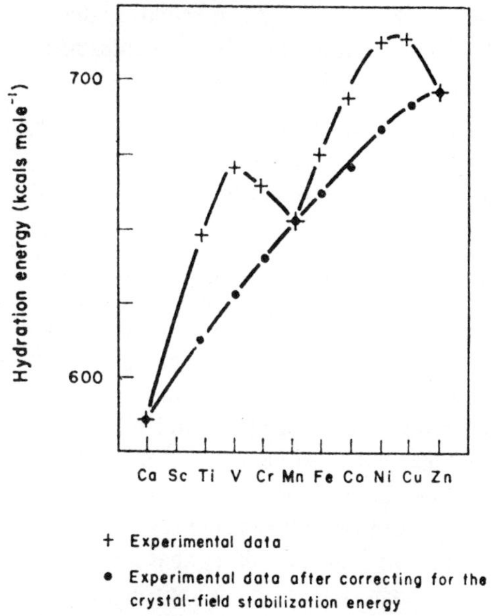

FIGURE 13.5. Experimental and corrected heats of hydration of divalent ions of the first transition series

† For a fuller discussion and for more examples, the reader is referred to GEORGE and MCCLURE, *Prog. Inorg. Chem.*, **1**, 381 (1959).

13.2 The detailed crystal-field theory of weak-field complexes

We now consider in more detail the effect of the crystal field on *d* orbitals. Crystal-field theory neglects any overlap between ligand and cation orbitals and considers only the effect of an electrostatic perturbation on the latter.

In calculating the energy levels of an isolated atom or ion in the Russell–Saunders scheme (page 115) one first considers the coulomb attraction between each electron and the nucleus, then the electron–electron repulsions, then spin-orbit coupling, this being the order of decreasing importance (figure 9.2). Crystal-field theory follows exactly the same pattern, but at an appropriate point in the development the crystal-field perturbation is inserted. For first row transition metal ions, this insertion takes place either before the electron-interaction stage (strong-field complexes) or before spin-orbit coupling (weak-field complexes). Even the weakest crystal field found in practice is strong compared with spin-orbit coupling, although this is not necessarily true for the other transition series and is certainly not true for the rare earths. The terms given by a d^n configuration in the Russell–Saunders scheme may be found by the methods described in chapter 9. The ground states are determined by Hund's rules and are listed in table 13.3.

TABLE 13.3. Ground state terms of the $d^1 \rightarrow d^9$ configurations

Configuration	d^1	d^2	d^3	d^4	d^5	d^6	d^7	d^8	d^9
Free-ion ground state	^2D	^3F	^4F	^5D	^6S	^5D	^4F	^3F	^2D

We shall now consider the effect of a weak field on these ground states. Since an electrostatic field does not interact directly with electron spin we need only consider how the crystal field removes the orbital degeneracy. Apart from the d^5 configuration, only *D* and *F* terms appear in table 13.3 and we can cover these by discussing the d^1 and d^2 cases; the rest, apart from d^5, will then have been dealt with.

A *D* state has five-fold orbital degeneracy, and an *F* state seven-fold. However, one can see from group theory that an orbital degeneracy of no more than three can persist in an octahedron (the $\mathbf{O}_h$ character table, given in table 13.4, shows no degeneracy greater than three) so that both *D* and *F* terms must split into at least two and three components respectively in an octahedral complex. Using group theory we shall first see how the *D* state splits up.

TABLE 13.4. The character table of the groups O_h

O_h	I	$6C_4$	$3C_2$	$6C_2'$	$8C_3$	i	$6S_4$	$3\sigma_h$	$6\sigma_d$	$8S_6$
A_{1g}	1	1	1	1	1	1	1	1	1	1
A_{1u}	1	1	1	1	1	−1	−1	−1	−1	−1
A_{2g}	1	−1	1	−1	1	1	−1	1	−1	1
A_{2u}	1	−1	1	−1	1	−1	1	−1	1	−1
E_g	2	0	2	0	−1	2	0	2	0	−1
E_u	2	0	2	0	−1	−2	0	−2	0	1
T_{1g}	3	1	−1	−1	0	3	1	−1	−1	0
T_{1u}	3	1	−1	−1	0	−3	−1	1	1	0
T_{2g}	3	−1	−1	1	0	3	−1	−1	1	0
T_{2u}	3	−1	−1	1	0	−3	1	1	−1	0

Clearly, the five components of a D state have the same symmetry properties as the five d orbitals since the symbols d and D both indicate an angular variation described by the spherical harmonics with $l = 2$. Any symmetry operation of the group transforms one of the components of the state into a mixture of the five. It follows that these five components form a basis for a representation of the group, and we want to know the component *I.R.*'s of this representation (page 103).

The character of the representation is the sum of the diagonal elements of the matrix which represents the operation of the group on the D states. This is given simply by the sum of the parts of each component which are unchanged by the operation. Let us take two easy ones to start with: the identity operation leaves all five components unchanged hence $\chi(I) = 5$; the inversion operator also leaves the sign of each component unchanged (d orbitals have g symmetry) hence $\chi(i) = 5$. The results of the other operations are more difficult to see, but we can get the answers by considering the general effect of a rotation by an angle φ on the states.

Suppose we want to find the effect of a rotation about a particular axis in space. There are many ways of choosing wave functions for the components of the D state (d orbitals) because all five components are degenerate. Clearly, we want to choose those wave functions which make this particular problem as simple as possible. This is done by arbitrarily taking the axis we are concerned with as the z axis and also taking the wave functions in complex form (3.11). The rotation we are considering only changes the reference point for the polar variable φ, and the five components depend on φ in the following manner (see 3.11)

$$e^{2i\varphi}, e^{i\varphi}, 1, e^{-i\varphi}, e^{-2i\varphi}.$$

If we rotate by an angle α these functions become

$$e^{2i(\varphi+\alpha)}, e^{i(\varphi+\alpha)}, 1, e^{-i(\varphi+\alpha)}, e^{-2i(\varphi+\alpha)},$$

hence we can say, for example,

$$\mathbf{R}_\alpha e^{2i\varphi} = e^{2i\alpha}.e^{2i\varphi}, \tag{13.1}$$

or uniting the effect on all five components in matrix form

$$\mathbf{R}_\alpha \begin{pmatrix} e^{2i\varphi} \\ e^{i\varphi} \\ 1 \\ e^{-i\varphi} \\ e^{-2i\varphi} \end{pmatrix} = \begin{pmatrix} e^{2i\alpha} & 0 & 0 & 0 & 0 \\ 0 & e^{i\alpha} & 0 & 0 & 0 \\ 0 & 0 & 1 & 0 & 0 \\ 0 & 0 & 0 & e^{-i\alpha} & 0 \\ 0 & 0 & 0 & 0 & e^{-2i\alpha} \end{pmatrix} \begin{pmatrix} e^{2i\varphi} \\ e^{i\varphi} \\ 1 \\ e^{-i\varphi} \\ e^{-2i\varphi} \end{pmatrix} \tag{13.2}$$

The character of the matrix which represents this operation on the components of the D states is therefore

$$e^{2i\alpha} + e^{i\alpha} + 1 + e^{-i\alpha} + e^{-2i\alpha} = \frac{\sin\frac{5\alpha}{2}}{\sin\frac{\alpha}{2}}. \tag{13.3†}$$

We can generalize this result to find the character for any state with a quantum number L whose $2L + 1$ components have a φ dependence $e^{iL\varphi}, e^{i(L-1)\varphi} + \ldots e^{-iL\varphi}$. The required character is

$$e^{iL\alpha} + e^{i(L-1)\alpha} + \cdots e^{-iL\alpha} = \frac{\sin(2L+1)\frac{\alpha}{2}}{\sin\frac{\alpha}{2}}. \tag{13.4}$$

We return now to the $\mathbf{O}_h$ group. To find the character appropriate to the operation C_4 on the D states we take the C_4 axis as the z axis and then from the formula (13.3) with $\alpha = 90°$ we find $\chi(C_4) = \sin 225°/\sin 45° = -1$. Likewise C_2 and C_2' (both $\alpha = 180°$) have characters 1, and lastly the C_3 operation ($\alpha = 120°$) (again considering the components defined with the C_3 axis as z axis) has a character $\chi(C_3) = \sin 300°/\sin 60° = -1$.

These characters are sufficient to establish that the component *I.R.*'s are of the form $E + T_2$, but they do not determine whether the suffixes should be g or u. Because the D states which concern us are derived from d^n

† This is proved by summing the left-hand side as a geometrical progression.

configurations, and because d orbitals are centrosymmetric, it follows that the component *I.R.*'s are $E_g + T_{2g}$. The D state derived from the p^2 configuration also gives rise to E_g and T_{2g} components in an octahedral ligand field, but those derived from the pd configuration give E_u and T_{2u} components.

Following the same method (problem 13.2) the F states split up under an octahedral field to give components $T_{1g} + T_{2g} + A_{2g}$. We now have to determine the relative energies of the components of the D and F terms. We start by considering the d^1 case. We already know how the orbitals split up, and this is summarized in table 13.5.

If the electron occupies the t_{2g} orbitals a ${}^2T_{2g}$ state results, if it occupies the e_g orbitals we get a ${}^2E_g$ state. We shall write these wave functions in complex form as follows $(l, m_l) = (2,2)$, $(2,1)$, $(2,0)$, $(2,-1)$, $(2,-2)$; and the real forms of the wave functions are then given as in table 13.5.

TABLE 13.5. Wave functions and energies from the d^1 configuration in an octahedral field

Real orbital	Wave function (see table 3.1)	Crystal-field energy
$d_{x^2-y^2}$	$\sqrt{\tfrac{1}{2}}\{(2,2) + (2,-2)\}$	$\tfrac{3}{5}\Delta$
d_{z^2}	$(2,0)$	
d_{xy}	$-i\sqrt{\tfrac{1}{2}}\{(2,2) - (2,-2)\}$	$-\tfrac{2}{5}\Delta$
d_{yz}	$-i\sqrt{\tfrac{1}{2}}\{(2,1) - (2,-1)\}$	
d_{zx}	$\sqrt{\tfrac{1}{2}}\{(2,1) + (2,-1)\}$	

The perturbation to the energies of these orbitals due to the crystal field is given by the following expressions (put $\mathcal{H}' = V_{\text{oct}}$ in 6.50)

$$\int d^*_{x^2-y^2} V_{\text{oct}}\, d_{x^2-y^2}\, \mathrm{d}v$$

$$= \frac{1}{2}\int [(2,2) + (2,-2)]^* V_{\text{oct}} [(2,2) + (2,-2)]\, \mathrm{d}v = \tfrac{3}{5}\Delta. \quad (13.5)$$

Similarly

$$\left.\begin{aligned}
&\int (2,0)^* V_{\text{oct}}(2,0)\,\mathrm{d}v = \tfrac{3}{5}\Delta \\
&\frac{1}{2}\int [(2,2)-(2,-2)]^* V_{\text{oct}}[(2,2)-(2,-2)]\,\mathrm{d}v = -\tfrac{2}{5}\Delta \\
&\frac{1}{2}\int [(2,1)-(2,-1)]^* V_{\text{oct}}[(2,1)-(2,-1)]\,\mathrm{d}v = -\tfrac{2}{5}\Delta \\
&\frac{1}{2}\int [(2,1)+(2,-1)]^* V_{\text{oct}}[(2,1)+(2,-1)]\,\mathrm{d}v = -\tfrac{2}{5}\Delta
\end{aligned}\right\},\qquad (13.6)$$

where, for example,

$$d_{xy}{}^* = \frac{\mathrm{i}}{\sqrt{2}}[(2,2)^* - (2,-2)^*]\dagger.$$

These integrals can be expanded (except the second, which does not need expanding). The first, for example, gives

$$\frac{1}{2}\left[\int (2,2)^* V_{\text{oct}}(2,2)\,\mathrm{d}v + \int (2,2)^* V_{\text{oct}}(2,-2)\,\mathrm{d}v \right.$$
$$\left. + \int (2,-2)^* V_{\text{oct}}(2,-2)\,\mathrm{d}v + \int (2,-2)^* V_{\text{oct}}(2,2)\,\mathrm{d}v\right] = \tfrac{3}{5}\Delta.$$

But if we either use the result that $(2,-2)^*=(2,2)$ (from 3.10) or note that $\int d^*_{x^2-y^2} V_{\text{oct}} d_{xy}\mathrm{d}v=0$ then

$$\int (2,2)^* V_{\text{oct}}(2,2)\,\mathrm{d}v + \int (2,-2)^* V_{\text{oct}}(2,2)\,\mathrm{d}v = \tfrac{3}{5}\Delta.$$

We are now in a position to calculate the effect of the octahedral field on the complex d orbitals. By combining expressions (13.5 and 6), we obtain,

$$\left.\begin{aligned}
&\int (2,2)^* V_{\text{oct}}(2,2)\,\mathrm{d}v = \tfrac{1}{10}\Delta \\
&\int (2,1)^* V_{\text{oct}}(2,1)\,\mathrm{d}v = -\tfrac{2}{5}\Delta \\
&\int (2,0)^* V_{\text{oct}}(2,0)\,\mathrm{d}v = \tfrac{3}{5}\Delta \\
&\int (2,-1)^* V_{\text{oct}}(2,-1)\,\mathrm{d}v = -\tfrac{2}{5}\Delta \\
&\int (2,-2)^* V_{\text{oct}}(2,-2)\,\mathrm{d}v = \tfrac{1}{10}\Delta
\end{aligned}\right\}\qquad (13.7)$$

† Since we are now working with complex functions we must always remember to take the complex conjugate in the first function of any integral of the type $\int\psi^* B\psi\,dv$.

$$\int (2,-2)^* V_{oct}(2,2)\,dv = \tfrac{1}{2}\Delta. \tag{13.8}$$

Note particularly the last integral; it is the only cross-term which is non-zero. The complete set of interaction integrals is most conveniently represented as a table or matrix (13.9)†

$$\begin{array}{r} \\ (2,2)^* \\ (2,1)^* \\ (2,0)^* \\ (2,-1)^* \\ (2,-2)^* \end{array} \begin{array}{c} \begin{array}{ccccc} (2,2) & (2,1) & (2,0) & (2,-1) & (2,-2) \end{array} \\ \begin{pmatrix} \frac{1}{10}\Delta & 0 & 0 & 0 & \frac{1}{2}\Delta \\ 0 & -\frac{2}{5}\Delta & 0 & 0 & 0 \\ 0 & 0 & \frac{3}{5}\Delta & 0 & 0 \\ 0 & 0 & 0 & -\frac{2}{5}\Delta & 0 \\ \frac{1}{2}\Delta & 0 & 0 & 0 & \frac{1}{10}\Delta \end{pmatrix} \end{array} \tag{13.9}$$

We have now expressed our crystal-field splittings in terms of complex orbitals rather than real. This is of no help in discussing the simple d^1 case (because of the cross-term between (2,2) and (2,−2)), but it will enable us to work out the energies in the d^2 case.

The eigenfunctions of the ^3F term arising from the d^2 configuration are readily obtained by the shift operator method described in chapter 9. The crystal-field calculation based on these is quite simple but is rather lengthy, so we shall give it in outline only. The two-electron F eigenfunctions may be labelled (3,3), (3,2), (3,1), (3,0), (3,−1), (3,−2) and (3,−3), in conformity with the nomenclature used for d^1.

The crystal-field perturbation does not act on spin so we can examine just one of the spin components (say $M_s = 1$). We shall give one example to show how the interaction integrals for these two-electron functions are evaluated. The function (3,3) can only arise from a combination of one electron in the d orbital (2,2) and one in (2,1). We can write this wave function in determinantal form.

$$(3,3) = |(2,2)(2,1)| = \sqrt{\tfrac{1}{2}}[(2,2)_1(2,1)_2 - (2,2)_2(2,1)_1], \tag{13.10}$$

where both electrons have α spin, and we label them by the subscripts 1 and 2. Then,

$$\int (3,3)^* V_{oct}(3,3)\,dv$$
$$= \frac{1}{2}\iint [(2,2)_1(2,1)_2 - (2,2)_2(2,1)_1]^* V_{oct}[(2,2)_1(2,1)_2,$$
$$-(2,2)_2(2,1)_1]\,dv_1\,dv_2 \tag{13.11}$$

† Because the name is readily comprehensible, we call integrals of the form (13.7) 'interaction integrals'. In the literature and in other texts the name 'matrix element' is used, an integral such as (13.8) being referred to as an 'off-diagonal matrix element'. The reason for this is obvious from an inspection of (13.9).

but since V_{oct} acts on each electron independently $V_{oct} = V_{oct}(1) + V_{oct}(2)$ and (13.11) is equal to

$$\int (2,1)_2{}^* V_{oct}(2)(2,1)_2 \, dv_2 + \int (2,2)_1{}^* V_{oct}(1)(2,2)_1 \, dv_1. \quad (13.12)$$

The values of these two integrals can be obtained from (13.7), so we have

$$\int (3,3)^* V_{oct}(3,3) \, dv = -\tfrac{2}{5}\Delta + \tfrac{1}{10}\Delta = -\tfrac{3}{10}\Delta. \quad (13.13)$$

Proceeding in this way, all of the perturbation energies can be evaluated. The results are collected together in the perturbation matrix (13.14).

$$\begin{array}{c} \\ (3,3)^* \\ (3,2)^* \\ (3,1)^* \\ (3,0)^* \\ (3,-1)^* \\ (3,-2)^* \\ (3,-3)^* \end{array} \begin{array}{c} \begin{array}{ccccccc} (3,3) & (3,2) & (3,1) & (3,0) & (3,-1) & (3,-2) & (3,-3) \end{array} \\ \begin{pmatrix} -\frac{3}{10}\Delta & 0 & 0 & 0 & \sqrt{\frac{3}{20}}\Delta & 0 & 0 \\ 0 & \frac{7}{10}\Delta & 0 & 0 & 0 & \frac{1}{2}\Delta & 0 \\ 0 & 0 & -\frac{1}{10}\Delta & 0 & 0 & 0 & \sqrt{\frac{3}{20}}\Delta \\ 0 & 0 & 0 & -\frac{3}{5}\Delta & 0 & 0 & 0 \\ \sqrt{\frac{3}{20}}\Delta & 0 & 0 & 0 & -\frac{1}{10}\Delta & 0 & 0 \\ 0 & \frac{1}{2}\Delta & 0 & 0 & 0 & \frac{7}{10}\Delta & 0 \\ 0 & 0 & \sqrt{\frac{3}{20}}\Delta & 0 & 0 & 0 & -\frac{3}{10}\Delta \end{pmatrix} \end{array}. \quad (13.14)$$

As in the d^1 case, functions differing by four in M_L are mixed, and the crystal-field interaction between (3,3) and (3, −1) is the same as that between (3, −3) and (3,1).

To obtain the final eigenfunctions and eigenvalues, we have to solve the secular equations (6.67)

$$\Sigma\, c_i(\mathscr{H}_{ik} - ES_{ik}) = 0, \text{ where } \mathscr{H} = V_{oct}.$$

If the matrix is rearranged as in (13.15) it is clear that only 2 × 2 secular determinants need to be solved.

$$\begin{array}{c} \\ (3,3)^* \\ (3,-1)^* \\ (3,-3)^* \\ (3,1)^* \\ (3,2)^* \\ (3,-2)^* \\ (3,0)^* \end{array} \begin{array}{c} \begin{array}{ccccccc} (3,3) & (3,-1) & (3,-3) & (3,1) & (3,2) & (3,-2) & (3,0) \end{array} \\ \begin{pmatrix} -\frac{3}{10}\Delta & \sqrt{\frac{3}{20}}\Delta & & & & & \\ \sqrt{\frac{3}{20}}\Delta & -\frac{1}{10}\Delta & & & & & \\ & & -\frac{3}{10}\Delta & \sqrt{\frac{3}{20}}\Delta & & & \\ & & \sqrt{\frac{3}{20}}\Delta & -\frac{1}{10}\Delta & & & \\ & & & & \frac{7}{10}\Delta & \frac{1}{2}\Delta & \\ & & & & \frac{1}{2}\Delta & \frac{7}{10}\Delta & \\ & & & & & & -\frac{3}{5}\Delta \end{pmatrix} \end{array}. \quad (13.15)$$

Carrying out this operation in the usual way we obtain the eigenfunctions and eigenvalues given in table 13.6.

TABLE 13.6. The components of $^3F(d^2)$ under O_h[a]

Eigenfunction	Energy	Symmetry (see table 13.4)
$\sqrt{\frac{3}{8}}(3,-1) + \sqrt{\frac{5}{8}}(3,3)$ $\sqrt{\frac{3}{8}}(3,1) + \sqrt{\frac{5}{8}}(3,-3)$ $(3,0)$	$-\frac{3}{5}\Delta$	$^3T_{1g}$
$\sqrt{\frac{5}{8}}(3,-1) - \sqrt{\frac{3}{8}}(3,3)$ $\sqrt{\frac{5}{8}}(3,1) - \sqrt{\frac{3}{8}}(3,-3)$ $\sqrt{\frac{1}{2}}[(3,2) + (3,-2)]$	$\frac{1}{5}\Delta$	$^3T_{2g}$
$\sqrt{\frac{1}{2}}[(3,2) - (3,-2)]$	$\frac{6}{5}\Delta$	$^3A_{2g}$

[a] We have evaluated wave functions for the case of both electrons having α spin. These are one component of each triplet state. The spin multiplicity is written as a prefix to the symmetry label.

It is convenient to show the data in tables 13.5 and 13.6 graphically. This is done in figure 13.6.

Figure 13.6 can readily be modified to include other configurations with D or F ground states. Consider the ^4F ground state of the d^3 configuration

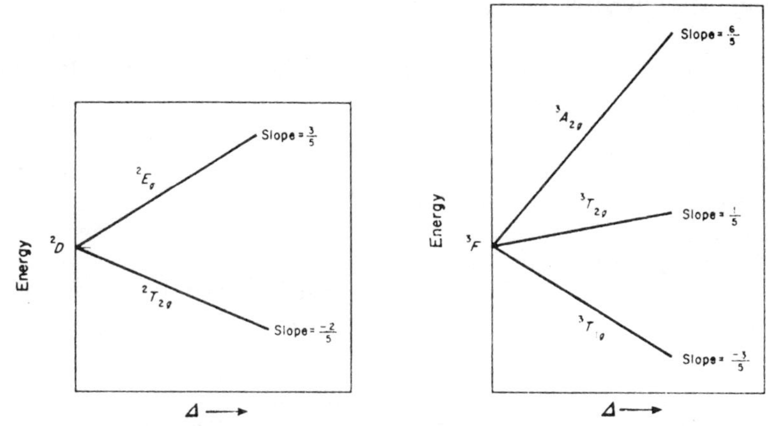

a) (b)

FIGURE 13.6. The splitting of D and F terms in an octahedral crystal field

as an example. The d^3 configuration differs from spherical symmetry (d^5) by two 'holes'. The splitting of the resulting $^4F(d^3)$ term will be exactly the same as for the $^3F(d^2)$ term except that V_{oct}, which repels electrons, will attract holes. Figure 13.6(b) has therefore to be *inverted* for $^4F(d^3)$. Analogous arguments show that figure 13.6(b) is correct for $^4F(d^7)$ but should be inverted for $^3F(d^8)$. Similarly, figure 13.6(a) is correct for $^5D(d^6)$ but has to be inverted for $^5D(d^4)$ and $^2D(d^9)$.

It remains to discuss the ground state of the d^5 configuration, which is ^6S. Since this is orbitally non-degenerate, it cannot be split by a crystal field.

There is an interesting consequence of our discussion of the weak-field $^3F(d^2)$ case. The ground state is stabilized by $-\frac{3}{5}\Delta$ with respect to the arbitrary zero. What occupation of t_{2g} and e_g orbitals is required to give this stabilization? From figure 13.2 we see that the occupation numbers must be $\frac{9}{5}$ in the t_{2g} and $\frac{1}{5}$ in the e_g orbitals $(\frac{1}{5} \times \frac{3}{5}\Delta - \frac{9}{5} \times \frac{2}{5}\Delta)$. In the weak-field d^2 case electron repulsion pushes a small amount of electron density into the unstable e_g orbital. It follows that in our earlier discussion of the magnetic properties of complex ions we were not quite correct in allocating two electrons to t_{2g} orbitals in the weak-field d^2 case. However, the small correction we deduce here is barely detectable in an experiment.

Non-integral occupation of t_{2g} and e_g orbitals is also found for weak-field $d^7(^4F)$ octahedral and $d^3(^4F)$ and $d^8(^3F)$ tetrahedral complexes.

13.3 Strong-field complexes

We now turn to a consideration of strong-field complexes. In order to do this we need to consider all of the terms which arise from a configuration, not just the ground term, because when Δ is large the ground state of the complex does not necessarily arise from the ground state of the free ion, and second-order perturbation energies become important.

Excited orbitally-degenerate terms of a configuration are split by a crystal field in just the same way as the ground term. Table 13.7 lists the terms arising from d^n configurations. Table 13.8 lists the components of all terms, up to thirteen-fold orbitally degenerate, in the presence of an octahedral crystal field.

TABLE 13.7. Terms arising from d^n configurations. $^2F(2)$ means that there are two ^2F terms

Configuration	Terms
d^1 d^9	^2D
d^2 d^8	^3F, ^3P, ^1G, ^1D, ^1S
d^3 d^7	^4F, ^4P, ^2H, ^2G, ^2F, $^2D(2)$, ^2P
d^4 d^6	^5D, ^3H, ^3G, $^3F(2)$, ^3D, $^3P(2)$, ^1I, $^1G(2)$, ^1F, $^1D(2)$, $^1S(2)$
d^5	^6S, ^4G, ^4F, ^4D, ^4P, ^2I, ^2H, $^2G(2)$, $^2F(2)$, $^2D(3)$, ^2P, ^2S

TABLE 13.8. Splitting of terms in an octahedral field

Term	Orbital degeneracy $=2L+1$	Components under O_h
S	1	A_{1g}
P	3	T_{1g}
D	5	$E_g + T_{2g}$
F	7	$A_{2g} + T_{1g} + T_{2g}$
G	9	$A_{1g} + E_g + T_{1g} + T_{2g}$
H	11	$E_g + 2T_{1g} + T_{2g}$
I	13	$A_{1g} + A_{2g} + E_g + T_{1g} + 2T_{2g}$

The eigenfunctions and energies of the components of each of these terms could be found by the method described earlier for the components of $^3F(d^2)$, but the results would only have significance if Δ were small. The reason for this is best illustrated by two examples. The terms arising from a d^5 configuration are sixteen in number, of which eleven are doublets (^2I, ^2H, $^2G(2)$, $^2F(2)$, $^2D(3)$, ^2P and ^2S); in the presence of an octahedral crystal field these split to give four $^2A_{1g}$, three $^2A_{2g}$, seven ^2E_g, eight $^2T_{1g}$ and ten $^2T_{2g}$ levels. In weak fields the ground state is $^6A_{1g}$, but one of the $^2T_{2g}$ levels becomes the ground state in strong fields. In order to determine how its energy depends on Δ, we apparently have to solve a 10×10 secular determinant and select the lowest root. Fortunately, as we shall see, the problem is greatly simplified when Δ is much greater than the electron repulsion energies—which is the strong-field limit. In the more common 'intermediate' field region, it is simplest to interpolate between the weak- and strong-field results. The d^5 case is, of course, a particularly difficult one. Much simpler are the d^2 and d^8 configurations, for here there is only one excited level to interact with the ground state. For d^2 and d^8 an upper component of the free-ion ground term (^3F) interacts with one component of an excited term (^3P), both of these having T_{1g} symmetry. This has important spectral consequences because the two components are close together. We shall consider the d^2 configuration in some detail.

We label the two $^3T_{1g}$ levels according to their parentage. In our earlier discussion of the weak-field case, we ignored the existence of the excited $^3T_{1g}(^3P)$ state. Since the ^3P level is not split by the crystal field (see table 13.8), the energy of this zeroth-order term is independent of Δ. Because $^3T_{1g}(^3F)$ and $^3T_{1g}(^3P)$ are of the same symmetry, there will be a cross-term

$$\int {}^3T_{1g}{}^*(^3P)V_{\text{oct}}{}^3T_{1g}(^3F)\,\mathrm{d}v \qquad (13.16)$$

which, following the procedure described on pages 235–6, can be shown to have the value $\frac{2}{5}\Delta$.

We now want to know the difference in energy between the two terms in the free ion. In terms of the Slater–Condon parameters (9.52) this is $15F_2 - 75F_4$, which we shall simply carry through the equations as a parameter X. Taking the energy of the ${}^3F$ term of the free ion as zero, the interaction integrals are given by the following matrix

$$\begin{array}{c} \\ {}^3T_{1g}({}^3F) \\ {}^3T_{1g}({}^3P) \end{array} \begin{array}{cc} {}^3T_{1g}({}^3F) & {}^3T_{1g}({}^3P) \\ \left(\begin{array}{c} -\frac{3}{5}\Delta \\ \frac{2}{5}\Delta \end{array} \right. & \left. \begin{array}{c} \frac{2}{5}\Delta \\ X \end{array} \right). \end{array} \tag{13.17}$$

The solutions of the secular determinant

$$\begin{vmatrix} -\frac{3}{5}\Delta - E & \frac{2}{5}\Delta \\ \frac{2}{5}\Delta & X - E \end{vmatrix} = 0 \tag{13.18}$$

are

$$E = \tfrac{1}{2}[X - \tfrac{3}{5}\Delta \pm (X^2 + \tfrac{6}{5}X\Delta + \Delta^2)^{\frac{1}{2}}]. \tag{13.19}$$

In the weak-field limit $X \gg \Delta$ we have, after a binomial expansion of the square root up to terms of order Δ, $E = X$ or $-\frac{3}{5}\Delta$. This is also the result obtained by ignoring the interaction between the ${}^3F$ and ${}^3P$ term. In the strong-field limit $\Delta \gg X$ we have, after expansion to terms of order X,

$$E = \frac{4X}{5} + \frac{\Delta}{5}, \quad \frac{X}{5} - \frac{4\Delta}{5}. \tag{13.20}$$

Figure 13.7 is the same as figure 13.6(b), modified to include the effect of the ${}^3P$ term. Both ${}^3T_{1g}$ levels are curved, the limiting slope of the lower level being $-\frac{4}{5}$, and that of the upper $\frac{1}{5}$. The ${}^3A_{2g}$ state crosses the upper ${}^3T_{1g}$ if Δ is large enough. For V^{3+}, where $X = 13{,}200\ \text{cm}^{-1}$ this occurs when $\Delta = 11{,}000\ \text{cm}^{-1}$. This phenomenon will be of importance when we discuss the electronic spectrum of $[V(H_2O)_6]^{3+}$.

Strong-field complexes differ in spin multiplicity from weak-field complexes because a component from an excited term of the free ion is rapidly stabilized as Δ increases, until it eventually becomes the ground state. For example, in the d^5 case a ${}^2T_{2g}({}^2D)$ component replaces ${}^6A_{1g}$ as the ground state. The latter has zero slope (being the only component of the orbitally non-degenerate ${}^6S$ term), whilst the former ultimately has a slope of -2. In the strong-field limit we no longer base our wave functions on the terms of the free ion, but on configurations built up from the crystal-field-split d orbitals (table 13.5). In this limit the ${}^2T_{2g}({}^2D)$ level corresponds to a $t_{2g}{}^5$ configuration that is, we have a 'hole' in one of the t_{2g} orbitals.

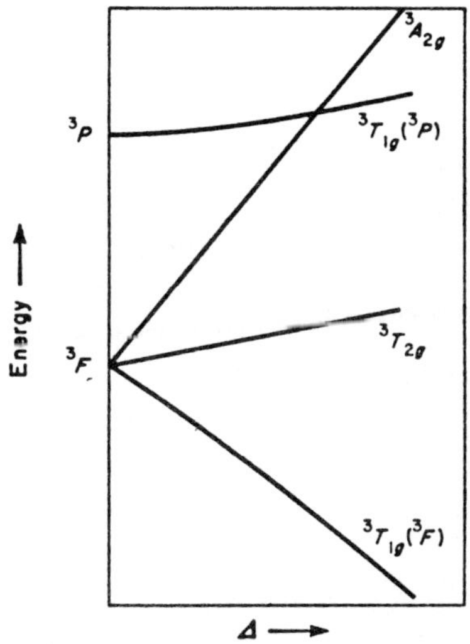

FIGURE 13.7. The effect of second-order interaction on the ground state of a d^2 octahedral complex

The variation of the energy of this level with Δ is obtained by allowing $-\frac{2}{5}\Delta$ for each electron in the t_{2g} orbitals, i.e. $5 \times (-\frac{2}{5}\Delta) = -2\Delta$.

A similar analysis holds for all other strong-field states, ground or excited. For example, d^7 gives rise to the ground configuration $t_{2g}{}^6e_g{}^1$, and this has an energy which varies with Δ as $6(-\frac{2}{5}\Delta) + \frac{3}{5}\Delta = -\frac{9}{5}\Delta$. Likewise, the slope of the ground state for d^2 is $-\frac{4}{5}\Delta$, which is the limit we found earlier (13.20).

As an example of a wave function in the strong-field limit we again take $t_{2g}{}^2$. The ground state is $^3T_{1g}$. If we consider the spin component with $M_s = 1$, then there are only three ways of allocating electrons to these three orbitals with this spin restriction, so that these must represent the three space components of the T_{1g} state. We can write them

$$|\phi_1\phi_2|, \quad |\phi_2\phi_3|, \quad |\phi_3\phi_1|,$$

where ϕ_1, ϕ_2 and ϕ_3 are any three independent functions for the t_{2g} orbitals, e.g. d_{xy} d_{yz} and d_{zx} or (2,1), (2,−1) and $\sqrt{\frac{1}{2}}\{(2,2) - (2,-2)\}$.

Following the method given in this and the last section crystal-field energy diagrams can be constructed for all of the d orbital configurations. Two typical cases are shown in figures 13.8 and 13.9†.

† Other cases may be found in GRIFFITH, *The Theory of Transition Metal Ions*, C.U.P., 1961.

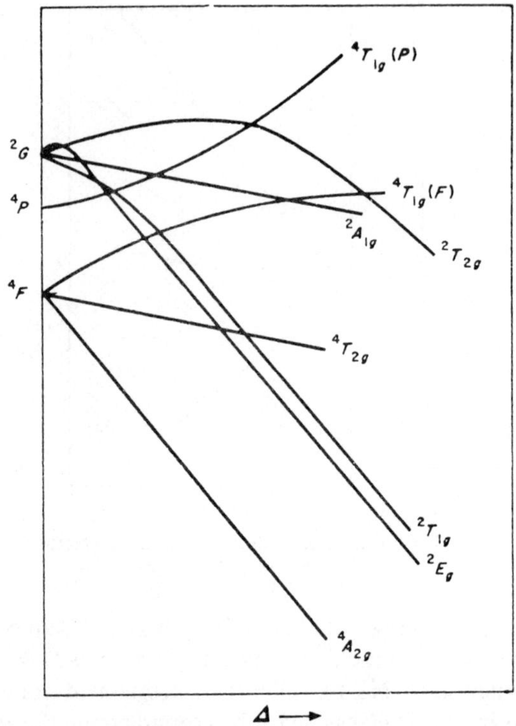

FIGURE 13.8. Crystal-field energy diagram for a d^3 configuration

13.4 Spin-orbit coupling

Although spin-orbit coupling is invariably weaker than the crystal-field interaction in the ions of the first transition series, it frequently has to be taken into account for a detailed understanding of the electronic spectra and magnetic properties of complexes.

We have already outlined the calculation of spin-orbit coupling energies in chapter 9 where we introduced the operator **L.S** (9.16, 9.19), and showed how this gave rise to the Landé interval rule. After the introduction of a crystal-field perturbation J and L are no longer good quantum numbers so that expression (9.20) can no longer be used. In order to calculate the spin-orbit coupling interactions, we must look in more detail at the form of the integrals involving the operator **L.S**. These are most easily evaluated using the following expansion

$$\begin{aligned} \mathbf{L}.\mathbf{S} &= \mathbf{L}_z\mathbf{S}_z + \mathbf{L}_x\mathbf{S}_x + \mathbf{L}_y\mathbf{S}_y \\ &= \mathbf{L}_z\mathbf{S}_z + \tfrac{1}{2}\{(\mathbf{L}_x + \mathrm{i}\mathbf{L}_y)(\mathbf{S}_x - \mathrm{i}\mathbf{S}_y) + (\mathbf{L}_x - \mathrm{i}\mathbf{L}_y)(\mathbf{S}_x + \mathrm{i}\mathbf{S}_y)\} \\ &= \mathbf{L}_z\mathbf{S}_z + \tfrac{1}{2}\{\mathbf{L}_+\mathbf{S}_- + \mathbf{L}_-\mathbf{S}_+\}, \qquad (13.21) \end{aligned}$$

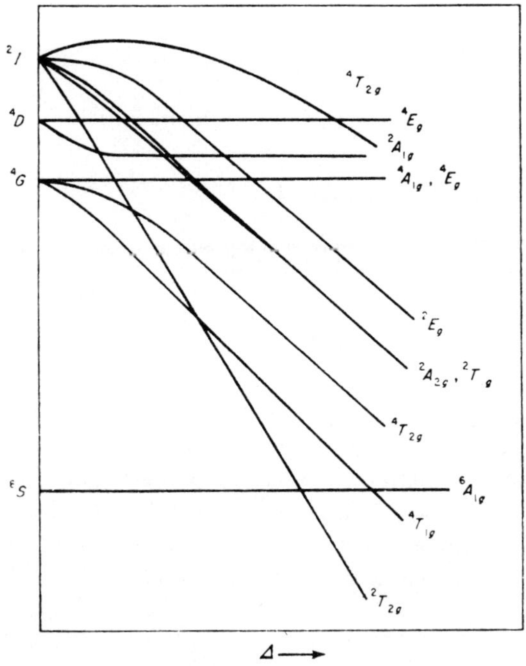

FIGURE 13.9. Crystal-field energy diagram for a d^5 configuration

where the shift operators $\mathbf{L}_+$ etc. are defined by (9.21).

We shall evaluate two integrals as examples†.

$$\text{(a)} \int (2,1,\alpha)^* \mathbf{L}\,.\,\mathbf{S}(2,1,\alpha)\,\mathrm{d}\tau$$

$$= \int (2,1,\alpha)^* \{\mathbf{L}_z\mathbf{S}_z + \tfrac{1}{2}(\mathbf{L}_+\mathbf{S}_- + \mathbf{L}_-\mathbf{S}_+)\}(2,1,\alpha)\,\mathrm{d}\tau$$

$$= \int (2,1,\alpha)^* \mathbf{L}_z\mathbf{S}_z(2,1,\alpha)\,\mathrm{d}\tau$$

$$= \int (2,1)^* \mathbf{L}_z(2,1)\,\mathrm{d}v \int \alpha^* \mathbf{S}_z\alpha\,\mathrm{d}s$$

$$= \left(1\,.\,\frac{h}{2\pi}\right)\left(\frac{1}{2}\cdot\frac{h}{2\pi}\right) = \frac{h^2}{8\pi^2}.$$

† We have included the spin wave functions in the integrals in an obvious manner.

$$\text{(b)} \int (2,1,\alpha)^*\mathbf{L}.\mathbf{S}(2,2,\beta)\,d\tau$$

$$= \int (2,1,\alpha)^*\{\mathbf{L}_z\mathbf{S}_z + \tfrac{1}{2}(\mathbf{L}_+\mathbf{S}_- + \mathbf{L}_-\mathbf{S}_+)(2,2,\beta)\,d\tau$$

$$= \int (2,1,\alpha)^*(\tfrac{1}{2}\mathbf{L}_-\mathbf{S}_+)(2,2,\beta)\,d\tau$$

$$= \frac{1}{2}\int (2,1)^*\mathbf{L}_-(2,2)\,dv \int \alpha^*\mathbf{S}_+\beta\,ds$$

$$= \frac{1}{2}\frac{h}{2\pi}[(2+2)(2-2+1)]^{\frac{1}{2}}\cdot\frac{h}{2\pi}[(\tfrac{1}{2}-\tfrac{1}{2}+1)(\tfrac{1}{2}+\tfrac{1}{2})]^{\frac{1}{2}} = \frac{h^2}{4\pi^2}.$$

The eigenvalues of the operator **L.S** for single d electron wave functions are given in matrix form below in units of $(h/2\pi)^2$. It should be noted that cross-terms (off-diagonal terms) only occur between functions with the same value of $M_J = M_L + M_S$. Many-electron spin-orbit coupling integrals can always be expressed as a sum of one-electron terms, using a technique similar to that which we adopted to evaluate two-electron integrals involving V_{oct}.

$$\begin{array}{c|cccccccccc|}
 & (2,2,\alpha) & (2,2,\beta) & (2,1,\alpha) & (2,1,\beta) & (2,0,\alpha) & (2,0,\beta) & (2,-1,\alpha) & (2,-1,\beta) & (2,-2,\alpha) & (2,-2,\beta) \\
(2,2,\alpha)^* & 1 & \cdot & \cdot & \cdot & \cdot & \cdot & \cdot & \cdot & \cdot & \cdot \\
(2,2,\beta)^* & \cdot & -1 & 1 & \cdot & \cdot & \cdot & \cdot & \cdot & \cdot & \cdot \\
(2,1,\alpha)^* & \cdot & 1 & \frac{1}{2} & \cdot & \cdot & \cdot & \cdot & \cdot & \cdot & \cdot \\
(2,1,\beta)^* & \cdot & \cdot & \cdot & -\frac{1}{2} & \sqrt{\frac{3}{2}} & \cdot & \cdot & \cdot & \cdot & \cdot \\
(2,0,\alpha)^* & \cdot & \cdot & \cdot & \sqrt{\frac{3}{2}} & 0 & \cdot & \cdot & \cdot & \cdot & \cdot \\
(2,0,\beta)^* & \cdot & \cdot & \cdot & \cdot & \cdot & 0 & \sqrt{\frac{3}{2}} & \cdot & \cdot & \cdot \\
(2,-1,\alpha)^* & \cdot & \cdot & \cdot & \cdot & \cdot & \sqrt{\frac{3}{2}} & -\frac{1}{2} & \cdot & \cdot & \cdot \\
(2,-1,\beta)^* & \cdot & \cdot & \cdot & \cdot & \cdot & \cdot & \cdot & \frac{1}{2} & 1 & \cdot \\
(2,-2,\alpha)^* & \cdot & \cdot & \cdot & \cdot & \cdot & \cdot & \cdot & 1 & -1 & \cdot \\
(2,-2,\beta)^* & \cdot & \cdot & \cdot & \cdot & \cdot & \cdot & \cdot & \cdot & \cdot & 1
\end{array} \quad (13.22)$$

Using these results, the number of spin-orbit components for the ground and lowest excited levels of weak-field octahedral complexes are found to be as shown in the following table.

	$^1A_{1g}$	$^6A_{1g}$	$^3A_{2g}$	$^4A_{2g}$	^2E_g	^5E_g	$^4T_{1g}$	$^5T_{1g}$	$^2T_{2g}$	$^3T_{2g}$	$^5T_{2g}$
Number of spin-orbit components	1	1	1	1	1	1	3	3	2	3	3

For example, the weak-field complexes of $Ni^{2+}(d^8)$ have a $^3A_{2g}$ ground state and $^3T_{2g}$ and $^3T_{1g}$ excited states. These are split by spin-orbit coupling into 1, 3 and 3 sub-levels respectively. If the splittings are large enough they may be observable spectroscopically, resulting in a splitting of one of the bands in the electronic absorption spectrum. A splitting has been observed in the ${}^3A_{2g} \rightarrow {}^3T_{1g}(F)$ transition of the $[Ni(H_2O)_6]^{2+}$ ion and has been attributed to spin-orbit coupling.

13.5 Molecular-orbital theory

Crystal-field theory suffers from the obvious limitation that it neglects the overlap between the ligand and transition metal orbitals, and if the ligand is a non-polar species (e.g. $CH_2{=}CH_2$) then this is likely to be more important than the electrostatic perturbation. In fact, even if one takes a realistic crystal-field potential for the polar ligands and ignores overlap effects it is difficult to obtain quantitatively correct answers from crystal-field theory. We turn now to MO theory which introduces a delocalization of electrons between the transition metal and ligand. When this is combined with crystal-field theory one obtains what is known as ligand-field theory.

We could just plunge in and form molecular orbitals by taking linear combinations of ligand and metal orbitals. This would only require the solution of, say, a 30 × 30 secular determinant. The problem is really much simplified by using group theory. We first take combinations of the metal orbitals which form a basis for an *I.R.* of the group; do the same for the ligand orbitals, and then combine ligand and metal orbitals which belong to the same *I.R.*

We have already seen how the *d* orbitals transform under the operations of the octahedral group. The same procedure can be used for other orbitals of this group, and other common symmetry groups of metal complexes. The data likely to be needed are collected in table 13.9.

TABLE 13.9. Transformation properties of atomic orbitals

	O_h	T_d	D_{3d}	D_{4h}	D_{2h}
s	a_{1g}	a_1	a_{1g}	a_{1g}	a_g
p_z	}	}	a_{2u}	a_{2u}	b_{1u}
p_x	} t_{1u}	} t_2	} e_u	} e_u	b_{3u}
p_y	}	}	}	}	b_{2u}
d_{z^2}	} e_g	} e	a_{1g}	a_{1g}	a_g
$d_{x^2-y^2}$	}	}	} e_g	b_{1g}	a_g
d_{xy}	}	}	}	b_{2g}	b_{1g}
d_{yz}	} t_{2g}	} t_2	} e_g	} e_g	b_{3g}
d_{zx}	}	}	}	}	b_{2g}

We now turn our attention to the ligand orbitals of σ symmetry (s orbitals and p orbitals pointing towards the cation). The first step is to determine the *I.R.*'s having these orbitals as basis. Using the method described earlier (asking how many orbitals are unchanged on each operation), we obtain the following characters of the representation.

I	$6C_4$	$3C_2$	$6C_2'$	$8C_3$	i	$6S_4$	$3\sigma_h$	$6\sigma_d$	$8S_6$
6	2	2	0	0	0	0	4	2	0

This reducible representation has components $A_{1g} + E_g + T_{1u}$.

We now use the method given in chapter 8 for obtaining the linear combinations of ligand orbitals (which we shall call ligand group orbitals) which transform according to these *I.R.*'s. We fix our attention on one ligand orbital and ask how it transforms under the forty-eight separate operations of the group. If we label the ligands A → F as shown in figure 13.10 and look at A, we obtain the first line in table 13.10. To obtain, for example, the t_{1u} ligand group orbitals, we multiply the orbitals under an operation by the t_{1u} character for that operation and then add the results together. These steps are shown in table 13.10.

TABLE 13.10. The determination of t_{1u} ligand group orbitals

	I	$6C_4$	$3C_2$	$6C_2'$	$8C_3$
A becomes	A	2A + B + C + E + F	A + 2D	2D + B + C + E + F	2B + 2C + 2E + 2F
T_{1u}	3	1	−1	−1	0
Multiply	3A	2A + B + C + E + F	−A − 2D	−2D − B − C − E − F	—
	i	$6S_4$	$3\sigma_h$	$6\sigma_d$	$8S_6$
A becomes	D	2D + B + C + E + F	2A + D	2A + B + C + E + F	2B + 2C + 2E + 2F
T_{1u}	−3	−1	1	1	0
Multiply	−3D	−2D − B − C − E − F	2A + D	2A + B + C + E + F	—
Add	8A − 8D				
Normalize	$\psi'(t_{1u}) = \sqrt{\tfrac{1}{2}}(A - D)$				

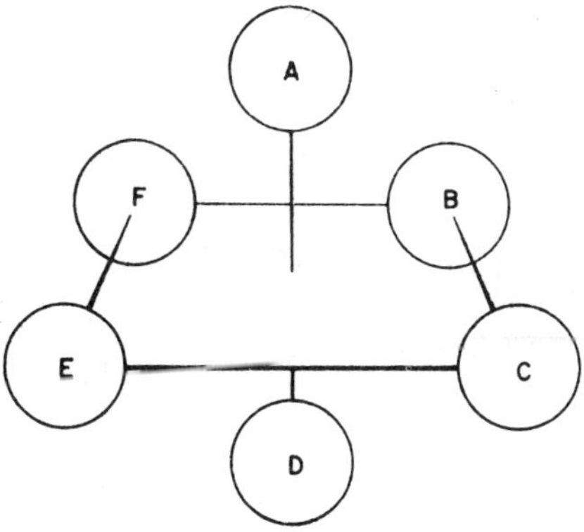

FIGURE 13.10. Labelling scheme for the ligand orbitals of an octahedral complex

The other t_{1u} ligand group orbitals are obtained by considering the transformation of B or E and C or F (D gives $\psi'(t_{1u})$ again) and are obviously

$$\psi''(t_{1u}) = \sqrt{\tfrac{1}{2}}(\text{B}-\text{E})$$

$$\psi'''(t_{1u}) = \sqrt{\tfrac{1}{2}}(\text{C}-\text{F}).$$

These orbitals are shown in figure 13.11, along with the atomic p orbitals. It is obvious that the two sets have identical symmetry, which is just what the common t_{1u} label tells us.

The a_{1g} and e_g ligand group orbitals are obtained in a similar way. For the latter set we can generate *three* orbitals, which, like d_{x^2}, d_{y^2} and d_{z^2}, are not linearly independent. But suitable linear combinations of two of these can be taken to form two which are independent (page 106). The a_{1g} and e_g ligand group orbitals are

$$\psi(a_{1g}) = \sqrt{\tfrac{1}{6}}(\text{A} + \text{B} + \text{C} + \text{D} + \text{E} + \text{F})$$

$$\psi'(e_g) = \tfrac{1}{2}(\text{B} - \text{C} + \text{E} - \text{F})$$

$$\psi''(e_g) = \sqrt{\tfrac{1}{12}}(2\text{A} - \text{B} - \text{C} + 2\text{D} - \text{E} - \text{F}).$$

These are shown in figure 13.12 along with their atomic-orbital counterparts.

We can now construct a molecular-orbital energy level diagram as in figure 13.13 for a complex of the first row transition series. This is highly qualitative. We have assumed that the main interaction is between the highest occupied orbitals of the ligand and the partly filled d and empty s and p orbitals of the cation. Further, we have assumed that the ligand orbitals all have a lower energy than the d orbitals of the cation. This is

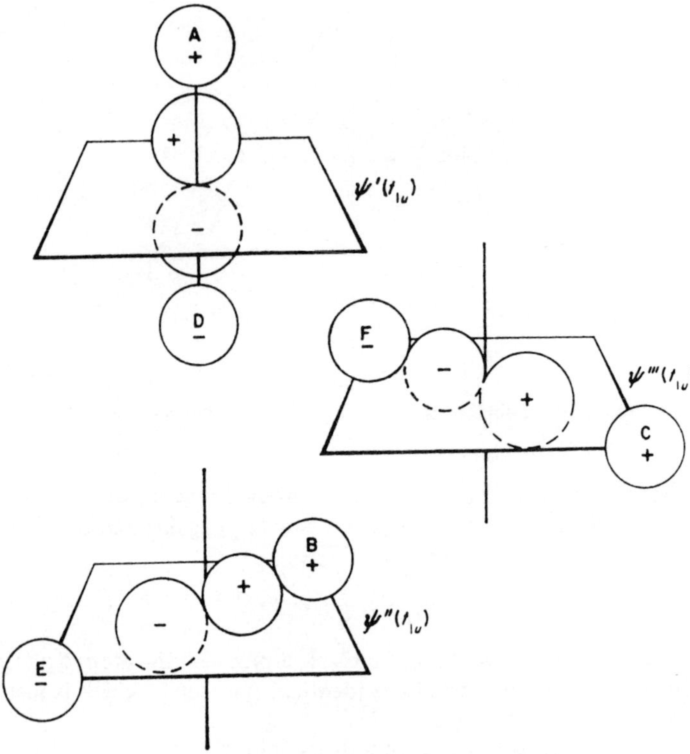

FIGURE 13.11. The t_{1u} ligand orbitals

supported by experiment, which shows that the molecular orbitals which contain unpaired electrons are mainly localized on the metal. The order of the antibonding t_{1u} and a_{1g} is not certain, but as we shall see, this order does not play any part in determining ground state symmetries. We have twelve electrons from the ligands plus any 3*d* electrons from the free ion to allocate to these orbitals. The lowest a_{1g}, t_{1u} and e_g orbitals will be fully occupied in all cases so we can simply consider how the extra electrons are distributed amongst the remaining orbitals which will be t_{2g} and e_g^* in the lowest energy state. This is just what we found with the electrostatic picture (figure 13.2), but we did not recognize the antibonding character of the upper level. Both the electrostatic and the molecular-orbital approaches focus our attention on the separation between the t_{2g} and e_g^* orbitals, which we have called Δ. Everything we said about the states of weak- and strong-field complexes in the section on crystal-field theory will therefore hold also in MO theory.

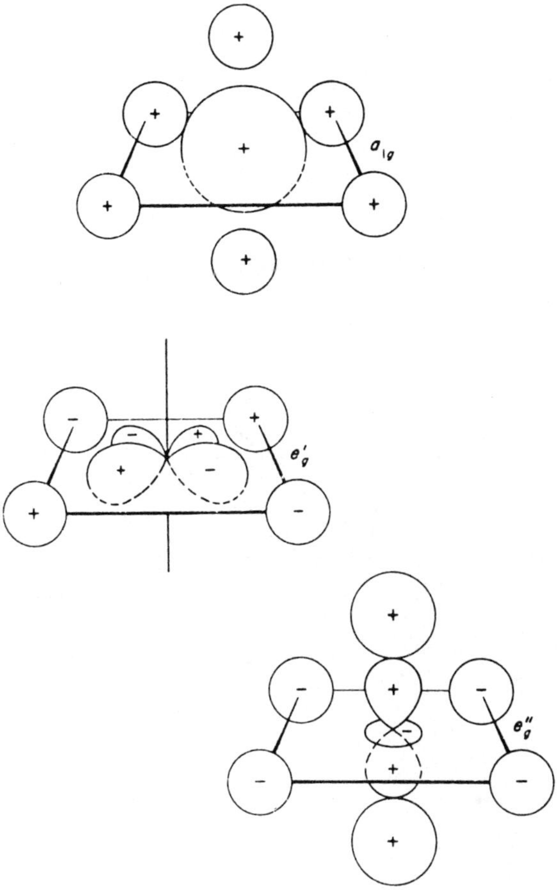

FIGURE 13.12. The ligand orbitals of a_{1g} and e_g symmetry

We noted that in molecular-orbital theory the t_{2g} orbitals are non-bonding, but if π bonding is important, this is no longer true. Figure 13.14 shows the ligand p orbitals, two on each ligand, which can be involved in π bonding.

These π orbitals can be compounded into ligand group orbitals which, in the $\mathbf{O}_h$ group, transform as $T_{1g} + T_{1u} + T_{2g} + T_{2u}$. The precise form of each of these ligand group orbitals could be determined by the same procedure that we used for the σ orbitals. We note that there are no orbitals of T_{1g} and T_{2u} symmetry present in figure 13.13, so these orbitals are non-bonding. Since our real concern is with the t_{2g} and e_g* orbitls, we may also ignore the t_{1u} ligand π orbitals. We are left with the problem of how the

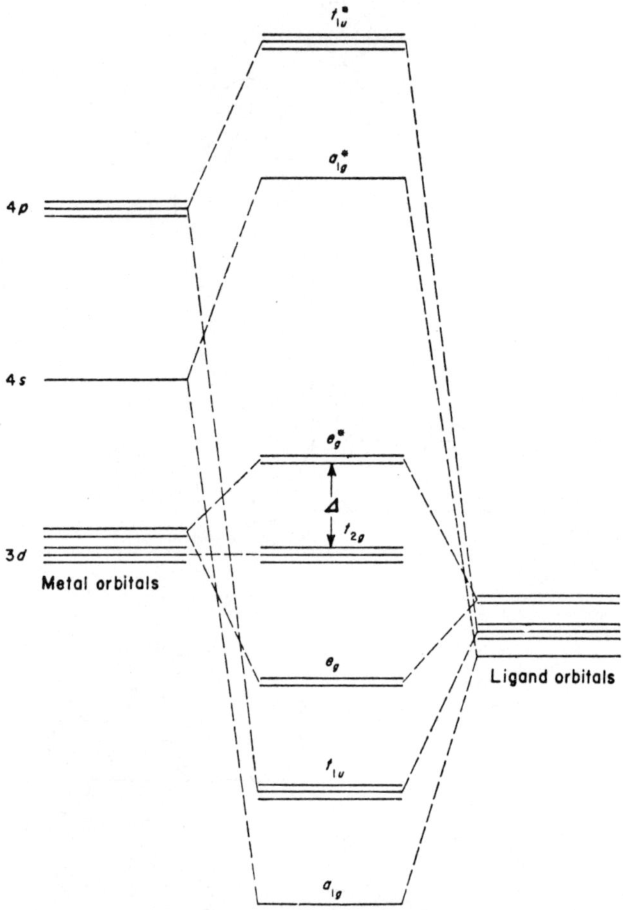

FIGURE 13.13. Molecular-orbital energy diagram for an octahedral complex. The asterisks represent antibonding orbitals

ligand group orbitals of symmetry t_{2g} interact with the non-bonding t_{2g} orbitals of figure 13.13. Two cases are likely to occur: (a) the ligand has filled π orbitals which have a lower energy than the 3*d*; (b) the ligand has empty π orbitals which have a higher energy than the 3*d*. The energy level diagrams in the two cases are shown in figure 13.15. The fluoride ion is an example of case (a), the cyanide ion an example of case (b). Particular notice should be taken of the different effect on Δ shown in figure 13.15. The increase of Δ with π bonding in case (b) probably explains why the cyanide ion heads the spectrochemical series (page 217).

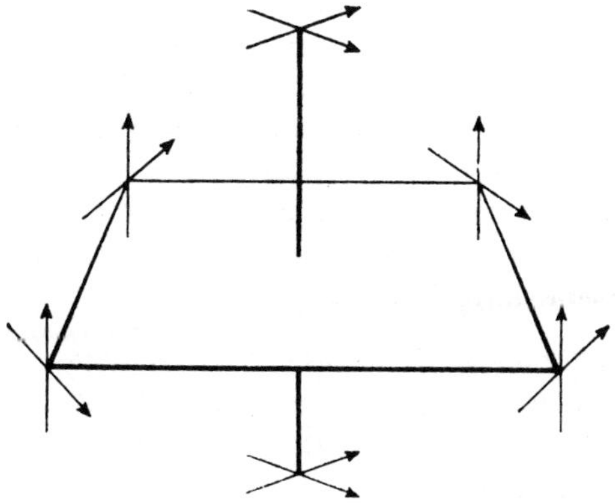

FIGURE 13.14. Ligand $p\pi$ orbitals. The arrows represent their directions, the head of the arrow being the positive end of the orbital

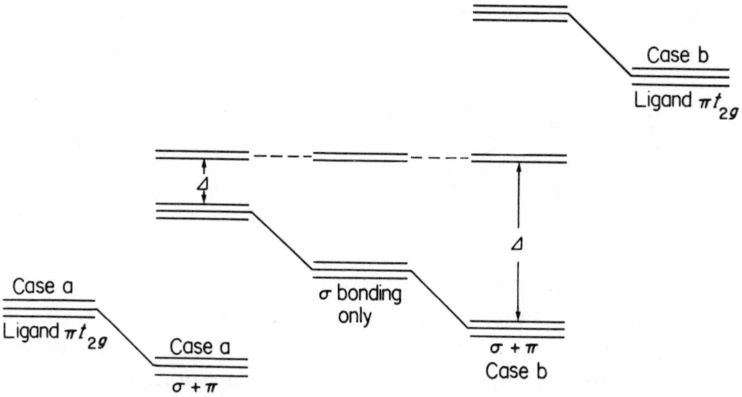

FIGURE 13.15. The effect of ligand orbitals on molecular-orbital energy levels: (a) ligand has filled π orbitals: (b) ligand has empty π orbitals

The problem of carrying out non-empirical calculations on transition-metal complexes is a most difficult one. There are several factors which contribute to this difficulty. Most important is the large number of electrons present in such molecules coupled with the fact that many do not have a closed-shell electronic structure. This means that there are large numbers of integrals to be evaluated (particularly electron repulsion integrals). Ultimately, the limiting factor is the size and sophistication of computer facilities; in practice the problem is beyond the vast majority of workers.

The hope must therefore be that it is possible to obtain worthwhile results from approximate calculations. However, the high polarity of most inorganic complex species means that the type of approximate calculation which has been so successful for organic molecules must be modified before being applied to inorganic systems. For example, the zero-overlap method has been adapted to complex ions. Such schemes have a common origin in the Roothaan equations (Section 10.7), differing in their subsequent simplifications.

A possible alternative is the use of calculations which parallel the Hückel approach to conjugated unsaturated organic molecules (Section 15.2) and dispense with an explicit Hamiltonian. These are often called *extended Hückel methods*.

Consider a diatomic complex ion $[M-L]^+$, in which we shall consider only a σ bond between the two atoms. The secular determinant which must be solved is of the form

$$\begin{vmatrix} H_{MM} - E & H_{ML} - S_{ML}E \\ H_{LM} - S_{LM}E & H_{LL} - E \end{vmatrix} = 0$$

Here, H_{MM} and H_{LL} are zeroth order energies of the relevant metal and ligand orbitals (for simplicity, we assume one orbital each), H_{ML} $(=H_{LM})$ is the interaction term and S_{ML} the σ overlap integral. The solutions of this equation are the energies of the σ bonding and antibonding orbitals. We could determine the magnitude of H_{MM}, say, by obtaining an explicit expression for it and evaluating the appropriate integrals. Alternatively, we could hope to employ an 'experimental' value for this quantity from ionization potential (or similar) data. This is the method adopted in extended Hückel calculations. However, it is impossible to obtain an experimental value for the off-diagonal matrix element, H_{ML}. This difficulty is commonly circumvented by relating it to the corresponding diagonal terms. The most common relationship assumed was first used in this context by Wolfsberg and Helmholz:

$$H_{ML} = FS_{ML}(H_{MM}+H_{LL}).$$

This equation has been found to hold reasonably well for small molecules.

The quantity F is a scale parameter which is commonly given a value of between 0·7 and 1·0. The main weakness of the method is that there is no one value of F which is widely successful and the results of a calculation are very sensitive to its magnitude.

The application of the method to octahedral and other complexes is similar to that discussed above, the only formal change being that the overlap integral S_{ML} is replaced by a group overlap integral G_{ML}. So, for the first interaction shown in Figure 13.12 the group overlap integral is

$$
\begin{aligned}
G_{ML}(a_{1g}) &= \int \psi_s(\text{metal})\, \psi\,(a_{1g})\, dv \\
&= \frac{1}{\sqrt{6}} \int \psi_s(\text{metal})\, (A+B+C+D+E+F) dv \\
&= \sqrt{6} \int \psi_s(\text{metal})\, A\, dv \\
&= \sqrt{6}\, S_{ML}\, (s,A).
\end{aligned}
$$

Group overlap integrals are always very simply related to diatomic overlap integrals.

Calculations of this type are usually carried out by an iterative scheme. This is because the values for the diagonal matrix elements (derived from ionization potential data) depend on the charges and electron configuration on the atoms and these in turn depend on the coefficients in the molecular orbitals. After each successive calculation, revised values for the diagonal and off-diagonal matrix elements are used in the next step of the iteration. This process is repeated until self-consistent charges and configurations (SCCC) are obtained.

13.6 Tetrahedral and planar complexes

We now turn to a discussion of the tetrahedral and square planar stereochemistries which are the only important high-symmetry structures other than the octahedron. Both the octahedron and tetrahedron are related to a cube, as shown in figure 13.16.

Following our discussion of an octahedron, we anticipate that the d orbitals in a tetrahedral field split into sets of two ($d_{x^2-y^2}$ and d_{z^2}) and three (d_{xy}, d_{yz} and d_{zx}). The character table for the group of a regular tetrahedron, $\mathbf{T}_d$, is given in table 13.11. It is instructive to compare this with the character table of the group $\mathbf{O}_h$ (table 13.4).

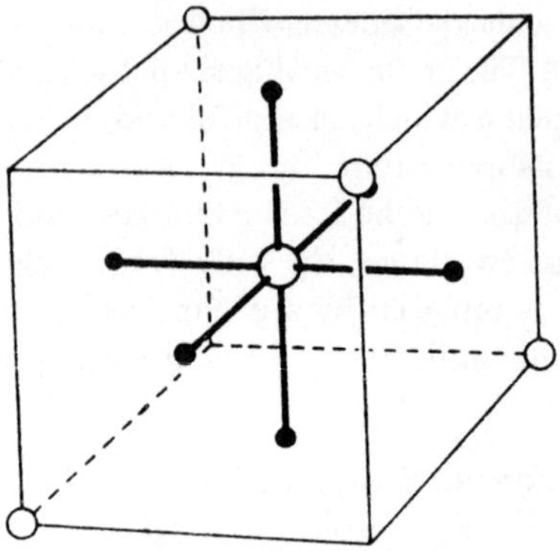

FIGURE 13.16. The relationship between the cube, octahedron and tetrahedron

TABLE 13.11. The T_d character table

T_d	I	$6S_4$	$3C_2$	$6\sigma_d$	$8C_3$
A_1	1	1	1	1	1
A_2	1	−1	1	−1	1
E	2	0	2	0	−1
T_1	3	1	−1	−1	0
T_2	3	−1	−1	1	0

The five d orbitals transform as $E + T_2$, the three p orbitals as T_2 (NOT T_1), and an s orbital as A_1. Since there is no inversion operation for the tetrahedron, p–d mixing can occur. This will be fairly small and for the moment we ignore it. In figure 13.17 are shown representative d orbitals of E and T_2 species. The lobes of $d_{x^2-y^2}$ are half a cube face

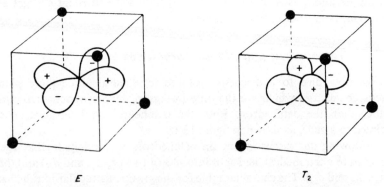

FIGURE 13.17. The representation of orbitals of E and T_2 species in a tetrahedral field

diagonal away from two ligands, those of d_{xy} are half a cube edge away from one. It seems probable that electron–ligand electrostatic repulsion is minimized if the electron occupies the e orbital $d_{x^2-y^2}$, although the splitting must be smaller than in the octahedral case. This is confirmed by calculation, the splitting being $\frac{4}{9}$ of that for the corresponding octahedron. It is most important to note that the t_{2g} orbitals have the lower energy in an octahedron, but the t_2 orbitals have the higher energy in a tetrahedron, i.e.

$$\Delta_{\text{tet}} = -\tfrac{4}{9}\Delta_{\text{oct}}.$$

In figure 13.18 the relationship between the octahedral and tetrahedral splittings is shown for the d^1 configuration. A similar pattern holds for the d^6 configuration; for the d^4 and d^9 configurations the tetrahedral and octahedral patterns are interchanged. A diagram such as figure 13.18 is frequently called an Orgel diagram.

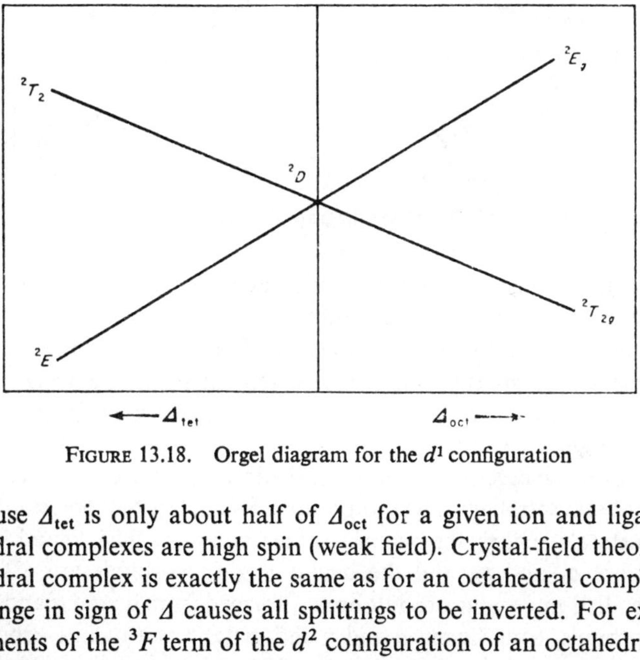

FIGURE 13.18. Orgel diagram for the d^1 configuration

Because Δ_{tet} is only about half of Δ_{oct} for a given ion and ligand, all tetrahedral complexes are high spin (weak field). Crystal-field theory for a tetrahedral complex is exactly the same as for an octahedral complex, but the change in sign of Δ causes all splittings to be inverted. For example, components of the ^3F term of the d^2 configuration of an octahedral complex are $^3T_{1g}$ (ground state), $^3T_{2g}$ (first excited state) and $^3A_{2g}$ (second excited state), but for a tetrahedral complex they are $^3A_2$ (ground), $^3T_2$ (first excited state) and $^3T_1$ (second excited state). This means that second-order interaction is larger for tetrahedral than for octahedral d^2 complexes, since the $^3T_1(^3F)$ and $^3T_1(^3P)$ levels are closer together in the former case.

The calculation of the effects of second-order interaction follows (13.17), the sign of terms involving Δ being changed to give energy levels in the strong-field limit at $4X/5 - \Delta/5$ and $X/5 + 4\Delta/5$. An Orgel diagram for the d^2 and related configurations is given in figure 13.19.

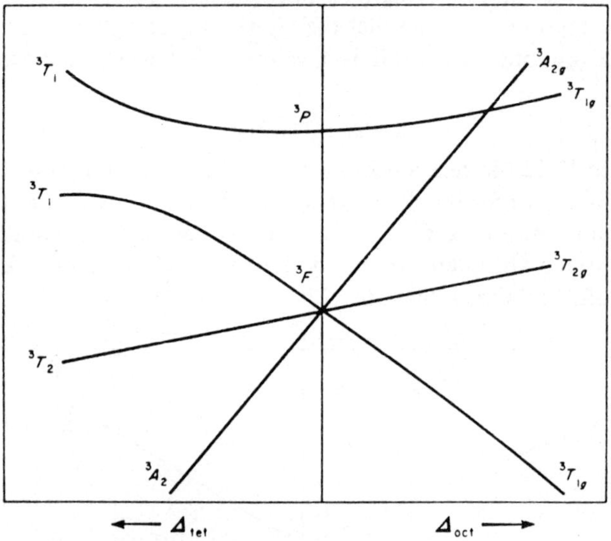

FIGURE 13.19. Orgel diagram for the d^2 configuration

As for an octahedral complex, the ^6S term of the d^5 configuration does not split in a tetrahedral ligand field.

Molecular-orbital theory is of considerable importance in the discussion of tetrahedral complexes, since in many of these the metal is in a high valence state (MnO_4^-, MnO_4^{2-}, CrO_4^{2-}, etc.) and considerable delocalization of ligand electrons onto the cation is anticipated. The technique described for the determination of octahedral ligand group orbitals can be used to obtain tetrahedral ligand group orbitals, but the inclusion of the π group orbitals makes the picture rather complex. The ligand σ group orbitals are a_1 and t_2, and the ligand π group orbitals are $e + t_1 + t_2$. Possible filled molecular orbitals are a_1, formed from ligand σ and metal s, two t_2 orbitals, from ligand σ and π and metal p and d, an e orbital from ligand π and metal d and the non-bonding t_1, which is entirely ligand π. Crystal-field theory suggests that the metal d orbitals of t_2 symmetry make a major contribution to the lowest antibonding t_2 molecular orbital (remembering that MO theory identified the e_g orbitals in an octahedral complex as antibonding). It also seems probable that the e molecular orbital is weakly antibonding. An energy level scheme which satisfies these conditions and which is generally accepted is shown in figure 13.20.

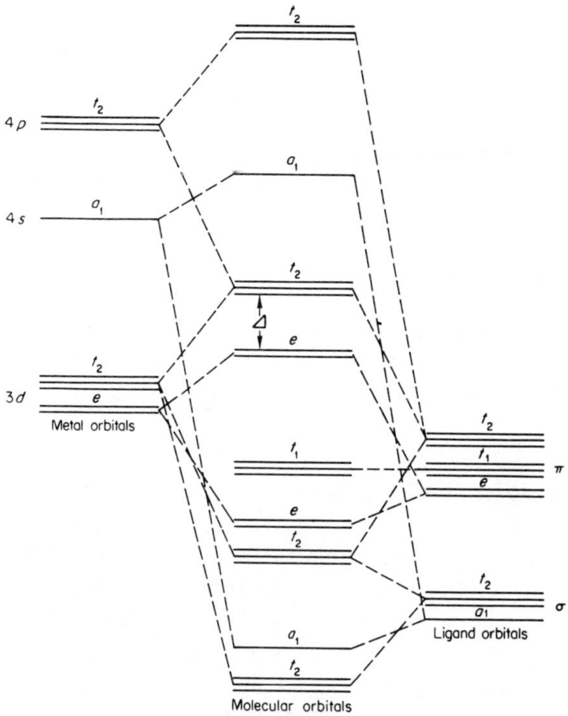

FIGURE 13.20. Molecular-orbital diagram for a tetrahedral complex including π bonding

Square planar complexes are best regarded as the limiting case of tetragonally distorted† octahedral complexes. If we take this distortion to be along the z axis then d_{xy} will no longer be degenerate with d_{yz} and d_{zx} although these remain degenerate with each other. The degeneracy between $d_{x^2-y^2}$ and d_{z^2} will also be relieved. Figure 13.21 shows the new splittings for a tetragonal extension and the correlation with the orbitals of a square planar complex.

There is no symmetry reason why δ' and δ'' should be equal. We therefore need two new parameters to discuss the splitting of d orbitals caused by a tetragonal ligand field. From a consideration of the interaction integrals of the tetragonal field it can be shown that $\delta' \simeq \delta''$.

† i.e. Distorted along a four-fold axis.

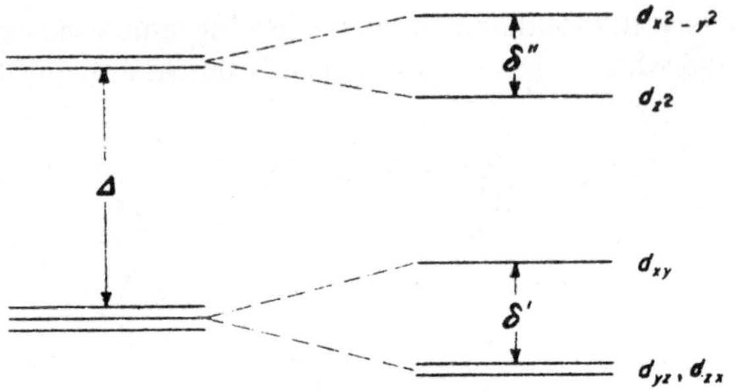

FIGURE 13.21. *d* orbital splitting for a tetragonal extension of an octahedron

The rule of 'average environment' is a useful aid in the interpretation of the spectra of complexes of lower than cubic symmetry. It is found that the value of Δ for MA_3B_3 is roughly the mean of the Δ's of MA_6 and MB_6; Δ for MA_2B_4 roughly $\frac{1}{3}\Delta(MA_6) + \frac{2}{3}\Delta(MB_6)$, and so on.

13.7 *The Jahn–Teller effect*

There is a general theorem due to Jahn and Teller which states that if a non-linear molecule is in an orbitally-degenerate state the system will distort so as to relieve that degeneracy. This is illustrated by the following example. Divalent copper is d^9, and in an octahedral complex the electron configuration would be $t_{2g}{}^6e_g{}^3$, which gives rise to a ^2E_g state. In this state the hole in the e_g orbitals is as likely to be in the d_{z^2} orbital as in the $d_{x^2-y^2}$. The Jahn–Teller theorem requires that the octahedron be distorted so that the degeneracy is removed. If this is done by a tetragonal extension along the z axis then the hole will be localized in $d_{x^2-y^2}$. If there is a tetragonal compression along the z axis the hole will be in d_{z^2}. The Jahn–Teller theorem does not tell us which of these will occur or if some other distortion will occur. Detailed calculations suggest that the tetragonal extension is the most likely, and this is supported by x-ray evidence that axial bonds in octahedral copper complexes are almost always longer than equatorial.

From group theory one can find out how the states appropriate to an octahedral group split up when the symmetry is lowered. This is illustrated for some important cases in table 13.12. What is not obvious at this point is why such splittings will lead to a more stable ground state.

Suppose we start with a regular octahedral complex in its equilibrium configuration. Any distortion of this configuration can be expressed in terms of symmetry distortions R which transform like *I.R.*'s of the group.

TABLE 13.12. Correlation between the *I.R.*'s of the octahedral, tetrahedral and lower symmetry groups

O_h	T_d	D_{4h}	D_{3d}	D_{2d}
A_{1g}	A_1	A_{1g}	A_{1g}	A_1
A_{2g}	A_2	B_{1g}	A_{2g}	B_1
E_g	E	$A_{1g} + B_{1g}$	E_g	$A_1 + B_1$
T_{1g}	T_1	$A_{2g} + E_g$	$A_{2g} + E_g$	$A_2 + E$
T_{2g}	T_2	$B_{2g} + E_g$	$A_{1g} + E_g$	$B_2 + E$
A_{1u}	A_2	A_{1u}	A_{1u}	B_1
A_{2u}	A_1	B_{1u}	A_{2u}	A_1
E_u	E	$A_{1u} + B_{1u}$	E_u	$A_1 + B_1$
T_{1u}	T_2	$A_{2u} + E_u$	$A_{2u} + E_u$	$B_2 + E$
T_{2u}	T_1	$B_{2u} + E_u$	$A_{1u} + E_u$	$A_2 + E$

these symmetry distortions being linear combinations of bond stretching and bending coordinates. For example, the A_{1g} symmetry coordinate is

$$R(A_{1g}) = \frac{1}{\sqrt{6}} \sum_{i=1}^{6} X_i,$$

where X_i is the change in length of the ith bond. If the Hamiltonian for the undistorted molecule is $\mathscr{H}$, then, using a Taylor expansion, that of the distorted molecule is

$$\mathscr{H}' = \mathscr{H} + \sum_k \left(\frac{\partial \mathscr{H}}{\partial R_k}\right) R_k + \text{terms in } R^2.$$

One now writes the wave functions of the distorted molecule as a linear combination of those for the undistorted molecule, and solves the secular determinant $|\mathscr{H}'_{rs} - ES_{rs}| = 0$. It is then found that there is at least one symmetry coordinate R_k which appears in the energy of the lowest state in linear form (or something that is equivalent to linear like $[R_{k_1}{}^2 + R_{k_2}{}^2]^{\frac{1}{2}}$, where R_{k_1} and R_{k_2} are components which together transform like a degenerate *I.R.* of the group). This means that superimposed on the normal quadratic form of the potential energy curve there will be a linear term in the displacement, so that the energy minimum no longer corresponds to the regular octahedral configuration.

It is only in the case of unequal occupation of the e_g orbitals that the Jahn–Teller effect is of recognized importance in the chemistry of transition metal complexes. Although unequal occupation of the t_{2g} orbitals must lead to a distortion, the non-bonding character of these orbitals suggests that any distortion of the complex will be small. In practice there are usually other mechanisms for relieving the degeneracy of these orbitals (spin-orbit coupling or crystal forces, for example).

If a Jahn–Teller distortion energy is of the same order of magnitude as a zero-point vibration energy, then there is no permanent distortion of the molecule. This situation is associated with a breakdown of the Born–Oppenheimer approximation and is referred to as the dynamic Jahn–Teller effect.

13.8 The electronic spectra of transition metal complexes

The visible and ultraviolet spectra of transition metal complexes show three types of band

(a) sharp, weak;
(b) broad, weak;
(c) broad, strong.

The first two of these are accounted for by crystal-field theory, the third requires ligand-field theory.

The weak bands are all associated with forbidden electronic transitions involving the excitation of an electron from one *d* orbital to another—a $g \rightarrow g$ transition (see page 125). The relative positions of these weak bands are readily correlated with energy level diagrams such as are shown in figures 13.6–13.9.

The strong broad bands which generally appear at the short wavelength end of the spectrum are due to transitions of an electron from one of the bonding molecular orbitals (largely located on the ligands) to either an e_g or t_{2g} orbital (largely located on the metal). These transitions are accompanied by a charge migration from the ligand to the metal and are called charge-transfer transitions (see chapter 18). They have not been greatly studied, although for some complexes a correlation has been found between peak position and the nature of the ligand.

The spectra of weak-field complexes are the easiest to interpret quantitatively. For the intermediate case, which occurs frequently in practice, second-order interactions are important and large secular determinants may have to be solved—for example, the 10×10 $^2T_2(d^5)$ determinant already mentioned†.

We shall discuss the spectra of two ions, $[Ti(H_2O)_6]^{3+}$ and $[V(H_2O)_6]^{3+}$ (d^1 and d^2 configurations respectively). The spectrum of $[Ti(H_2O)_6]^{3+}$ is shown in figure 13.22. It contains one weak peak at 20,300 cm^{-1}, ascribed to the $^2T_{2g} \rightarrow {}^2E_g$ transition, and this is the value of Δ in wavenumbers. In most octahedral complexes of Ti^{3+} this peak carries a shoulder, sometimes

† The interaction matrices are given by TANABE and SUGANO, *J. Phys. Soc. Japan*, **9**, 753 (1954) and by MCCLURE, *Solid State Phys.* **9**, 399 (1959), and one has only to substitute the numbers appropriate to the ion under consideration. The solution of these large secular determinants is a job for a computer.

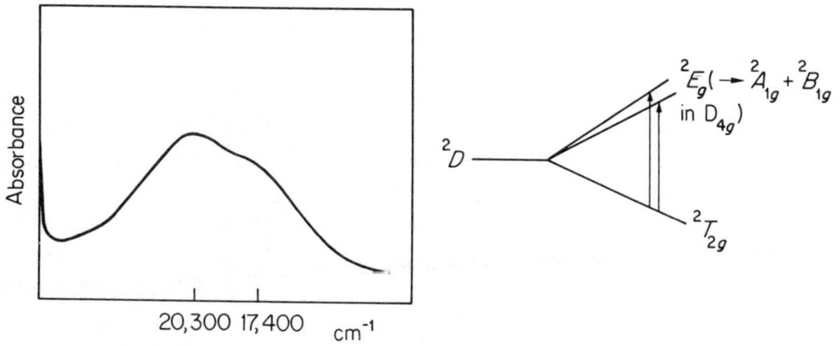

FIGURE 13.22. The absorption spectrum of $[Ti(H_2O)_6]^{3+}$ and its interpretation

almost resolvable as a separate peak. This is probably because the excited level shows a Jahn–Teller distortion since there is only one electron in the e_g* orbitals. $[V(H_2O)_6]^{3+}$ has the spectrum shown in figure 13.23. It consists of two weak peaks at 17,100 and 25,200 cm^{-1} assigned to ${}^3T_{1g} \rightarrow {}^3T_{2g}$ and ${}^3T_{1g} \rightarrow {}^3T_{1g}$ transitions.

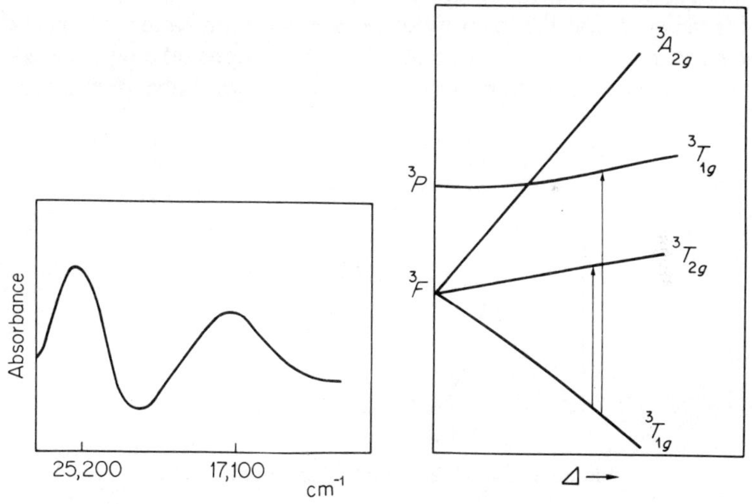

FIGURE 13.23. The absorption spectrum of $[V(H_2O)_6]^{3+}$ and its interpretation

Using the formulae derived earlier (page 229 and table 13.6), we have

$$E({}^3T_{1g} \rightarrow {}^3T_{1g}) = (X^2 + (\tfrac{6}{5})\Delta X + \Delta^2)^{\frac{1}{2}} = 25{,}200\ \text{cm}^{-1},$$

$$E({}^3T_{1g} \rightarrow {}^3T_{2g}) = -\tfrac{1}{2}X + (\tfrac{1}{2}\Delta) + \tfrac{1}{2}(X^2 + (\tfrac{6}{5})\Delta X + \Delta^2)^{\frac{1}{2}} = 17.100\ \text{cm}^{-1}$$

which have solutions $\Delta = 18{,}406\ \text{cm}^{-1}$ and $X = 9{,}406\ \text{cm}^{-1}$. This value of X is to be compared with that of the ^3F–^3P separation of the free ion which is $13{,}200\ \text{cm}^{-1}$. The agreement between these two figures is a measure of the validity of the crystal-field approximations. The predicted $^3T_{1g} \rightarrow {}^3A_g$ transition is at $34{,}595\ \text{cm}^{-1}$ where it is obscured by a strong charge-transfer band.

It is seen from figure 13.8 that for a d^3 configuration three of the excited states have about the same dependence on Δ as the ground state whereas others have a very different dependence on Δ. Since Δ is modulated by thermal vibrations of the ligands (a compressed complex will have a larger, and a stretched complex a smaller Δ), we expect that the transitions such as $^4A_{2g} \rightarrow {}^2E_g$, whose energies are almost independent of Δ, will give sharper bands than, say, $^4A_{2g} \rightarrow {}^4T_{2g}$, whose energy is very sensitive to Δ. This phenomenon is well illustrated by the crystal spectrum of ruby (Cr_2O_3, Al_2O_3), which is given in figure 13.24. All sharp bands are spin forbidden, because the ground and excited states have the same occupation of e_g and t_{2g} orbitals (this is why their energies have the same dependence on Δ) and only differ in their spin multiplicity. However, the converse is not true; not all spin-forbidden transitions give rise to sharp bands.

It is believed that the intensity of weak absorption bands is largely due to d–p mixing. If d–p mixing occurs $d \rightarrow d$ transitions take on some $d \rightarrow p$ and $p \rightarrow d$ character, both of which are allowed transitions for the free ion

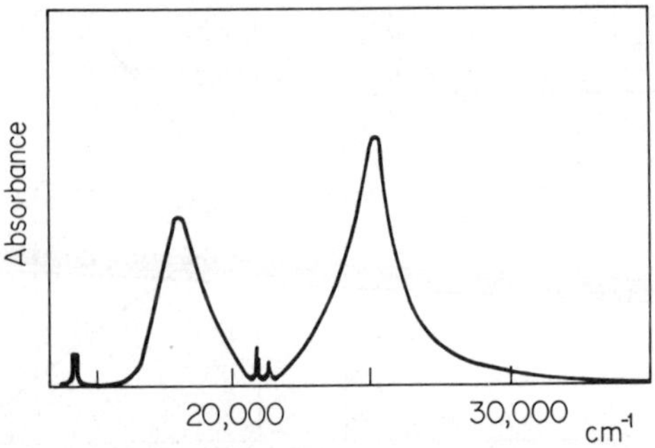

FIGURE 13.24. The absorption spectrum of ruby

and the complex. In tetrahedral complexes, for example, the p and three of the d orbitals transform as T_2 so that mixing is allowed. Tetrahedral complexes tend to have much more intense absorption than octahedral—the manganate ion is a good example of a tetrahedral complex with a strong absorption. Another useful comparison is $[CoCl_4]^{2-}$ (deep blue) and $[Co(H_2O)_6]^{2+}$ (pale pink). This still does not explain the spectra of strictly octahedral complexes, for which *d–p* mixing is not possible. However, our model is defective because it takes no account of the vibrations of the molecule. Some of these vibrations reduce the symmetry and in this reduced symmetry *d–p* mixing is allowed. The intensity of an absorption band should really be calculated by averaging over all configurations taken up by the molecule during the vibrations.

Transitions between states of different spin multiplicity are forbidden in the absence of spin-orbit coupling (page 124). In the presence of spin-orbit coupling S ceases to be a good quantum number, and transitions between states of different spin multiplicity give rise to weak absorption bands.

There are other physical properties which can shed much light on the nature of metal complexes. Firstly, magnetic measurements, both of bulk susceptibility and of electron-spin resonance, can give direct evidence on the symmetry of the ground state. Secondly, a study of the absorption of polarized light by a single crystal, and, for optically active complexes, a study of the circular dichroism or optical rotary dispersion spectra, will give added information about excited states. Although these techniques are extremely important, we feel they are rather outside the scope of this book.

13.9 Conclusions

It has been the object of this chapter to describe the techniques of ligand-field theory rather than to give a critical appraisal of it. Crystal-field theory has obvious deficiencies all based on the fact that it allows for no delocalization of the *d* electrons. On the other hand, a non-empirical MO theory is completely out of the question with the techniques at present available, and the qualitative results obtained from MO theory parallel those obtained from crystal-field theory.

Because of the inherent simplicity of crystal-field theory it is useful to ask how it should be changed to allow for electron delocalization. The answer is that three quantities need to be modified. The first are the electron repulsion integrals. These need to be reduced because covalency spreads out the electrons and therefore reduces their repulsion. In practice a reduction of up to 30% of the free-ion value may be required, as we saw in the calculation in the V^{3+} spectrum. Ligands can be arranged in a series, the Nephelauxetic (cloud-expanding) series, such that the electron repulsion parameters decrease from left to right, as covalency increases:

$$F^- > H_2O > NH_3 > SCN^- > Cl^- > CN^-.$$

Secondly, the spin-orbit coupling constant is reduced below the free-ion value. This is to be expected if an electron spends some of its time in ligand orbitals where coupling between its spin and the (metal) orbital angular momentum will be small. This reduction may also be of the order of 30%. Thirdly, in a detailed calculation of magnetic susceptibility the orbital contribution to the magnetic moment (9.10) must be reduced, for the same reason that the spin-orbit coupling is reduced. These reductions increase the empirical nature of crystal-field calculations, but do not detract from the great success which the theory has had in correlating a vast amount of spectral, magnetic, thermodynamic and kinetic data.

Bibliography

ORGEL, *An Introduction to Transition Metal Chemistry*, Methuen and Wiley, 1960.
BALLHAUSSEN, *Introduction to Ligand Field Theory*, McGraw-Hill, 1962.
GRIFFITH, *The Theory of Transition Metal Ions*, C.U.P., 1961.
MOFFITT and BALLHAUSEN, *Ann. Rev. Phys. Chem.*, **7**, 107 (1956).
LOW, *Paramagnetic Resonance in Solids*, Academic Press, 1960.

Problems 13

13.1 The complex ion $[FeF_6]^{3-}$ is colourless. How many unpaired electrons would you expect it to have?

13.2 Show that the *F* states arising from a d^n configuration split up in an octahedral field to give T_{1g}, T_{2g} and A_{2g} components.

13.3 Write down the eigenfunctions of the ground state for d^8 and d^9 octahedral complexes in terms of atomic state wave functions (cf. table 13.6) and in terms of real *d* orbitals.

13.4 Assign the peaks in the visible spectra of $[Mn(H_2O)_6]^{2+}$ given below.

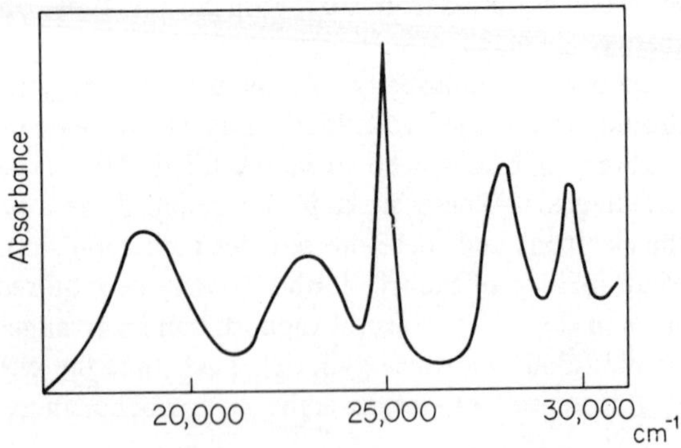

FIGURE 13.25. The absorption spectrum of $[Mn(H_2O)_6]^{2+}$

13.5 Prove, by considering the eigenfunctions, the assertion on page 229 that the ^3P level of a d^2 configuration is not split by a crystal field. The eigenfunctions of the ^3P components having $M_s = 1$ are as follows

$$M_L = 1 \quad \sqrt{\frac{2}{5}}\,\big|(2,2)(2,-1)\big| - \sqrt{\frac{3}{5}}\,\big|(2,1)(2,0)\big|$$

$$M_L = 0 \quad \sqrt{\frac{4}{5}}\,\big|(2,2)(2,-2)\big| - \sqrt{\frac{1}{5}}\,\big|(2,1)(2,-1)\big|$$

$$M_L = -1 \quad \sqrt{\frac{2}{5}}\,\big|(2,1)(2,-2)\big| - \sqrt{\frac{3}{5}}\,\big|(2,0)(2,-1)\big|$$

13.6 Derive the T_{2g} eigenfunctions formed from the ligand π orbitals in an octahedral complex.

Chapter 14

The electronic structure of electron-deficient molecules

14.1 The structure of diborane

A chemist usually implies several things when he refers to a molecule as 'covalent'. For example, the volatility of the compound is roughly related to its molecular weight; the atoms within the molecule are held in a more or less fixed orientation one to another by highly directional valence forces; each pair of bonded atoms has at least two electrons to form a bond between them and may have more. There are some molecules which by these criteria are covalent, except that they do not appear to comply with the requirement of a minimum of two electrons between each pair of bonded atoms. The best known example is diborane, B_2H_6, which requires at least seven bonds, and therefore fourteen electrons, to hold it together. In fact, only twelve valence electrons are available. Compounds of this type are called *electron deficient*, this term being used when there are less than $2(n-1)$ valence electrons, n being the number of atoms†.

All of the known boron hydrides are electron deficient as too are most of the anions derived from them. Although we shall spend the larger part of this chapter discussing electron-deficient molecules containing boron, most of the elements in group 3, and some in group 2, of the periodic table also form electron-deficient compounds. A less restrictive definition of electron deficiency is that there are less than $2m$ valence electrons where m is the number of formal bonds. Into this category fall all compounds in which an unsaturated organic molecule is bonded to a transition metal (e.g. ferrocene). These will be discussed in the last chapter of this book.

Experiment shows that the diborane molecule has four (terminal) hydrogen and two boron atoms in a plane, and two (bridge) hydrogen

† n atoms need at least $n-1$ bonds to hold them together and this requires at least $2(n-1)$ electrons in the usual electron-pairing scheme.

atoms, one above and one beneath this plane as shown in figure 14.1†. The molecule belongs to the $\mathbf{D}_{2h}$ symmetry group whose character table is given in table 14.1. This structure apparently requires eight two-electron bonds, that is, a total of sixteen electrons. There have been several different

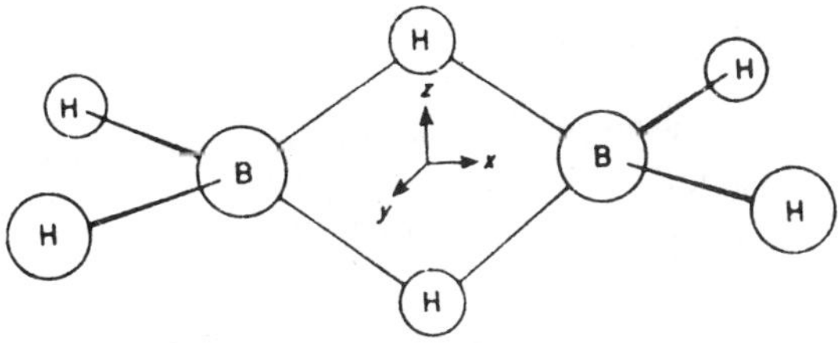

FIGURE 14.1. The coordinate axes for B_2H_6

theories to explain the bonding in this molecule‡. We shall just give the MO picture which is generally agreed to be the most satisfactory.

TABLE 14.1. The character table for the $\mathbf{D}_{2h}$ group

D_{2h}	I	$C_2(z)$	$C_2(y)$	$C_2(x)$	i	$\sigma(xy)$	$\sigma(zx)$	$\sigma(yz)$
A_g	1	1	1	1	1	1	1	1
B_{1g}	1	1	−1	−1	1	1	−1	−1
B_{2g}	1	−1	1	−1	1	−1	1	−1
B_{3g}	1	−1	−1	1	1	−1	−1	1
A_u	1	1	1	1	−1	−1	−1	−1
B_{1u}	1	1	−1	−1	−1	−1	1	1
B_{2u}	1	−1	1	−1	−1	1	−1	1
B_{3u}	1	−1	−1	1	−1	1	1	−1

In the MO calculation we take linear combinations of the six hydrogen $1s$ orbitals and the $2s$ and $2p$ orbitals of the boron atoms. The calculation is simplified by first combining these atomic orbitals to give group orbitals, which transform as *I.R.*'s of the group. We need then only take linear combinations of group orbitals belonging to the same *I.R.* An equivalent procedure, but one for which it is a little easier to see how the molecular orbitals bind the bridge atoms together, involves hybridization of the boron orbitals as a first step. We shall go over this in detail.

The terminal H—B—H angle is close to 120° which suggests that these bonds are formed by overlap of hydrogen $1s$ orbitals and the sp^2 hybrids of

† For a review of the evidence see LONGUET-HIGGINS, *Quart. Rev. (London)*, **11**, 121 (1957).

‡ See LONGUET-HIGGINS, *loc. cit.*

the boron atom. If these two hybrids are sp^2 there remains $(1/3)s$ and $(5/3)p$ character to be divided between the other two hybrids which point towards the bridge hydrogens. Since these two orbitals must each have an equal amount of s and p character they must have wave functions of the form

$$\sigma_1 = \sqrt{\tfrac{1}{6}}s + \sqrt{\tfrac{5}{6}}(ap_x + bp_z)$$

$$\sigma_2 = \sqrt{\tfrac{1}{6}}s + \sqrt{\tfrac{5}{6}}(ap_x - bp_z).$$

By symmetry, p_y cannot contribute to the wave functions. The coefficients a and b are determined by the orthogonality condition

$$\int \sigma_1\sigma_2\,dv = \tfrac{1}{6} + \tfrac{5}{6}(a^2 - b^2) = 0,$$

$$a^2 - b^2 = -\tfrac{1}{5},$$

and the normalization requirement

$$a^2 + b^2 = 1.$$

It follows that

$$a = \sqrt{\tfrac{2}{5}}, \qquad b = \sqrt{\tfrac{3}{5}}.$$

The orbital $(\sqrt{2}p_x + \sqrt{3}p_z)/\sqrt{5}$ makes an angle of $\tan\vartheta = \sqrt{3/2} = 50°46'$ with the x axis. The hybrids σ_1 and σ_2 therefore make an angle of 101°32′ with each other, which fits in very well with the bridge H—B—H angle of 96°36′.

Suppose we now label† the orbitals involved in the bridge bonds as in figure 14.2.

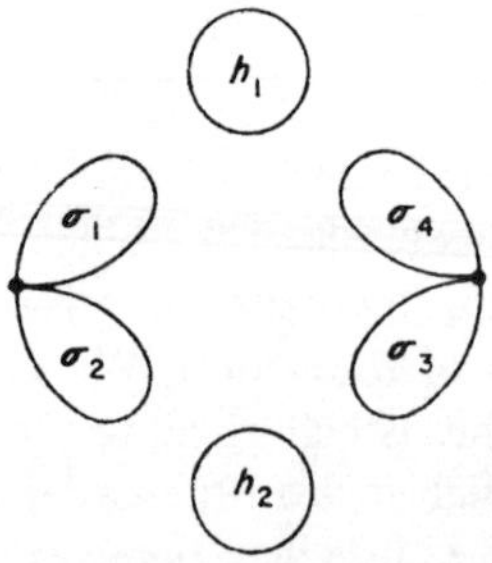

FIGURE 14.2. Orbitals forming the bridge bonds in B_2H_6

We first form these into group orbitals following the method given in chapter 8. These are shown below.

† It should be noted that lower case group theory symbols are used to label orbitals and upper case for states (see page 216).

Symmetry species	
$\sqrt{\frac{1}{2}}(h_1 + h_2)$	A_g
$\sqrt{\frac{1}{2}}(h_1 - h_2)$	B_{1u}
$\frac{1}{2}(\sigma_1 + \sigma_2 + \sigma_3 + \sigma_4)$	A_g
$\frac{1}{2}(\sigma_1 + \sigma_2 - \sigma_3 - \sigma_4)$	B_{3u}
$\frac{1}{2}(\sigma_1 - \sigma_2 + \sigma_3 - \sigma_4)$	B_{2g}
$\frac{1}{2}(\sigma_1 - \sigma_2 - \sigma_3 + \sigma_4)$	B_{1u}

The two a_g orbitals will now give rise to a bonding and an antibonding molecular orbital, and the same for the b_{1u}. The b_{3u} and b_{2g} orbitals are approximately non-bonding, although, since they are antisymmetric to reflection in the *yz* plane, they have a small B—B antibonding character due to the overlap of the hybrids σ_1 and σ_4, σ_2 and σ_3. If these two orbitals are expanded back into *s* and *p* atomic orbitals then b_{2g} is found to be entirely made up of *p* orbitals and the b_{3u} of sp^2 hybrids. This means that of the two non-bonding orbitals b_{3u} will have the lower energy. Of the two bonding orbitals the a_g is H—H bonding but the b_{1u} is H—H antibonding. It follows that the a_g orbital will have the lower energy. The same situation will hold for the antibonding orbitals. The results are summarized in figure 14.3. There are only four electrons available to form the bridge bonds, so these occupy the two bonding molecular orbitals a_g and b_{1u}.

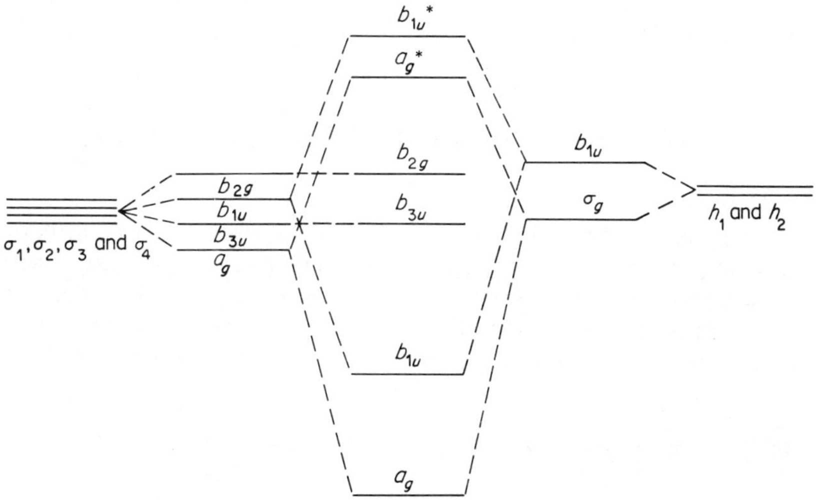

FIGURE 14.3. The molecular-orbital energy scheme for the bridge bonds in B_2H_6

The a_g and b_{1u} bonding molecular orbitals have the form

$$\psi(a_g) = c_1(h_1 + h_2) + c_2(\sigma_1 + \sigma_2 + \sigma_3 + \sigma_4)$$

$$\psi(b_{1u}) = c_3(h_1 - h_2) + c_4(\sigma_1 - \sigma_2 - \sigma_3 + \sigma_4)$$

and are shown in figure 14.4.

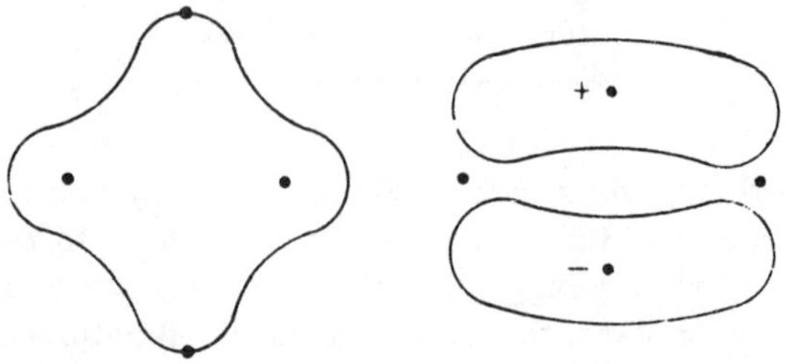

FIGURE 14.4. The bonding molecular orbitals for the bridge bonds in B_2H_6

These orbitals transform as *I.R.*'s of the group. Equivalent orbitals (page 186) are obtained if we take linear combinations of $\psi(a_g)$ and $\psi(b_{1u})$. These equivalent orbitals are localized on either side of the bridge and may be called three-centred bonding orbitals (figure 14.5).

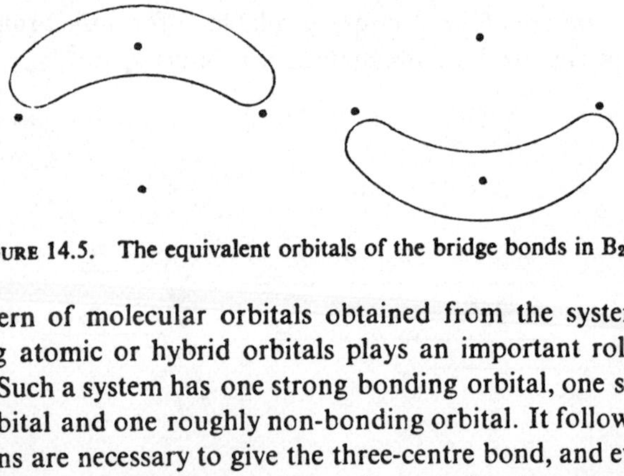

FIGURE 14.5. The equivalent orbitals of the bridge bonds in B_2H_6

The pattern of molecular orbitals obtained from the system of three overlapping atomic or hybrid orbitals plays an important role in boron chemistry. Such a system has one strong bonding orbital, one strong anti-bonding orbital and one roughly non-bonding orbital. It follows that only two electrons are necessary to give the three-centre bond, and even if there were additional electrons they would not increase the strength of the bond.

14.2 *Three-centre orbitals between boron atoms*

We have already met one sort of three-centre orbital, involving two boron and one hydrogen atom. There are two other types of importance in boron chemistry, both involving three boron atoms. One of the orbitals is nodeless, the other contains one planar node. Both originate from a situation in which atoms A and C have hybrid orbitals roughly pointing towards atom B, and atom B has either a hybrid orbital σ *or* a p orbital, as shown in

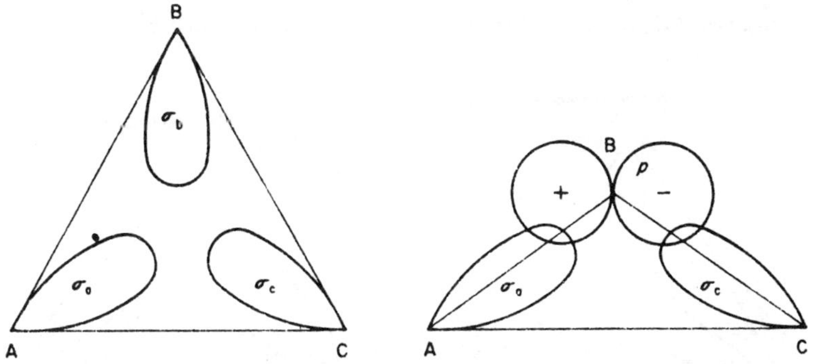

FIGURE 14.6. Three-centre interactions in boron compounds

figure 14.6. If atoms A and C are equivalent then in both cases we can form symmetry combinations of σ_a and σ_c as follows

$$\sigma_1 = \sigma_a + \sigma_c$$
$$\sigma_2 = \sigma_a - \sigma_c,$$

σ_1 can, by symmetry, interact with σ_b but not with p_b, σ_2 can interact with p_b but not with σ_b.

Case (*a*) We get molecular orbitals

$$\psi_3 = c_3(\sigma_a + \sigma_c) - c_4(\sigma_b)$$
$$\psi_2 = \sigma_a - \sigma_c$$
$$\psi_1 = c_1(\sigma_a + \sigma_c) + c_2(\sigma_b),$$

where the coefficients $c_1 \ldots c_4$ are all positive. ψ_1 is A—C, A—B and B—C bonding since it has no nodes; ψ_2 is A—C antibonding; ψ_3 is A—C bonding but A—B and B—C antibonding. Clearly, only ψ_1 is a strongly bonding orbital. If all three atoms are equivalent ψ_2 and ψ_3 are degenerate and equally antibonding.

Case (*b*) We get molecular orbitals

$$\psi_3 = c_3(\sigma_a - \sigma_c) - c_4 p_b$$
$$\psi_2 = \sigma_a + \sigma_c$$
$$\psi_1 = c_1(\sigma_a - \sigma_c) + c_2 p_b.$$

ψ_1 is A—C antibonding but A—B and B—C bonding; ψ_2 is A—C bonding; ψ_3 is A—B, B—C and A—C antibonding. In this case, however, the overlap between orbitals σ_a and σ_c is small and so the A—B and B—C bonding and antibonding character of the orbitals dominates the situation. Again there is only one strong bonding orbital ψ_1. The two three-centre bonding orbitals for cases (a) and (b) are shown in figure 14.7.

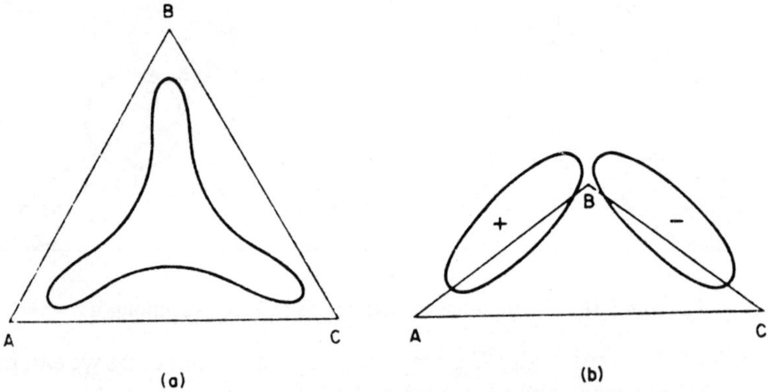

FIGURE 14.7. Three-centre bonding orbitals in boron compounds

It is conventional to represent these two bonding situations by

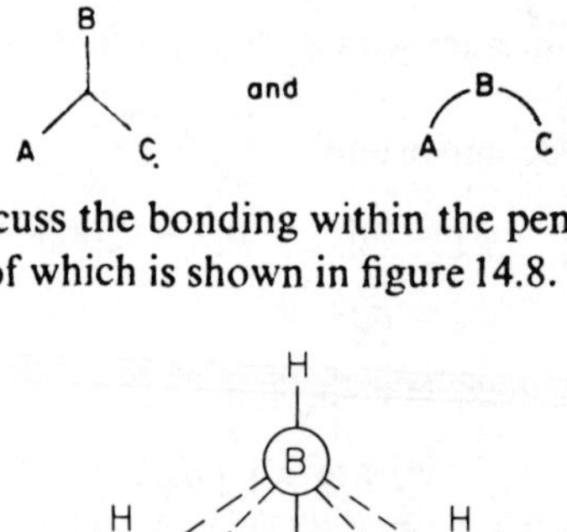

We now briefly discuss the bonding within the pentaborane-11 molecule B_5H_{11}, the structure of which is shown in figure 14.8. The molecule consists

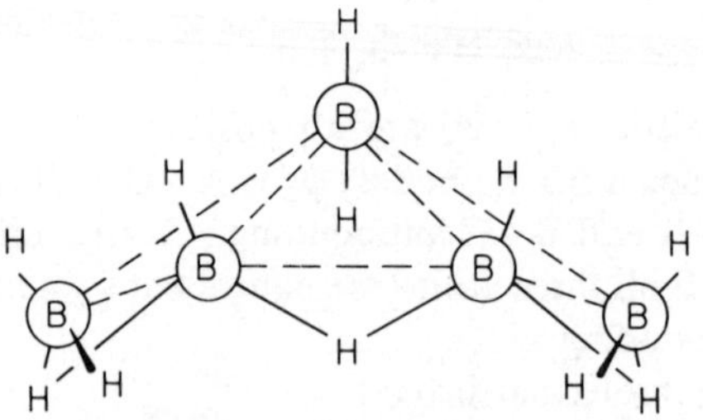

FIGURE 14.8. The structure of B_5H_{11}

of a plane of four boron atoms, two having two terminal hydrogen atoms and two having one. The apical boron carries two hydrogen atoms which we shall take to be terminal (one hydrogen may be involved to some extent with bonding in the basal plane). In this molecule, eighteen two-electron bonds appear to be necessary but only twenty-six valence electrons are available.

We have chosen this particular hydride because it can be described using either of the three-centre orbitals shown in figure 14.7 depending on the hybridization which is used for the apical boron atom. The H—B—H angle at this atom is 116°, with an error of about ±9° and this is compatible with sp^2 or sp^3 hybridization.

The two descriptions are shown diagrammatically in figure 14.9, in which the molecule is viewed from above.

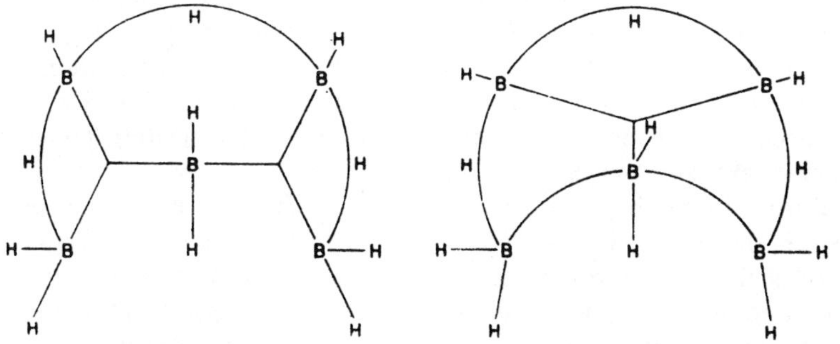

(a) (b)

FIGURE 14.9. The bonding schemes for B_5H_{11}
(a) assuming sp^3 hybridization of the apical boron atom.
(b) assuming sp^2 hybridization of the apical boron atom.

In either model there are thirteen bonding molecular orbitals (eight terminal B—H, three bridging $B\overset{H}{\frown}B$, and two three-centred bonds involving boron atoms only). These are filled by the twenty-six available electrons. There is no evidence at the moment to indicate which is the better of the two representations of the molecule.

14.3 Boron polyhedra

The borides formed by many electropositive elements contain some interesting examples of electron-deficient structures. CaB_6, for example, contains six boron atoms in the form of an octahedron, each corner of the octahedron being bonded to an adjacent octahedron so that the boron atoms form a giant anion which extends throughout the crystal. The calcium ions are packed between the boron octahedra to give a sodium chloride-type lattice.

A discussion of the molecular orbitals of the B_6 octahedra follows closely our discussion of the ligand molecular orbitals of an octahedral complex (chapter 13). Two main alterations have to be made. Firstly, we must include the 2*s* orbital on each boron; these transform as the ligand σ lone-pair orbitals in an octahedral complex (i.e. $A_{1g} + E_g + T_{1u}$). Secondly, we have to consider overlap between the boron atoms. The boron 2*s* and $2p\sigma$ orbitals mix to give one set of $A_{1g} + E_g + T_{1u}$ orbitals, directed outwards from the B_6 octahedron and responsible for the bonds between octahedra, and another set pointing inwards towards the centre of the octahedron. Including the twelve boron $2p\pi$ orbitals, we have A_{1g}, E_g, $2T_{1u}$, T_{1g}, T_{2g} and T_{2u} orbitals available for bonding within each octahedron. MO calculations† show that, of these orbitals, only the A_{1g}, T_{2g} and T_{1u} are bonding, a total of 7. The bonds between octahedra are normal two-electron bonds, so the structure requires a total of $7 \times 2 + (6 \times 2)/2 = 20$ electrons to completely fill all bonding molecular orbitals. This is the number available in a B_6^{2-} anion.

Boron forms B_{12} icosahedra in boron carbide, B_4C, and in the $(B_{12}H_{12})^{2-}$ anion. This structure is important because the structures of most of the other boron hydrides and hydride anions are related to it. If boron atoms are hypothetically removed from the $(B_{12}H_{12})^{2-}$ anion and the edges 'sewn up' with either terminal or bridging hydrogen atoms, the structures of most boron hydrides can be systematically generated.

An icosahedron is shown in figure 14.10. There are 120 symmetry operations which will convert this figure into itself. This is the order of the group. These operations fall into ten classes; so it follows that the *I.R.*'s in the character table must have degeneracies greater than three-fold (since even if they were all three-fold degenerate $10 \times 3^2 = 90$ which is less than the order of the group (page 97)). In fact there are two singly-degenerate representations (A_g and A_u), four triply-degenerate (T_{1g}, T_{1u}, T_{2g} and T_{2u}), two four-fold degenerate (U_g and U_u), and two five-fold degenerate representations (V_g and V_u). We shall not attempt to derive the molecular orbitals appropriate to this figure, for this is a rather lengthy process, but shall merely quote the results‡. As before, we find that 2*s* and $2p\sigma$ orbitals on the boron transform together so that we may take a linear combination of them, one mixed orbital pointing towards the centre of the icosahedron, the other directed outwards. In $(B_{12}H_{12})^{2-}$ the latter orbitals are used in forming B—H electron-pair bonds; in $(B_4C)_n$, which is best represented by $(C_3^{2+}—B_{12}^{2-})_n$, six orbitals are involved in bonding with other icosahedra and six bond with the C_3 groups which the structure also contains.

† LONGUET-HIGGINS and ROBERTS, *Proc. Roy. Soc.*, **224A**, 336 (1954).
‡ LONGUET-HIGGINS and ROBERTS, *Proc. Roy. Soc.*, **230A**, 110 (1955).

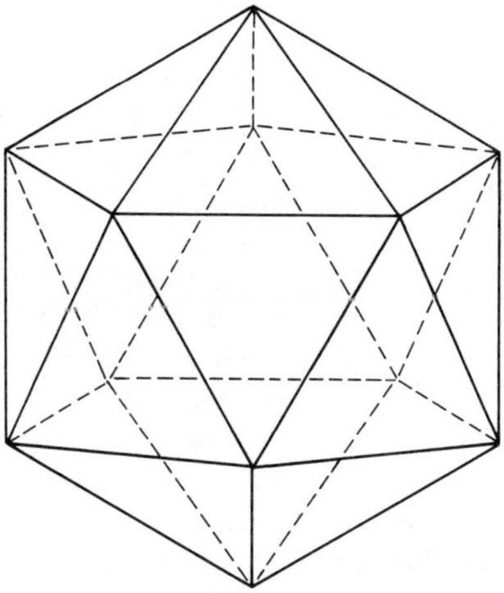

FIGURE 14.10. The icosahedron

Within the icosahedron, there are thirteen bonding molecular orbitals, A_g, T_{1u}, V_g and U_u. These require twenty-six electrons to fill them, so that each icosahedron needs thirty-eight electrons to give a closed-shell structure (twelve being needed for the external bonds). This condition is fulfilled by $(B_{12}H_{12})^{2-}$ and $(B_4C)_n$ where there are $3 \times 12 + 2 = 38$ electrons available.

Despite the presence of three-fold symmetry axes in the icosahedron, its structure cannot be described in terms of three-centre bonds in a straightforward way. This is because each boron atom has five nearest neighbours and one cannot obtain five independent hybrids from its $2s$ and $2p$ atomic orbitals. Thus, although a three-centred bond description leads to an energy level pattern for the thirteen bonding molecular orbitals similar to that obtained by a more detailed treatment, seven antibonding combinations of three-centred orbitals result also.

Closely related to the boron hydrides are the so-called *carboranes*. These are boron hydrides where a B—H (terminal) unit has been replaced by a carbon atom. So, corresponding to $B_{12}H_{12}^{2-}$ there are three corresponding carboranes $B_{10}C_2H_{10}^{2-}$, differing in the relative positions of the two carbon atoms in the icosahedron.

The electronic structures of the carboranes resemble those of the corresponding boron hydrides. In particular, a three-centred bond approach may be employed. Differences of stability and reactivity exist, however.

14.4 *Other electron-deficient structures*

The occurrence of simple electron-deficient compounds is largely limited to alkyl and hydride derivatives of the second and third group elements, although a few halogen compounds such as B_4Cl_4 (a tetrahedron of boron atoms with a chlorine bonded radially outwards to each) and B_8Cl_8, are known.

Aluminium trimethyl is a dimer with methyl bridges; its structure and bonding resemble that of diborane. However, BMe_3 is a monomer and there appears to be no convincing explanation for this, particularly since $BeMe_2$ (like BeH_2) is polymeric. In these latter molecules each beryllium atom is tetrahedrally surrounded by bridging groups.

A molecule which has been regarded as electron deficient is the so-called 'platinum tetramethyl', believed to be a tetramer with bridging methyl groups on the basis of incomplete X-ray evidence. It has recently been shown that the compound is, in fact, tetrameric trimethyl platinum hydroxide, with bridging hydroxyl groups, a molecule which is not electron deficient.

14.5 *A critique of electron deficiency*

In all of the molecules which we have discussed, electron deficiency arose only because we insisted that there be a two-electron bond between bonded atoms. A more sophisticated discussion revealed in every case that there was no electron deficiency in the sense of empty bonding orbitals. *In these molecules there are just enough electrons to fill all the bonding orbitals.* Any molecule which had empty bonding orbitals would be a powerful oxidizing agent—and this is by no means one of the properties of the molecules that we have considered.

It is always possible to relieve electron deficiency, hypothetically at least, by invoking ionic structures. For example, diborane is not electron deficient in the 'protonated double bond' model. If this structure were at

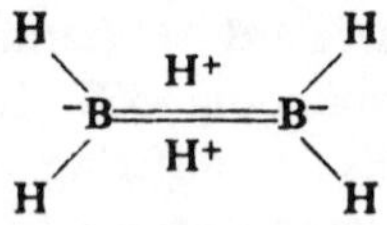

all close to the true picture, diborane should be a strong dibasic acid. In fact, it has no acidic properties. The converse is also true; ionic crystals are not classed as electron deficient because we choose to ignore any covalent contribution to the bonding.

In summary, a necessary condition for a molecule to be classed as electron deficient is that its structure can only be understood in terms of delocalized molecular orbitals. It is usually possible to give an adequate description of such molecules using only three-centre orbitals, although the delocalization of the molecular orbitals may be greater than this.

Problems 14

14.1 The molecule B_3H_9 has not been prepared. Is it likely to be stable?

14.2 Investigate the relative energies of the orbitals discussed in case (a) page 261 with variation of the angle ABC.

Chapter 15

π-*Electron theory of organic molecules*

15.1 Introduction

Organic chemists divide hydrocarbons into three types: saturated, unsaturated and aromatic. Saturated hydrocarbons are relatively inert substances, while unsaturated and aromatic hydrocarbons undergo a wide variety of reactions. It is the presence of electrons in π orbitals in the latter two classes which is responsible for their greater reactivity.

In chapter 10, orbitals of diatomic molecules were classified by their angular momentum about the internuclear axis: σ with zero and π with one unit of angular momentum. In polyatomic molecules the same terms are used when they do not refer to angular momentum but to the spatial characteristics of the orbitals. A π orbital has one nodal plane containing the bond axis while a σ orbital has no nodal plane. A π orbital of a molecule like ethylene therefore looks like a π orbital of oxygen taken in real form. Saturated hydrocarbons can be considered to be built up only from σ orbitals whereas unsaturated and aromatic hydrocarbons (e.g. ethylene and benzene) contain both types of orbital. Both benzene and ethylene are planar molecules, their σ orbitals are symmetric and their π orbitals antisymmetric to reflection in the molecular plane (cf. figures 5.15, 5.16).

In terms of VB theory the π bonds in an aromatic compound cannot be described by a single valence structure. Benzene will be some combination of the two Kekulé structures

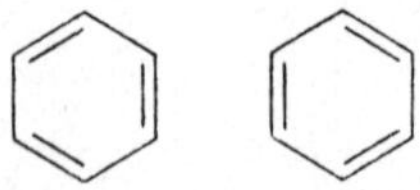

1 2

together with smaller contributions from the higher-energy 'Dewar' structures†. These five 'structures' represent the possible independent

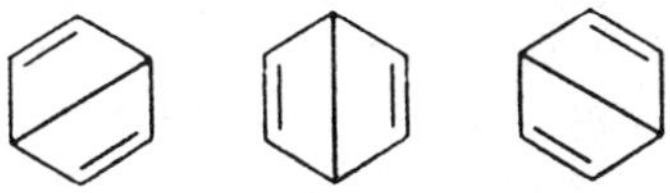

3 4 5

perfect-pairing schemes for benzene (page 204). The eigenfunction for the system may be represented as

$$\Psi = c_1\Psi_1 + c_2\Psi_2 + c_3\Psi_3 + c_4\Psi_4 + c_5\Psi_5.$$

The coefficients c_1 and c_2 are equal on the one hand and c_3, c_4 and c_5 are equal on the other.

If just a single Kekulé structure represented the ground state of benzene, the π-bond energy would be about the same as that of three ethylenes. However, the wave function which is a mixture of all five structures gives a lower energy than this, the extra stabilization being called the 'resonance energy'. For benzene the two Kekulé structures contribute about 80% of the resonance energy that can be calculated using all five structures. For a more complete calculation ionic structures of the type 6 and 7 must also be considered, although the contribution these species make to the ground state is extremely small.

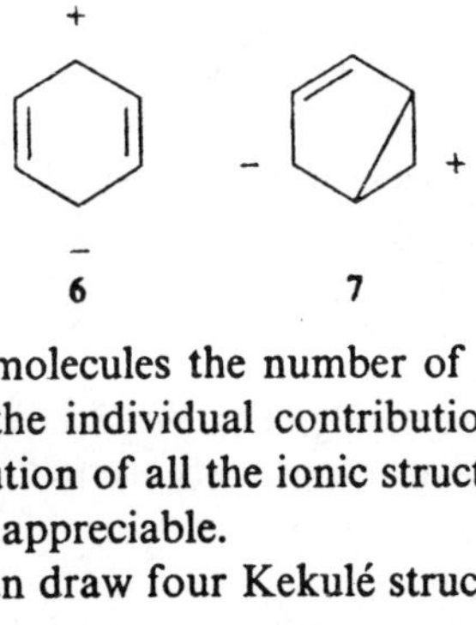

6 7

For larger aromatic molecules the number of ionic structures is much greater and, although the individual contribution of each remains very small, the total contribution of all the ionic structures to the ground state wave function becomes appreciable.

For anthracene we can draw four Kekulé structures.

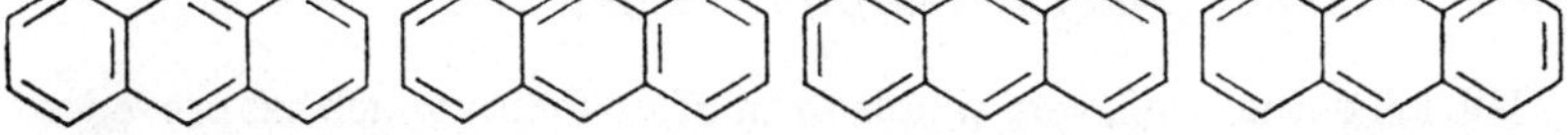

† We emphasize that these structures (which represent VB electron-pairing schemes) are all based on a regular planar hexagon for the carbon skeleton. A puckered compound bicyclo[2.2.0]hexadiene has been synthesized and called 'Dewar Benzene'. This is not the same as the planar 'Dewar' structure represented above.

Although these are not equivalent to each other, they have the same energy within the perfect-pairing approximation (12.31). It follows that the ground state must have an equal weighting of all four structures (provided interactions with higher energy structures are ignored.) By taking the average of the four Kekulé structures of anthracene we get the fractional double-bond character of each bond. The appreciable double-bond character of the 1,2 bond is in accord with chemical evidence and with the observed bond lengths (cf. ethylene 1·34 Å, benzene 1·40 Å and ethane 1·54 Å).†

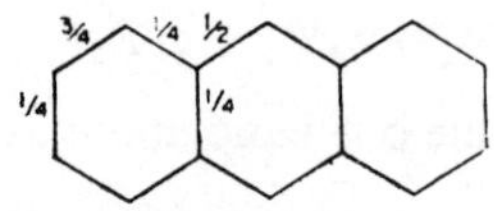

Double-bond character

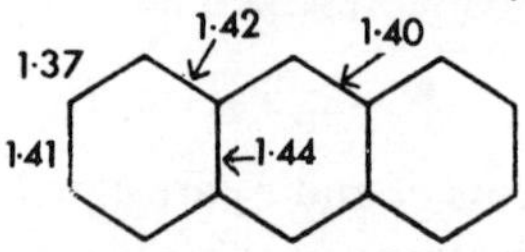

Observed bond lengths in Å

Valence-bond calculations become extremely tedious for large molecules because of the large number of structures that have to be taken into account (see section 12.5). For this reason most workers today use the MO method on this type of compound.

Molecular-orbital theory is particularly suitable for conjugated olefins and aromatic compounds, since delocalized orbitals are the principal feature of the model. Each unsaturated carbon atom in a molecule of this type has one $2p$ atomic orbital with π symmetry (ϕ_ν). Using the LCAO approximation the molecular orbitals will then be of the form (10.27)

$$\psi = \sum_\nu c_\nu \phi_\nu. \tag{15.1}$$

The coefficients and energies of these orbitals are obtained by solving the secular equations (10.30)

$$\sum_\nu c_\nu (\mathrm{H}_{\mu\nu} - ES_{\mu\nu}) = 0. \tag{15.2}$$

We have already discussed the significance of the one-electron Hamiltonian which comes into this expression (sections 10.7 and 12.4), and we shall return to this point in section 15.8.

15.2 Hückel molecular-orbital theory

Hückel developed a very simple form of MO theory which has proved to be extremely valuable for correlating the properties of unsaturated organic molecules. The secular equations are simplified by making the following approximations:

† CRUICKSHANK, *Acta Cryst.* **9**, 915 (1956).

1. Zero overlap is assumed even between neighbouring atomic orbitals i.e. $S_{\mu\nu} = 0$ if $\mu \neq \nu$. We can always take normalized atomic orbitals such that $S_{\mu\mu} = 1$.
2. $\mathbf{H}_{\mu\mu}$ is assumed to be the same for each atom, it is given the symbol α and is called the *coulomb integral.*
3. $\mathbf{H}_{\mu\nu}$ is a constant called the *resonance integral* if atom μ is bonded directly to atom ν which is represented by $(\mu \to \nu)$; it is given the symbol β.
4. $\mathbf{H}_{\mu\nu} = 0$ if atom μ is not bonded directly to atom ν.

With these approximations equations (15.2) become

$$c_\mu(\alpha - E) + \sum_{\nu \to \mu} c_\nu \beta = 0, \tag{15.3}$$

or by writing

$$x = \frac{\alpha - E}{\beta} \tag{15.4}$$

we have

$$x c_\mu + \sum_{\nu \to \mu} c_\nu = 0. \tag{15.5}$$

The simplest molecule we can deal with is ethylene. Labelling the carbon atoms 1 and 2, the secular equations are (cf. problem 10.6)

$$\mu = 1, \quad x c_1 + c_2 = 0,$$
$$\mu = 2, \quad c_1 + x c_2 = 0.$$

These two equations only have a non-trivial solution if the determinant of the multipliers of c is zero (page 78), i.e.

$$\begin{vmatrix} x & 1 \\ 1 & x \end{vmatrix} = x^2 - 1 = 0.$$

This has solutions

$$x = \pm 1,$$

and hence $E = \alpha + \beta$ or $E = \alpha - \beta$.

Substituting $x = -1$ back into the secular equations we have

$$c_1 = c_2.$$

Applying the normalization condition we have

$$c_1^{\,2} + c_2^{\,2} = 1.$$

There are no terms in $c_1 c_2$ because of the zero-overlap approximation, hence we arrive at the solution

$$\psi_1 = \frac{1}{\sqrt{2}}(\phi_1 + \phi_2), \qquad E_1 = \alpha + \beta.$$

Similarly, by substituting $x = 1$ back into the secular equations we have the second solution

$$\psi_2 = \frac{1}{\sqrt{2}}(\phi_1 - \phi_2), \qquad E_2 = \alpha - \beta.$$

The resonance integral β is a negative quantity, being roughly the energy of an electron density $\phi_\mu\phi_\nu$ in the field of the nuclei shielded by the σ electrons. ψ_1 has energy less than an isolated carbon atom (α in the Hückel approximation) and is therefore bonding; and ψ_2 has energy greater than an isolated carbon atom and is therefore antibonding. The shapes of the orbitals are given in figure 10.3.

A much more interesting example is butadiene.

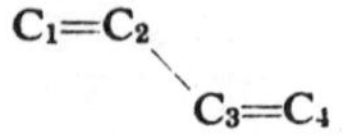

The secular equations are†

$$\begin{aligned}
(\mu = 1) \quad & xc_1 + c_2 && = 0 \\
(\mu = 2) \quad & c_1 + xc_2 + c_3 && = 0 \\
(\mu = 3) \quad & c_2 + xc_3 + c_4 && = 0 \\
(\mu = 4) \quad & c_3 + xc_4 && = 0,
\end{aligned} \tag{15.6}$$

and the secular determinant is

$$\begin{vmatrix} x & 1 & 0 & 0 \\ 1 & x & 1 & 0 \\ 0 & 1 & x & 1 \\ 0 & 0 & 1 & x \end{vmatrix} = 0. \tag{15.7}$$

Expanding this gives

$$x^4 - 3x^2 + 1 = 0,$$

which has solutions

$$x = \pm 1{\cdot}62, \quad x = \pm 0{\cdot}62.$$

If we take one of these solutions ($x = -1{\cdot}62$) and substitute it back into the secular equations (15.6) we have

from $\mu = 1$ $\quad c_2 = 1{\cdot}62\, c_1$;

from $\mu = 2$ $\quad c_3 = -c_1 + 1{\cdot}62 c_2 = 1{\cdot}62 c_1$;

from $\mu = 4$ $\quad c_3 = 1{\cdot}62\, c_4$, i.e. $c_4 = c_1$.

† With this level of sophistication there is no difference between the secular equations for *cis*- and *trans*-butadiene.

From the normalization condition,

$$c_1^2 + c_2^2 + c_3^2 + c_4^2 = 1,$$

we have

$$c_1 = \pm 0{\cdot}37,$$

but since the overall sign of a wave function is unimportant we can take the positive solution, hence

$$\psi_1 = 0{\cdot}37\phi_1 + 0{\cdot}60\phi_2 + 0{\cdot}60\phi_3 + 0{\cdot}37\phi_4; \qquad E_1 = \alpha + 1{\cdot}62\beta.$$

In a similar way, the remaining three molecular orbitals can be obtained

$$\psi_2 = 0{\cdot}60\phi_1 + 0{\cdot}37\phi_2 - 0{\cdot}37\phi_3 - 0{\cdot}60\phi_4; \qquad E_2 = \alpha + 0{\cdot}62\beta,$$
$$\psi_3 = 0{\cdot}60\phi_1 - 0{\cdot}37\phi_2 - 0{\cdot}37\phi_3 + 0{\cdot}60\phi_4; \qquad E_3 = \alpha - 0{\cdot}62\beta,$$
$$\psi_4 = 0{\cdot}37\phi_1 - 0{\cdot}60\phi_2 + 0{\cdot}60\phi_3 - 0{\cdot}37\phi_4; \qquad E_4 = \alpha - 1{\cdot}62\beta.$$

These orbitals are illustrated diagramatically in figure 15.1.

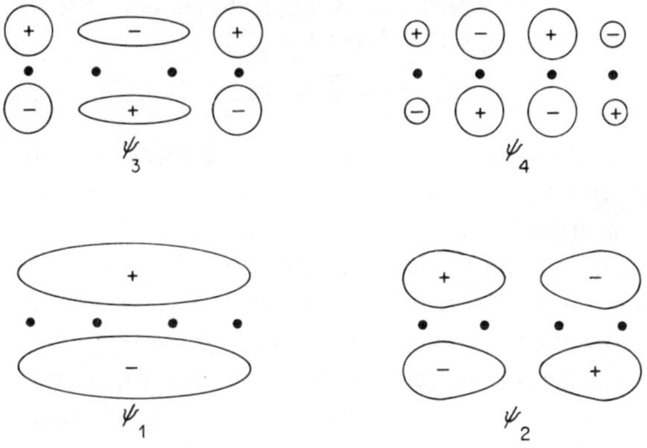

FIGURE 15.1. The Hückel orbitals of butadiene

$\psi_1 \to \psi_4$ is the order of increasing energy, so that in the ground state the four π electrons fill the bonding orbitals ψ_1 and ψ_2. The ground state configuration is therefore $\psi_1^2\psi_2^2$ and the ground state wave function is represented by the Slater determinant $|\psi_1\bar{\psi}_1\psi_2\bar{\psi}_2|$ (see section 7.1).

Although we have shown in section 10.7 that the sum of the orbital energies does not equal the total electronic energy, since it does not take into account the details of electron repulsion, nevertheless, we may reasonably expect that there is a rough correlation between the two. For this reason the total Hückel energy of a molecule is a useful quantity. The sum of the Hückel energies of the four electrons in the ground state of butadiene is $2(\alpha + 1{\cdot}62\beta) + 2(\alpha + 0{\cdot}62\beta) = 4\alpha + 4{\cdot}48\beta$. For two localized ethylenic double bonds we would have $4\alpha + 4\beta$, so $0{\cdot}48\beta$ is the stabilization energy resulting from the delocalization of the electrons†. This is analogous to the resonance energy in VB theory.

We shall see in the next chapter that electron densities play an important part in the interpretation of organic reactions. An electron in an orbital $\psi_r = \sum_\mu c_{r\mu}\phi_\mu$ has a density distribution $\sum_\mu c_{r\mu}{}^2\phi_\mu{}^2$ (neglecting all overlap density terms such as $\phi_\mu\phi_\nu$). We can say that $c_{r\mu}{}^2$ is a measure of the charge at atom μ contributed by this electron. If we sum over all occupied orbitals (2 electrons in each) we get the total *π-electron charge* at atom μ

$$q_\mu = 2 \sum_{r \text{ occ}} c_{r\mu}{}^2. \tag{15.8}$$

In the case of butadiene we have $q_1 = 2c_{11}{}^2 + 2c_{21}{}^2 = 2(0{\cdot}37)^2 + 2(0{\cdot}60)^2 = 1{\cdot}00$, and $q_2 = 2c_{12}{}^2 + 2c_{22}{}^2 = 2(0{\cdot}60)^2 + 2(0{\cdot}37)^2 = 1{\cdot}00$. By symmetry, atoms 3 and 4 have the same charge as 2 and 1 respectively, so that *the π-electron charge is unity at each carbon atom.*

A second quantity of importance is the *π-electron bond order* (sometimes called the mobile bond order) defined by

$$p_{\mu\nu} = 2 \sum_{r \text{ occ}} c_{r\mu}c_{r\nu}. \tag{15.9}$$

If atom μ is directly bonded to atom ν this is a measure of the π-electron density in the bond. For butadiene we have $p_{12} = 2c_{11}c_{12} + 2c_{21}c_{22} = 2(0{\cdot}37 \times 0{\cdot}60) + 2(0{\cdot}60 \times 0{\cdot}37) = 0{\cdot}89$ and $p_{23} = 2c_{12}c_{13} + 2c_{22}c_{23} = 0{\cdot}45$; $p_{34} = p_{12}$ by symmetry. The π-electron bond order of ethylene is 1, so that the closer a bond order is to 1 the greater the double-bond character of the bond. We see that, according to Hückel theory, the central bond in butadiene has a much smaller double-bond character than the outer bonds, which corresponds to the usual representation of the molecule by a single valence structure with no double bond in the middle.

It is found experimentally that the π-electron bond order and the bond length are related to one another as shown in figure 15.2. The only compounds for which the bond orders are determined by symmetry and are not dependent on the Hückel approximations are graphite (0·525), benzene (0·667) and ethylene (1·0). Taking the bond length in these three compounds as 1·421, 1·397 and 1·344 respectively, the bond length is found to be linearly related to the bond order by the expression

$$R\,(\text{Å}) = 1{\cdot}50 - 0{\cdot}16p. \tag{15.10}$$

† This is only true if β for ethylene is the same as β for butadiene. This means that the delocalization energy is relative to the π-electron energy of two ethylenes having the same bond length as the average bond length in butadiene.

Figure 15.2 shows that this relation fits the data for other compounds quite well. Several non-linear relationships between R and p have been proposed, but when one considers that the uncertainty in the measured bond lengths is usually about 0·02 Å it is clear that the experimental data are insufficiently accurate to confirm these non-linear relationships.

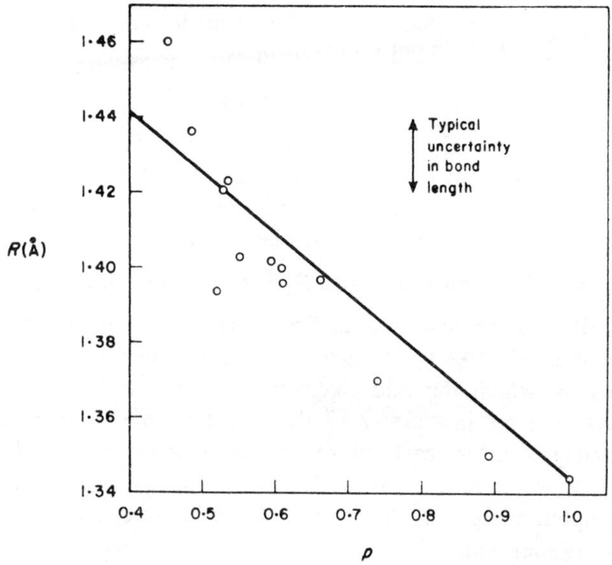

FIGURE 15.2. The bond-order bond-length correlation

The energy of a Hückel orbital is

$$
\begin{aligned}
E_r &= \int \psi_r \mathbf{H} \psi_r \, \mathrm{d}v \\
&= \int \Big(\sum_{\mu} c_{r\mu}\phi_{\mu}\Big)\mathbf{H}\Big(\sum_{\nu} c_{r\nu}\phi_{\nu}\Big)\, \mathrm{d}v \\
&= \sum_{\mu} c_{r\mu}{}^2\alpha + 2\sum_{\nu\to\mu} c_{r\mu}c_{r\nu}\beta. \qquad (15.11)
\end{aligned}
$$

The total Hückel energy is therefore obtained by summing over twice the occupied orbitals

$$
\begin{aligned}
\varepsilon &= 2\sum_{r} E_r = 2\sum_{r}\Big(\sum_{\mu} c_{r\mu}{}^2\alpha + 2\sum_{\nu\to\mu} c_{r\mu}c_{r\nu}\beta\Big) \\
&= \sum_{\mu} q_{\mu}\alpha + 2\sum_{\nu\to\mu} p_{\mu\nu}\beta. \qquad (15.12)
\end{aligned}
$$

We thus see how the total energy is related to the π-electron charges and the bond orders. For open-shell configurations the definitions of q and p can be generalized as follows

$$q_\mu = \sum_r n_r c_{r\mu}^2; \quad p_{\mu\nu} = \sum_r n_r c_{r\mu} c_{r\nu},$$

where n_r is the number of electrons in orbital ψ_r.

There is one further quantity that has been found to be useful, this is the *free-valence index* defined by Coulson as

$$F_\mu = n_{\max} - n_\mu, \tag{15.13}$$

where n_μ is the sum of the π-electron bond orders of all the bonds coming from atom μ and $n_{\max}$ is the maximum value of such bond orders (usually taken as $\sqrt{3}$, the value appropriate to the central atom in the hypothetical molecule $((CH_2)_3C)$ (see problem 15.1). For butadiene we have

$$F_1 = F_4 = 1{\cdot}73 - 0{\cdot}89 = 0{\cdot}84 \text{ and } F_2 = F_3 = 1{\cdot}73 - 0{\cdot}89 - 0{\cdot}45 = 0{\cdot}39.$$

This quantity can be regarded as the MO equivalent of Thiele's residual affinity. The usual value for an aromatic carbon atom is $F \simeq 0{\cdot}4$ and in a free radical in which the odd electron is almost confined to one atom $F \approx 1$. We shall see in chapter 17 that the free-valence index correlates approximately with the reactivity of an atom in a molecule to free-radical attack.

Our complete picture of the π electrons in butadiene is summarized in the following diagram.

$q=1$		$q=1$		$q=1$		$q=1$
CH_2	$p=0{\cdot}89$	CH	$p=0{\cdot}45$	CH	$p=0{\cdot}89$	CH_2
$F=0{\cdot}84$		$F=0{\cdot}39$		$F=0{\cdot}39$		$F=0{\cdot}84$

There are two important points to be noted about the butadiene orbitals. Firstly, they occur in pairs with energies $\alpha \pm x\beta$, and secondly, the coefficients of the paired orbitals are either the same or simply change in sign. These are general properties of so-called *alternant* hydrocarbons. Alternant hydrocarbons are conjugated molecules in which the carbon atoms can be divided into two sets, starred and unstarred, such that no two members of the same set are bonded together (in effect this means that the only conjugated hydrocarbons which are non-alternants are cyclic compounds with odd-membered rings such as azulene). In an odd-alternant, that is one

Typical alternants

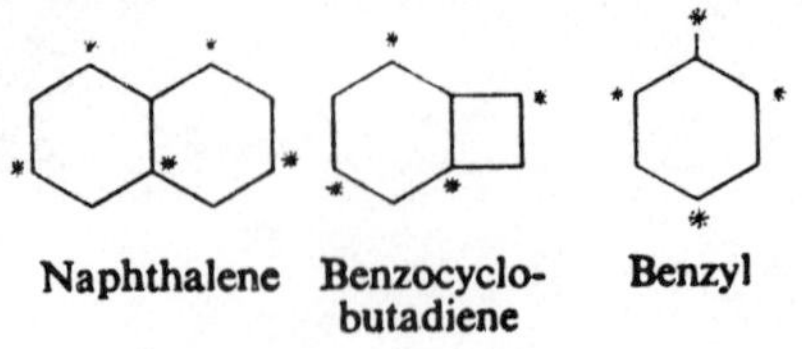

Naphthalene Benzocyclobutadiene Benzyl

with an odd number of conjugated atoms, like benzyl, the more numerous set is taken to be starred.

The general form of the Hückel secular equations is (15.5)

$$xc_\mu + \sum_{\nu \to \mu} c_\nu = 0.$$

But for an alternant if μ is a starred atom then ν must be an unstarred atom; thus if x_r, $c_{r\mu}$ and $c_{r\nu}$ define one solution of these equations then

$$x_r c_{r\mu} + \sum_{\nu \to \mu} c_{r\nu} = 0. \qquad (15.14)$$

But also

$$(-x_r)c_{r\mu} + \sum_{\nu \to \mu} (-c_{r\nu}) = 0,$$

hence the orbital with $-x_r$, $c_{r\mu}$ and $-c_{r\nu}$ must also be a solution of the secular equations. That is, if ψ_r is a bonding orbital of energy $\alpha - x\beta$, then there is also an antibonding orbital ψ_{-r} (say) of energy $\alpha + x\beta$. Moreover, if the bonding orbital has the wave function

$$\psi_r = \sum_\mu^* c_{r\mu}\phi_\mu + \sum_\nu^\circ c_{r\nu}\phi_\nu, \qquad (15.15)$$

where the summations over the starred and unstarred atoms are represented by $\sum^*$ and $\sum^\circ$ respectively, then the paired antibonding orbital has the wave function:

$$\psi_{-r} = \sum_\mu^* c_{r\mu}\phi_\mu - \sum_\nu^\circ c_{r\nu}\phi_\nu. \qquad (15.16)$$

To get the antibonding orbital we just change the coefficients of the unstarred atoms in the bonding orbital.

There is another property of *neutral* alternant hydrocarbons we have already found for butadiene, and this is that the π-electron charge (defined by 15.8) is unity at each atom. To prove this, we note that for any orthonormal set of molecular orbitals

$$\sum_r c_{r\mu}{}^2 = 1. \qquad (15.17)$$

But because of the pairing properties of the orbitals of an alternant

$$\sum_{r\,\text{occ}} c_{r\mu}{}^2 = \sum_{r\,\text{unocc}} c_{r\mu}{}^2.$$

Hence

$$2\sum_{r\,\text{occ}} c_{r\mu}{}^2 = 1. \qquad (15.18)$$

It follows that alternant hydrocarbons should be non-polar, in agreement with the observation that molecules like phenanthrene have no measurable dipole moment; in contrast, non-alternant hydrocarbons such as azulene or fulvene have appreciable dipole moments.

In a similar way we find that for an alternant the mobile bond order between two starred atoms, or between two unstarred atoms is zero. Because the Hückel coefficients $c_{r\mu}$ form an orthogonal matrix we have

$$\sum_r c_{r\mu}\, c_{r\nu} = 0 \quad (\mu \neq \nu), \tag{15.19}$$

and if we divide this into sums over occupied and unoccupied orbitals we obtain

$$\sum_{r\ occ.} c_{r\mu}\, c_{r\nu} + \sum_{r\ unocc.} c_{r\mu}\, c_{r\nu} = 0 \tag{15.20}$$

However, from the pairing property of the orbitals (15.15) and (15.16) both summations are equal if μ and ν are both starred or both unstarred atoms, and they must therefore both be equal to zero. The mobile bond orders between two atoms of the same set are therefore zero.

If an alternant hydrocarbon has an odd number of atoms, it has an odd number of Hückel orbitals. Since we have proved that every bonding molecular orbital ψ_r with energy $(\alpha - x\beta)$ has a corresponding anti-bonding orbital ψ_{-r} with energy $(\alpha + x\beta)$, in an odd alternant there must be a *non-bonding orbital* with energy α, $(x = 0)$. We shall later need to know the coefficients of such non-bonding orbitals and they can be found very easily. We see from the general Hückel secular equation (15.5) that when $x = 0$, the sum of the coefficients around any one atom is zero. Since only starred atoms are directly bonded to unstarred atoms either the starred or the unstarred coefficients are zero. It can easily be seen that it is the smaller set that has the zero coefficients and this is usually called the unstarred set. If we take a polymethine chain as an example and call the non-bonding orbital coefficient of the terminal carbon atom a, then the remaining coefficients can be represented as follows

$$\begin{array}{ccccccccccc} a & & 0 & & -a & & 0 & & (-1)^{k-2}a & 0 & (-1)^{k-1}a \\ C & - & C & - & C & - & C & - \;\ldots\; & C & - \; C \; - & C \end{array} \tag{15.21}$$

From the normalization condition we have $\Sigma a^2 = 1$. Hence the numerical value of the coefficients of the non-bonding orbital in an odd alternant hydrocarbon is given by $a = k^{-\frac{1}{2}}$, where $2k-1$ is the total number of conjugated carbon atoms. We shall see later that this technique can be used when the molecule also contains rings of atoms.

15.3 The use of symmetry for determining Hückel orbitals

The method of finding the Hückel molecular orbitals outlined for butadiene can be followed for any conjugated molecule. But for large molecules the calculations, although not difficult, can become exceedingly tedious without the use of a computer. However, if the molecule has symmetry then this can be used to simplify the calculation: in effect to factorize the secular determinant. For example, the Hückel orbitals of naphthalene arise from the solution of a 10×10 determinant; but using symmetry we are faced with evaluating nothing more than a 3×3.

Naphthalene belongs to the group $\mathbf{D}_{2h}$ (see table 14.1). Half of the symmetry species of this group are symmetric to reflection in the yz plane and the other half are antisymmetric. The σ molecular orbitals will belong exclusively to the former and the π orbitals to the latter. If we restrict our attention to the π orbitals, then these are distinguished one from another by their behaviour under the three two-fold rotations of the group. It follows that we can separate the π molecular orbitals into different symmetry species by considering only the sub-group of operations I, C_2^z, C_2^y, C_2^x: this is the $\mathbf{D}_2$ group whose character table is as follows.

TABLE 15.1. THE Character table of $\mathbf{D}_2$

$\mathbf{D}_2$	I	C_2^z	C_2^y	C_2^x
A_1	1	1	1	1
B_1	1	1	−1	−1
B_2	1	−1	1	−1
B_3	1	−1	−1	1

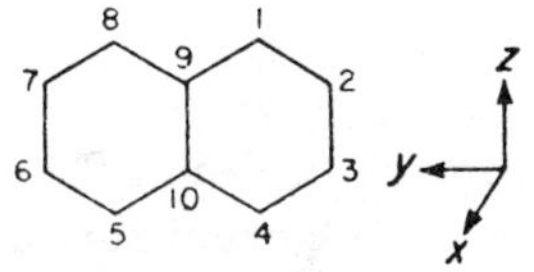

To find the symmetry species of the π molecular orbitals in this sub-group we follow the procedure used before (pages 102–104), and ask how many of the ten p_x atomic orbitals are left unchanged (taking −1 for a change of sign) by the operations of the group. This gives us the following representation of the group with these ten atomic orbitals as basis.

	I	C_2^z	C_2^y	C_2^x
Γ	10	−2	0	0

We now break this representation down into *I.R.*'s by taking the scalar product of this with the *I.R.*'s of the $\mathbf{D}_2$ group (treating the sets of numbers as components of a four-dimensional vector)

$$\Gamma . A_1 = 8, \quad \Gamma . B_1 = 8, \quad \Gamma . B_2 = 12, \quad \Gamma . B_3 = 12.$$

Dividing these numbers by the order of the group (4) gives us the number of molecular orbitals which transform as each of the *I.R.*'s: $2A_1 + 2B_1 + 3B_2 + 3B_3$. In the full $\mathbf{D}_{2h}$ group of the molecule these become $2A_u + 2B_{1g} + 3B_{2g} + 3B_{3u}$.

Let us now look at the form of the Hückel orbitals with these symmetry properties; the following diagrams may be readily obtained from an examination of the character table.

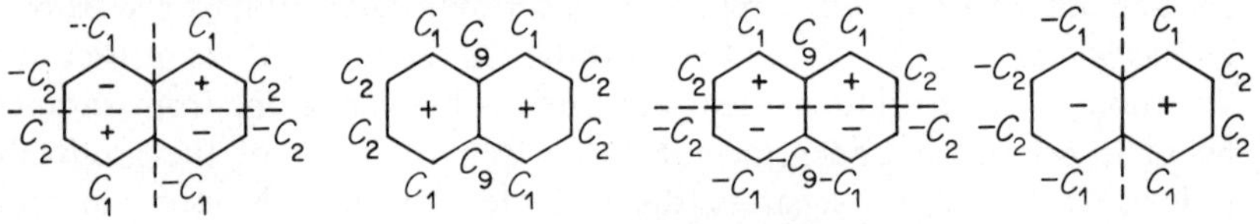

A_u $\quad$ B_{3u} $\quad$ B_{2g} $\quad$ B_{1g}

The A_u species, for example, must have xy and xz as nodal planes since it is symmetric under C_2^y and C_2^z (remember we are dealing with p_x atomic orbitals all of which are antisymmetric to a reflection in the yz plane). This means that there are only two unknown coefficients, c_1 and c_2 for an orbital of A_u symmetry. It can be seen that *the number of times a symmetry species occurs in the set of ten Hückel orbitals is equal to the number of different unknown Hückel coefficients that can exist for each species.* This means we could have arrived at the number of molecular orbitals of each species by simply writing down diagrams for each of these species and taking the number of unknown coefficients for each case.

We now want the secular determinants appropriate to each symmetry species. The general secular equations are ten in number, but they fall into *three* different categories since there are only three different types of atom in the molecule. For example we have (15.5 and figure accompanying table 15.1)

$$xc_9 + c_1 + c_8 + c_{10} = 0, \qquad (15.21a)$$

and there is a similar equation involving xc_{10}. The other two categories are

$$xc_1 + c_2 + c_9 = 0, \qquad (15.21b)$$

$$xc_2 + c_1 + c_3 = 0. \qquad (15.21c)$$

If we now make use of the symmetries of the molecular orbitals these three equations are sufficient to determine the wave functions and energies.

For example, the B_{3u} species has $c_1 = c_8 = c_4 = c_5$, $c_2 = c_3 = c_6 = c_7$, $c_9 = c_{10}$ hence the equations (15.21) become in this case

$$xc_1 + c_2 + c_9 = 0$$
$$c_1 + (x+1)c_2 = 0$$
$$2c_1 + (x+1)c_9 = 0, \qquad (15.22)$$

which have solutions given by

$$\begin{vmatrix} x & 1 & 1 \\ 1 & x+1 & 0 \\ 2 & 0 & x+1 \end{vmatrix} = 0.$$

On expansion this gives

$$(x+1)(x^2 + x - 3) = 0,$$

from which one obtains

$$x = 1{\cdot}303, \quad -1, \quad -2{\cdot}303.$$

Taking the solution $x = -1$ and substituting it back into the secular equations (15.22) we find

$$c_1 = 0;\ c_2 = -c_9,$$

and by symmetry $c_4 = c_5 = c_8 = 0$, $c_3 = c_6 = c_7 = c_2$ and $c_9 = c_{10}$. Applying the normalization condition $6c_2{}^2 = 1$ and remembering that the overall sign of the wave function is unimportant we have the orbital.

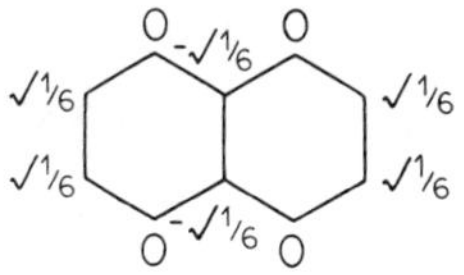

The coefficients of the other B_{3u} orbitals are obtained in a similar manner.

The molecular orbitals of the other symmetry species are obtained in a similar manner. The secular determinants are as follows

$$\begin{matrix} c_1 \\ c_2 \\ c_9 \end{matrix}\begin{vmatrix} x & 1 & 1 \\ 1 & x-1 & 0 \\ 2 & 0 & x-1 \end{vmatrix} \qquad \begin{matrix} c_1 \\ c_2 \end{matrix}\begin{vmatrix} x & 1 \\ 1 & x-1 \end{vmatrix} \qquad \begin{matrix} c_1 \\ c_2 \end{matrix}\begin{vmatrix} x & 1 \\ 1 & x+1 \end{vmatrix}.$$

$$B_{2g} \qquad\qquad A_u \qquad\qquad B_{1g}$$

The coefficients of the bonding orbitals and their Hückel energies are listed in table 15.2.

TABLE 15.2. Coefficients of bonding orbitals and Hückel energies

		c_1	c_2	c_9	$x = (\alpha - E)/\beta$
(A_u)	ψ_1	0·4253	0·2629	0	−0·618
(B_{3u})	ψ_2	0	0·4082	−0·4082	−1·000
(B_{2g})	ψ_3	0·3996	0·1735	0·3470	−1·303
(B_{1g})	ψ_4	0·2628	0·4253	0	−1·618
(B_{3u})	ψ_5	0·3006	0·2307	0·4614	−2·303

Naphthalene is an alternant, so to obtain the corresponding antibonding orbitals one changes the signs of the coefficients of every other atom (B_{3u} bonding becomes B_{2g} antibonding, B_{1g} bonding becomes A_u antibonding, and vice versa).

15.4 Cyclic conjugated polyolefins and Hückel's 4n+2 rule

The secular equations for a cyclic polyolefin take the form (15.5)

$$c_{\mu-1} + xc_\mu + c_{\mu+1} = 0, \tag{15.23}$$

where $\mu - 1$, μ and $\mu + 1$ are successive atoms on the ring. In seeking a general solution for the coefficient c_μ we must look for a function that is periodic in t (the number of atoms in the ring). As a trial solution we take $c_\mu = \sin k\mu$ where $k = 2\pi r/t$, r being an integer. This trial solution satisfies the boundary condition that $c_\mu = c_{t+\mu}$. Substituting this trial solution into (15.23) we get

$$\sin k(\mu - 1) + x \sin k\mu + \sin k(\mu + 1) = 0. \tag{15.24}$$

Since $\sin A + \sin B = 2 \sin \frac{1}{2}(A + B) \cos \frac{1}{2}(A - B)$ we have, after combining the first and last terms in (15.24)

$$2 \sin k\mu \cos k + x \sin k\mu = 0. \tag{15.25}$$

Hence $x = -2 \cos k$, that is

$$E_r = \alpha + 2\beta \cos k = \alpha + 2\beta \cos \frac{2\pi r}{t} \tag{15.26}$$

If there is an even number of atoms in the ring then possible solutions are

$$r = 0 \cdots \frac{t}{2} \text{ (t is even).} \tag{15.27}$$

If there is an odd number of atoms then

$$r = 0 \cdots \frac{t-1}{2} \; (t \text{ is odd}). \tag{15.28}$$

Our trial solution is clearly satisfactory. However, $c_\mu = \cos k\mu$ would be an equally reasonable choice; we now have

$$\cos k(\mu - 1) + x \cos k\mu + \cos k(\mu + 1) = 0. \tag{15.29}$$

Since $\cos A + \cos B = 2 \cos \frac{1}{2}(A + B) \cos \frac{1}{2}(A - B)$, this leads to

$$2 \cos k\mu \cos k + x \cos k\mu = 0, \tag{15.30}$$

which gives the same energies as (15.26).

There are thus two acceptable solutions for equation (15.23) for all values of r except for $r = 0$, or $r = t/2$ for an even chain, when the sine solution is zero. This means that the lowest orbital in a cyclic polyolefin is non-degenerate but that all the higher orbitals are doubly degenerate, except the top orbital for t even. This result can be represented geometrically as in figure 15.3. To obtain the Hückel energy levels for a cyclic system of

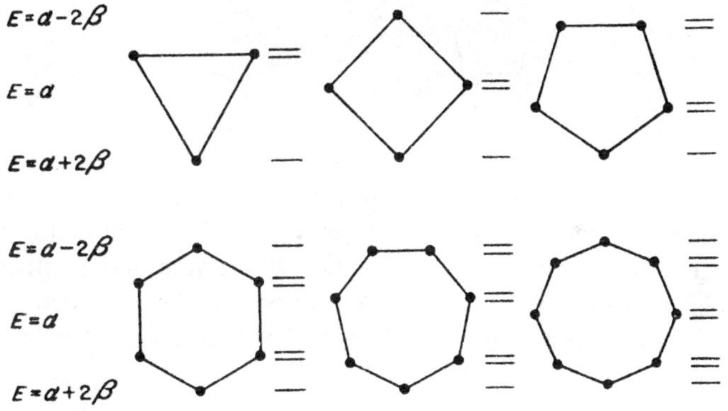

FIGURE. 15.3. The energy level scheme for cyclic polyenes

t atoms, we inscribe a t-sided regular polygon in a circle radius 2β such that one corner is the lowest point. Each remaining corner corresponds to a single Hückel orbital, the energy of which is given in units of β by the distance from the horizontal diameter of the circle (problem 15.5).

Because each orbital can contain two electrons, only cyclic unsaturated molecules having $(4n + 2)$ π electrons will have a closed-shell configuration and therefore be chemically stable. Compounds with $(4n + 1)$ π electrons, will be free radicals, and compounds with $4n$ π electrons will have a triplet ground state.

The relationship between this rule and aromaticity has produced rather a lot of muddled thinking. Benzene is the archetype of an aromatic compound, and the aromaticity of a compound should be measured by the extent to which its properties resemble those of benzene. There has been a tendency to call compounds aromatic which have no chemical resemblance to benzene at all, just because they satisfy the $4n + 2$ rule. The exceptional stability of the benzene molecule is partly due to its internal angle being 120°, which imposes no strain on the sp^2 hybridized carbon atoms.

The cyclopentadienyl anion $C_5H_5^-$, radical $C_5H_5^{\bullet}$, and cation $C_5H_5^+$ all need at least five equivalent VB structures to make the carbon atoms equivalent. If we adopt a naive approach and identify stability with number of structures then we would expect them all to be equally stable†. However, Hückel theory shows that only the cyclopentadienyl anion will have a closed shell and experimentally this is the only stable form of the five-membered ring which is found (e.g. dimethyl cyclopentadiene dicarboxylate forms a stable anion I in aqueous alkali). Similarly, the seven-membered ring has a stable cation, the tropylium cation II.

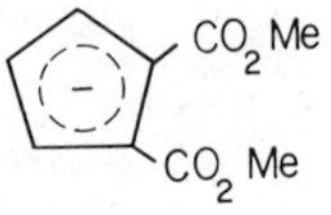

I

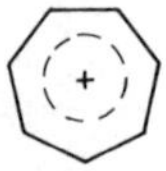
II

However, it must be emphasized that neither of these ions show any of the characteristic chemical properties of benzene and it is therefore not very illuminating to describe them as aromatic. On the other hand, ferrocene III (see section 18.4) and tropolone IV, which can be regarded as being derived from these ions, could well be called aromatic since they have many of the characteristic chemical properties of benzene.

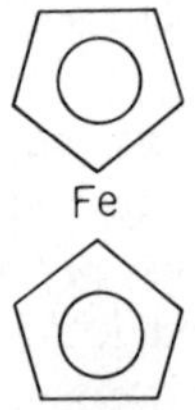

III

IV

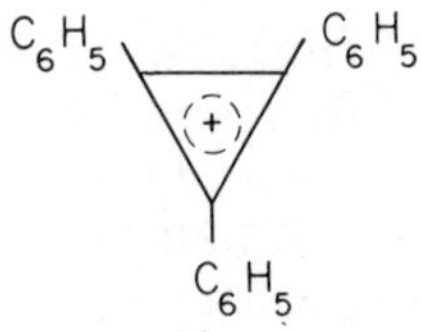

V

† It has been found that when the interaction between these structures is examined in detail, VB theory does give the same result as Hückel theory (FISCHER and MURRELL, *Theoret. Chim. Acta*, **1**, 463 (1963).

The cyclopropenyl cation with two π electrons ($n = 0$) represents a somewhat similar example. Derivatives of this ion such as V are relatively stable, whereas derivatives of the corresponding anion or radical are not.

It is of interest to consider what happens when $n > 1$. These compounds are known as annulenes. More detailed theory shows that in large rings a distortion into long and short C—C bonds is likely to occur even if they have $(4n + 2)$ π electrons†. If n is not large the molecules cannot adopt the planar configuration, which is necessary for π-electron stabilization, because of the introduction of bond angle strain and the repulsion of non-bonded atoms. Thus $C_{10}H_{10}$ VI has not yet been prepared and $C_{14}H_{14}$ VII which has, is very unstable.

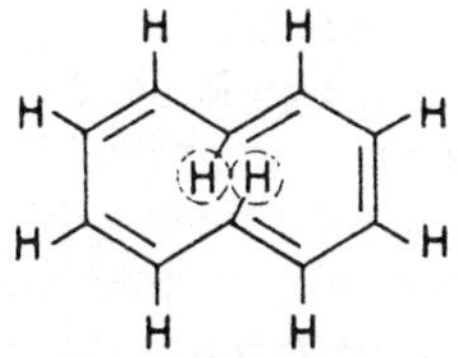

VI

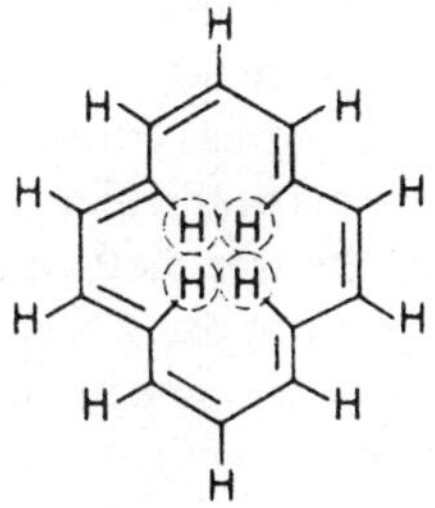

VII

The ring system containing 18 carbon atoms ($n = 4$) VIII has been prepared and has been found to be more stable than the annulene containing

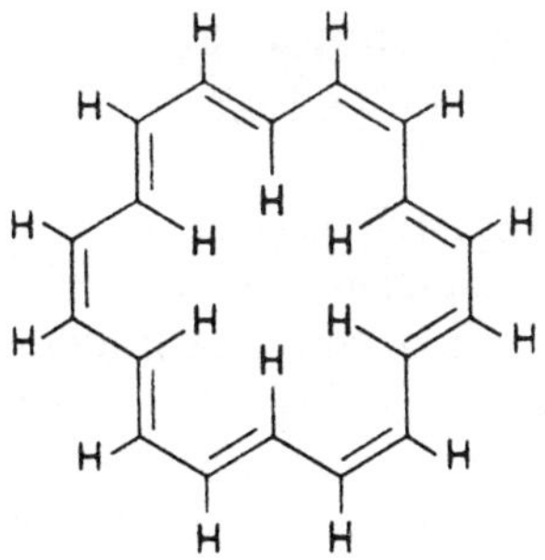

VIII

† Longuet-Higgins and Salem, *Proc. Roy. Soc.*, **A251**, 172 (1959).

20 carbon atoms. It will undergo typical aromatic substitution reactions under rather special conditions. However, the molecule is in no sense as stable as benzene, presumably because of the repulsion of the hydrogen atoms inside the ring.

Hückel theory predicts that square cyclobutadiene C_4H_4 should have a triplet ground state, and present experimental evidence supports this view since all attempts to make its simple derivatives have resulted in dimers. A naive VB approach suggests the molecule should be stable since it has two equivalent valence structures, the same number as benzene. Cyclooctatetraene would also be a triplet molecule if it were planar, but it exists in a non-planar form and has the properties of an olefin. Its dinegative ion $C_8H_8^{2-}$ has, however, been shown to have a delocalized π-electron system and is presumably planar.

The most convincing evidence for the delocalization of π electrons in cyclic molecules comes from their high diamagnetism. In the presence of a magnetic field perpendicular to the plane of the molecule the π electrons will circulate like free electrons in a superconducting ring. This produces a magnetic field opposing the applied field and a resulting diamagnetism. The effect of this diamagnetism is clearly seen in nuclear magnetic resonance spectroscopy by a large magnetic deshielding of any proton outside the ring and a large magnetic shielding of protons inside the ring. However, we emphasize that such a physical property is not by itself sufficient grounds for describing a molecule as aromatic.

15.5 Aromaticity and anti-aromaticity

Some authors have applied Hückel's $4n + 2$ rule to polycyclic systems. It will be clear from its derivation that there can be no justification for doing this, and in many cases it does not give correct predictions. Another suggestion has been that the rule may be applied to the perimeter of the molecule. According to this picture pyrene (I) with a perimeter containing 14 π electrons should show aromatic properties in spite of the presence of 16 π electrons all told. In fluoranthene (II), on the other hand, the perimeter contains 15 π electrons. In practice, both these compounds show fully aromatic properties in the chemical sense, and in order to account for this fluoranthene has to be considered as two units with separate perimeters of ten and six π electrons joined by two single bonds.

The Hückel delocalization energy is not always a reliable guide of aromaticity or of molecular stability, as is shown by table 15.3. The first four compounds are stable, typically aromatic molecules; on the other hand, heptalene has been synthesized but is an extremely unstable molecule showing no aromatic properties, and pentalene has so far resisted

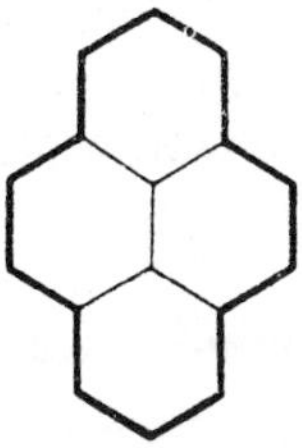

I

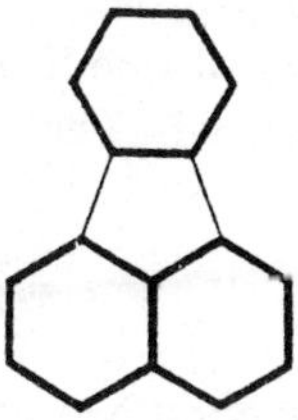

II

TABLE 15.3. Delocalization energies

	Delocalization energy (units of β)
Benzene	2·00
Naphthalene	3·68
Azulene III	3·36
Biphenylene IV	4·51
Pentalene V	2·46
Heptalene VI	3·62

all attempts at synthesis. By comparison with benzene, there is nothing in the delocalization energy to show why the last two molecules should be so unstable.

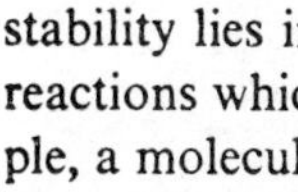

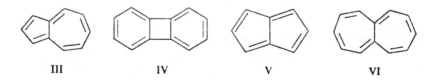

III

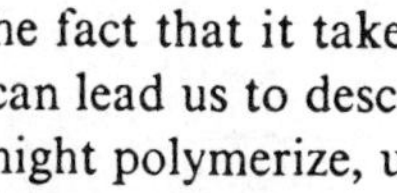

IV

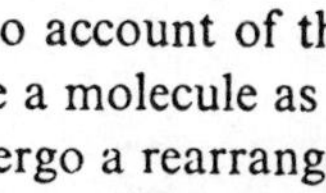

V

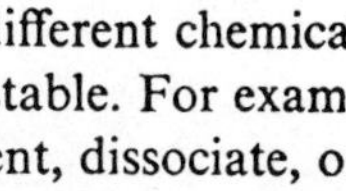

VI

The limitation of the Hückel delocalization energy as a criterion of stability lies in the fact that it takes no account of the different chemical reactions which can lead us to describe a molecule as unstable. For example, a molecule might polymerize, undergo a rearrangement, dissociate, or be oxidized by air, any of which would be sufficient to label it as unstable. Yet none of these processes are necessarily correlated with the Hückel delocalization energy. In the case of heptalene, the highest occupied orbital is one having $x = 0$ (a non-bonding orbital) and the molecule should therefore be easily oxidized.

It is in many cases possible to estimate the delocalization energy of a molecule by a perturbation calculation. As a simple example we will examine the effect of bond formation between two positions in the same alternant, hexatrienyl. The total π-electron energy of the molecule is given by (15.12)

$$\varepsilon = \sum_{\mu} q_{\mu}\, \alpha_{\mu} + 2 \sum_{\mu \to \nu} p_{\mu\nu}\, \beta_{\mu\nu}$$

If we form a new bond between atoms ρ and σ, the energy change is given by $2p_{\rho\sigma}\beta_{\rho\sigma}$. Now from hexatrienyl we can formally construct benzene by joining atoms 1 and 6, or fulvene by joining atoms 1 and 5.

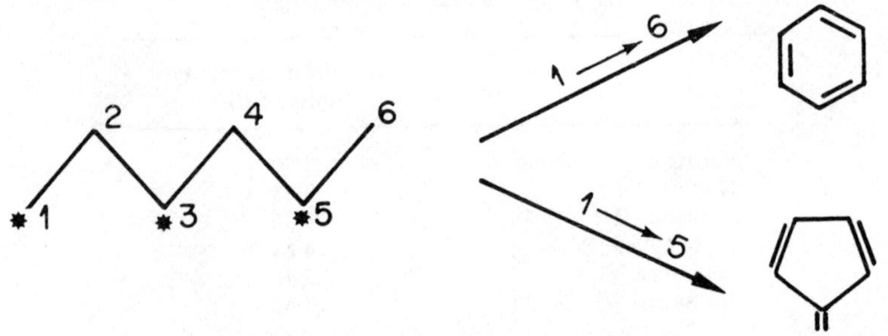

In making benzene we join a starred atom to an unstarred atom and p_{16} is positive representing a gain in π-electron energy. In making fulvene however, we joined two starred atoms and we have shown in (15.20) that the mobile bond order between such atoms is zero. In other words the π-electron energy of fulvene is the same as that of hexatriene according to this simple approach.

To go further with this approach we will now consider the union of two odd alternant radicals R and S

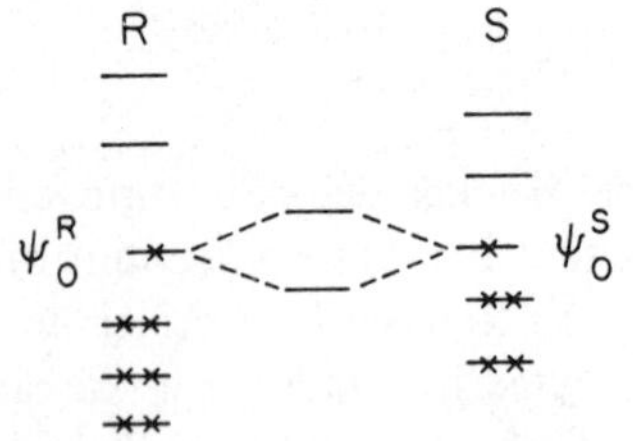

FIGURE 15.4. The interaction between the orbitals of two odd alternants.

The completely filled orbitals of R and S can interact with one another but this will not lead to any net stabilization of the system (see problem 15.6). Stabilization will arise from the interaction of the non-bonding and *anti-bonding* orbitals of one fragment with the non-bonding and *bonding* orbitals of the other. The most important interaction is that between the non-bonding orbitals, which gives rise to two molecular orbitals of RS, $\sqrt{\tfrac{1}{2}}(\psi_o^R \pm \psi_o^S)$. The stabilization of the lower of these two molecular orbitals relative to the non-bonding energy is given by

$$\int \psi_o^R \, \mathbf{H} \, \psi_o^S \, dv = \sum_{\nu \to \mu} c_{o\mu}^R \, c_{o\nu}^S \, \beta_{\mu\nu} \tag{15.31}$$

This expression holds for the interaction of R^- and S^+ (or R^+ and S^-) as well as for the two radicals, since in each case there are two electrons which occupy non-bonding orbitals in the fragments, but a bonding orbital in RS. This treatment will always underestimate the energy of RS because it ignores all interactions except that between the non-bonding orbitals.

Let us now apply this approach to a particular example. We shall estimate the π-electron energy of a cyclic molecule RS which we can consider to be made up of two odd alternant radicals R and S. The energy of the cyclic compound will then be given approximately by

$$E_{RS} \approx E_R + E_S + 2\beta\left(\sum_{\nu \to \mu} c_{o\mu}^R \, c_{o\nu}^S\right) \tag{15.32}$$

where $c_{o\mu}^R$ and $c_{o\nu}^S$ are the coefficients of the non-bonding orbitals of R and S respectively for atoms μ (of R) and ν (of S) which are joined together. The summation is over all $\mu \to \nu$ bonds. We can consider benzene as made up of two allyl radicals. The total Hückel energy of the allyl radical is $3\alpha + 2{\cdot}82\,\beta$ and the coefficients of the non-bonding orbital at the starred positions are $\pm\sqrt{\tfrac{1}{2}}$ (from the procedure on p. 290). Then

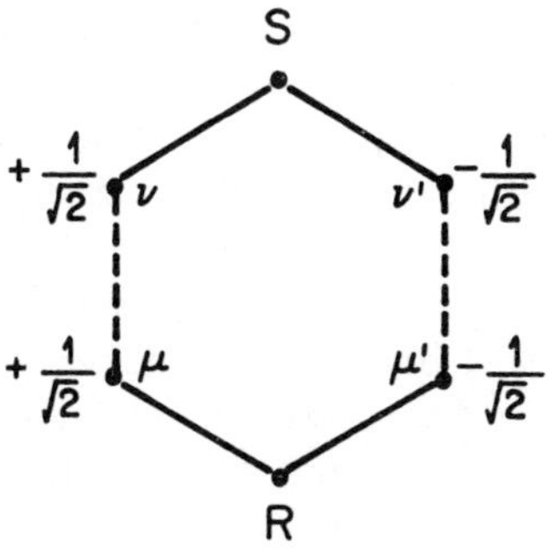

from (15.32) we have

$$E_{RS} \approx (3\alpha+2{\cdot}82\beta)+(3\alpha+2{\cdot}82\beta)+2\beta[(\sqrt{\tfrac{1}{2}})(\sqrt{\tfrac{1}{2}})+(-\sqrt{\tfrac{1}{2}})(-\sqrt{\tfrac{1}{2}})]$$
$$\approx 6\alpha+7{\cdot}63\beta$$

We could instead regard benzene made up of a single methine unit and a pentadienyl fragment. We can obtain the coefficients

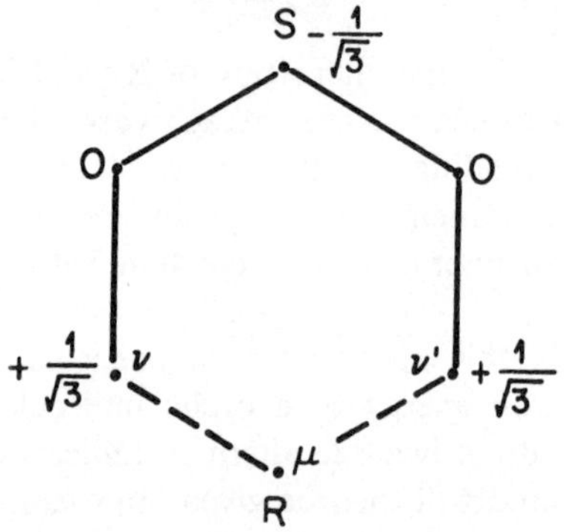

of the non-bonding orbital from the procedure on p. 290. We now have :—

$$E_{RS} \approx \alpha+(5\alpha+5{\cdot}46\beta)+2\beta(1\times\sqrt{\tfrac{1}{3}}+1\times\sqrt{\tfrac{1}{3}})$$
$$\approx 6\alpha+7{\cdot}77\beta$$

We can compare these two approximate energies for benzene with the value of $6\alpha+8\beta$ we obtain for a complete Hückel calculation on benzene. Apart from using this method to obtain approximate π-electron energies we shall see in chapter 17 that it is of considerable value in calculating a reactivity index called the Dewar Number.

At present we wish to use this approach to estimate the 'aromaticity' of a cyclic conjugated molecule C_rH_r. We shall do this by seeing whether the π-electron energy of the cyclic compound is greater or less than that of the corresponding linear polyene C_rH_{r+2}. We shall consider the cyclic annulene as made up of a polymethine chain joined at both ends to a single carbon fragment, whereas we shall regard the linear polyene as made up of the same polymethine chain joined to the single carbon fragment at one end only. Let us put the total number of conjugated atoms in the chain equal to $(2k-1)$. Then following exactly the same procedure as before we have,

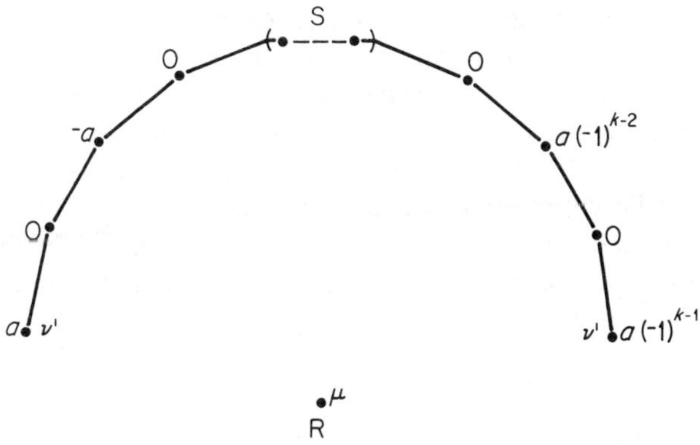

for the linear olefin in which the atom μ is joined to only one end of S,

$$E_{RS} \approx E_S + \alpha + 2a\beta$$

whilst for the cyclic annulene in which atom μ is joined to both ν and ν' of S we have

$$E_{RS} \approx E_S + \alpha + 2a\beta[1+(-1)^k].$$

The difference in π-electron energy between the linear polyene and the cyclic annulene is given by

$$\mathcal{E}_{(annulene)} - \mathcal{E}_{(polyene)} \approx 2a\beta(-1)^k.$$

Therefore, if k is even, the π-electron energy ('resonance energy') of the annulene is greater than that of the polyene, i.e. the annulene is *aromatic*. If k is odd the π-electron energy of the annulene is less than that of the polyene and this is sometimes referred to as *anti-aromaticity*. If k is even the annulene will have $(4n+2)$ π-electrons whereas if k is odd it will have $(4n)$ π-electrons, i.e. we have obtained the same result as Hückel's rule derived in the previous section.

We can extend the argument to cyclic polymethine rings C_rH_r (where r is odd) and compare their π-electron energies with those of the corresponding linear polyene radical or ion. We find the cyclic polymethines with $(4r+1)$ atoms are 'aromatic' (i.e. have a greater π-electron energy than the linear compound) if they are an anion and 'anti-aromatic' (i.e. a lower π-electron energy) if they are a cation. Similarly, a cyclic polymethine with $(4r+3)$ atoms is 'anti-aromatic' if it is an anion and 'aromatic' if it is a cation. In each case the 'aromatic' species contains $(4n+2)$ π-electrons.

This perturbation theory approach has two advantages over the Hückel $(4n+2)$ rule. Firstly it shows that complete delocalization of π-electrons in $4n$-annulenes causes an increase in energy, i.e. these molecules will exist with alternate single and double bonds in so far as they can. The second advantage of the perturbation treatment is that it can be extended to polycyclic compounds, about which the original Hückel rule says nothing. We employ exactly the same technique as before; we break down the polycyclic compound into a polycyclic odd alternant and a single carbon fragment. We compare the π-electron energy of the odd alternant simply joined to the single carbon unit, with the system formed by joining the odd atom to give an additional ring

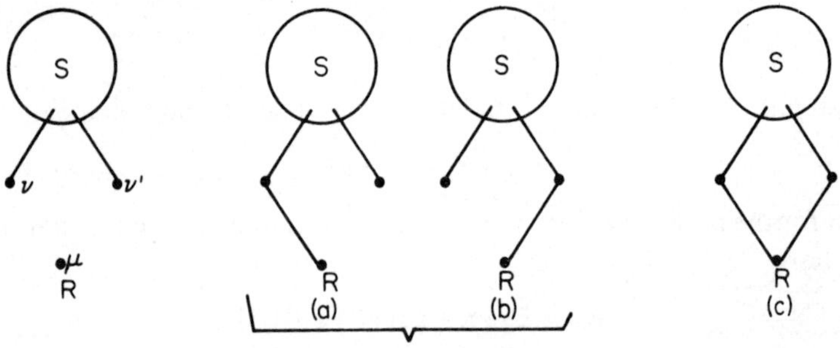

The change in π-electron energy in forming (a) or (b) is $2\,c_{0\nu}\beta$ and $2\,c_{0\nu'}\beta$ respectively. The change in π-electron energy in forming (c) is $2\beta\,(c_{0\nu}+c_{0\nu'})$. The system with the extra ring (c), will have greater π-electron energy (i.e. will have more resonance energy) than either (a) or (b) if $c_{0\nu}$ and $c_{0\nu'}$ have the same sign, on the other hand the system will be less stable than either (a) or (b) if the coefficients $c_{0\nu}$ and $c_{0\nu'}$ are of opposite sign. Now it can be shown that the signs depend on the size of the new ring formed. If it has 4r+2 members the signs are the same, if it has 4r members the signs are opposite.

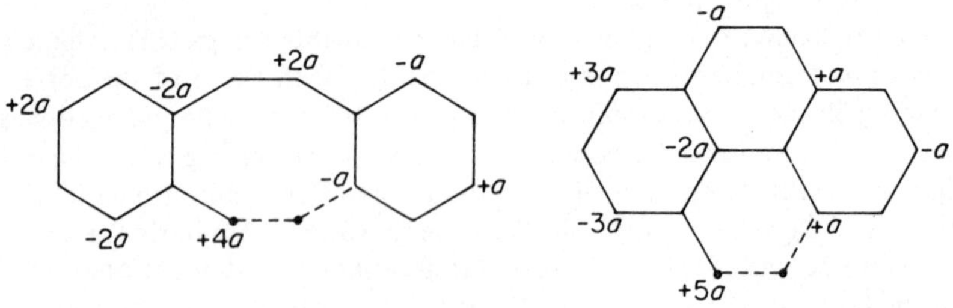

Both dibenzo-cyclo-octatetraene and pyrene contain 16 π-electrons, but pyrene is 'aromatic' while dibenzo-cyclo-octatraene is not.

The same treatment can be extended to systems with two odd membered rings fused together as in azulene. The transannular bond joins two unstarred carbon atoms. This bond therefore makes no first order contribution to the π-electron energy and azulene, according to this picture, is a distorted [10]-annulene with a transannular 'single' bond.

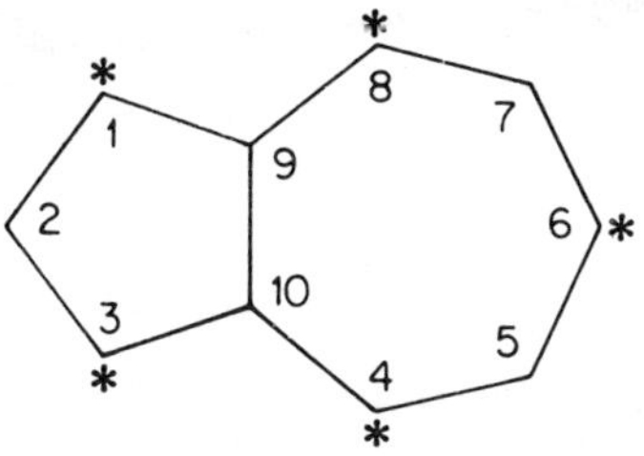

15.6 *Non-classical structures*

In order to explain a wide variety of chemical phenomena it has become customary to invoke so-called 'non-classical' intermediates; that is, species which cannot be represented by Kekulé-type structures of single, double and triple bonds. A typical example is given by the hydrolysis of cholesteryl halides (I). Kinetic studies provide evidence that this reaction involves the 'non-classical' carbonium ion (II) which in turn reacts to the two products (III) and (IV). The fact that two products are formed could be explained by assuming tautomerism between two isomeric carbonium ions, but this would not account for the exceptionally rapid hydrolysis of (I).

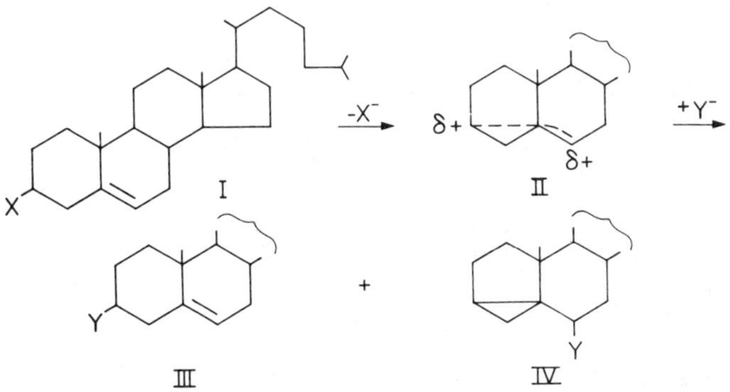

We therefore ask, can such delocalization occur and if it does, will it be stabilizing? This type of structure, sometimes called a homoallylic cation, differs from the allyl cation by the interposition of a methylene group (figure 15.5). We can see from the diagrams that we can treat the homoallylic cation by Hückel theory if we introduce a value β' for the

Hückel molecular-orbital levels

$-\sqrt{2}\beta, 0, +\sqrt{2}\beta$ $\qquad$ $-\sqrt{(\beta^2+\beta'^2)}, 0, +\sqrt{(\beta^2+\beta'^2)}$

FIGURE 15.5. The allyl (a) and homoallyl (b) cations

resonance integral between atoms 2 and 4. Winstein and Simonetta estimated $\beta'/\beta = 0{\cdot}30$ on the basis of calculated overlap integrals, so obtaining the small, but significant value of $0{\cdot}09\beta$ as the stabilizing energy due to delocalization†. They showed that a slight reduction in ϑ from the tetrahedral angle gives an increased value of the delocalization energy which more than offsets the σ-bond strain energy, and calculated that if $\beta = -20$ kcal mole^{-1}, the maximum net stabilization is about 7 kcal mole^{-1}.

The homoallyl cation is but one of many such non-classical carbonium ions which have been proposed. Another, for which there appears good experimental evidence, is the 7-norbornenyl cation (V).

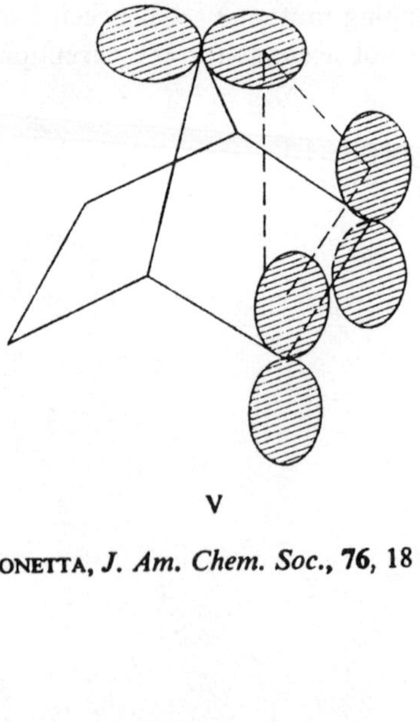

V

† WINSTEIN and SIMONETTA, *J. Am. Chem. Soc.*, **76**, 18 (1954).

Another type of non-classical carbonium ion is often suggested as an intermediate for molecular rearrangements of the pinacol type.

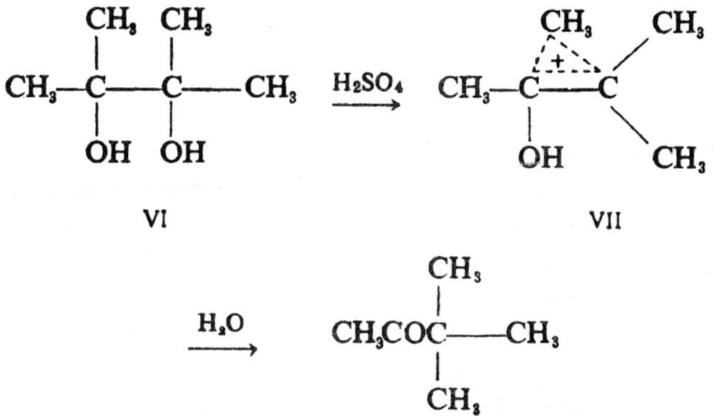

VI VII VIII

The carbonium ion VII can be regarded as a modified cyclopropenyl cation. Again the calculated resonance energy of such a species will depend on values chosen for the coulomb integral of the migrating group and for the resonance integrals of the 'dotted' bonds. Using normal values for α and β, the secular equations for this system leave only a single negative root. Hückel theory is thus able to account for the fact that this type of 1,2 shift is much commoner among reactions involving carbonium ions than those involving carbanions or radicals. However, the introduction of a heteroatom, such as a positively charged quaternary nitrogen atom, lowers the energy of all three molecular orbitals, so that in such a case the migration of an alkyl group can occur through a carbanion intermediate (e.g. the Stevens rearrangement)†.

15.7 Heteroatomic molecules

In discussing Hückel theory so far we have only dealt with systems containing carbon atoms. The introduction of a nitrogen or oxygen atom into an unsaturated system will require new values for both the coulomb

† STEVENS *et al.*, *J. Chem. Soc.*, 3193 (1928).

and resonance integrals. Changes in α and β are usually expressed in terms of α_C and β_{C-C} with benzene as the reference substance

$$\alpha_X = \alpha_C + h_X \beta_{C-C} \tag{15.33}$$

$$\beta_{C-X} = k_{C-X} \beta_{C-C}. \tag{15.34}$$

It is difficult to arrive at definite values of the parameters h_X and k_{C-X} because, as we shall see in the next section, the Hückel parameters depend on what experiment is being studied. The approximate magnitudes of h_X are known, for α_X must be related to the electronegativity of atom X. It has been suggested that h_X is linearly related to the difference between the electronegativity of X and C. One must distinguish between such cases as the nitrogen in pyridine which donates one electron to the π system and the nitrogen in pyrrole which donates two electrons. Generally acceptable values† are as follows.

$h_{O-} = 2{\cdot}0$	$k_{C-O} = 0{\cdot}8$
$h_{O=} = 1{\cdot}0$	$k_{C=O} = 1{\cdot}0$
$h_{N-} = 0{\cdot}5$	$k_{C-N} = 0{\cdot}8$
$h_{N=} = 1{\cdot}5$	$k_{C=N} = 1{\cdot}0$

Once the values of h_X and k_{C-X} are decided upon, the calculation of the coefficients of the secular equation is carried out as before. For example, the secular determinant for formaldehyde $H_2C{=}O$ is as follows

$$\begin{vmatrix} x & 1 \\ 1 & x+1 \end{vmatrix} = 0 \qquad x = 0{\cdot}62, \qquad x = -1{\cdot}62,$$

† STREITWEISER, *Molecular Orbital Theory for Organic Chemists*, chapter 4, Wiley, 1961.

which gives $E_1 = \alpha + 1{\cdot}62\beta$; $E_2 = \alpha - 0{\cdot}62\beta$. The bonding molecular orbital is close in energy to the oxygen atomic orbital and the antibonding molecular orbital is close in energy to the carbon atomic orbital. This is illustrated in figure 15.6.

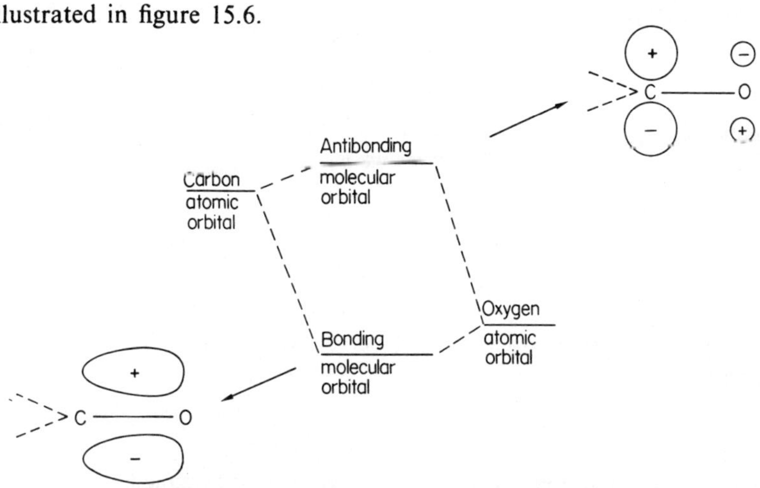

FIGURE 15.6. The Hückel orbitals of the carbonyl group

An alternative method of tackling this type of problem is to consider the introduction of the heteroatom as producing a perturbation on the molecular orbitals and energies of the isoelectronic hydrocarbon. Thus the molecular orbitals of formaldehyde could be written as a linear combination of the molecular orbitals of ethylene. Some useful expressions relating to electron density changes on heterosubstitution have been derived on this basis.

From (15.12) one can obtain an expression for the complete differential of ε

$$d\varepsilon = \sum_{\mu} q_\mu \, d\alpha_\mu + 2\sum_{\mu\nu} p_{\mu\nu} \, d\beta_{\mu\nu} + \sum_{\mu} \alpha_\mu \, dq_\mu + 2\sum_{\mu\nu} \beta_{\mu\nu} \, dp_{\mu\nu}. \qquad (15.35)$$

However, a variation of q_μ and $p_{\mu\nu}$ with constant α_μ and $\beta_{\mu\nu}$ will make no contribution to $d\varepsilon$ because the orbital energies have been obtained by a method which satisfies the variation theorem, that is, ε is already a minimum with respect to variation of the coefficients $c_{r\mu}$ and hence to a variation of q_μ and $p_{\mu\nu}$. It follows from (15.35) that

$$(\partial\varepsilon/\partial\alpha_\nu) = q_\nu, \quad (\partial\varepsilon/\partial\beta_{\mu\nu}) = 2p_{\mu\nu}. \qquad (15.36)$$

Hence

$$\frac{\partial q_\nu}{\partial \alpha_\mu} = \frac{\partial^2 \varepsilon}{\partial\alpha_\mu \partial\alpha_\nu} = \frac{\partial q_\mu}{\partial \alpha_\nu} \equiv \pi_{\mu,\nu}. \qquad (15.37)$$

That is, a change in α_μ produces the same perturbation to the π-electron charge at atom ν as a similar change in α_ν would produce at atom μ. The derivative $(\partial q_\mu/\partial \alpha_\nu) = (\partial q_\nu/\partial \alpha_\mu)$ is called the *mutual polarizability of atoms* μ and ν and is given the symbol $\pi_{\mu,\nu}$†.

We can also ask how q_ν is affected by a change in resonance integral $\beta_{\mu\kappa}$, and this quantity

$$\pi_{\nu,\mu\kappa} = \frac{\partial q_\nu}{\partial \beta_{\mu\kappa}} = \frac{\partial^2 \varepsilon}{\partial \beta_{\mu\kappa}\partial \alpha_\nu} \tag{15.38}$$

is called the atom–bond polarizability. Further we can ask how the bond orders are affected by changes in α and β

$$\pi_{\mu\nu,\kappa} = \frac{\partial p_{\mu\nu}}{\partial \alpha_\kappa} = \frac{1}{2}\frac{\partial^2 \varepsilon}{\partial \alpha_\kappa \partial \beta_{\mu\nu}}; \qquad \pi_{\mu\nu,\kappa\rho} = \frac{\partial p_{\mu\nu}}{\partial \beta_{\kappa\rho}} = \frac{1}{2}\frac{\partial^2 \varepsilon}{\partial \beta_{\kappa\rho}\partial \beta_{\mu\nu}}, \tag{15.39}$$

and these are called bond–atom and bond–bond polarizabilities. The first of all these quantities $\pi_{\mu,\nu}$ is by far the most important. An expression for $\pi_{\mu,\nu}$ may be derived using first-order perturbation theory (6.52) (cf. problem 15.4)

$$\pi_{\mu,\nu} = \frac{\partial q_\nu}{\partial \alpha_\mu} = 4 \sum_{r\,\text{occ}} \sum_{s\,\text{unocc}} \frac{c_{r\mu}c_{r\nu}c_{s\mu}c_{s\nu}}{E_r - E_s}, \tag{15.40}$$

where the summations are over occupied and unoccupied orbitals respectively. It can be proved that for an alternant hydrocarbon $\pi_{\mu,\nu}$ is positive if μ and ν are both starred or both unstarred and negative if one is starred and one unstarred. This means that in an alternant hydrocarbon its value is positive or negative according as the atoms are separated by an odd or an even number of bonds. In a neutral alternant hydrocarbon all the electron densities are unity, so if we replace one carbon atom by a hetero-atom X with coulomb integral α_X, the altered values of q_ν throughout the molecule are given to first-order by

$$q_\nu = 1 + (\partial q_\nu/\partial \alpha_\mu)_0 \delta\alpha_\mu = 1 + \pi_{\mu,\nu}(\alpha_X - \alpha_C). \tag{15.41}$$

The mutual polarizabilities for benzene are (in units of β^{-1}) *ortho* $(\pi_{1,2}) = -17/108$, *meta* $(\pi_{1,3}) = +1/108$ and *para* $(\pi_{1,4}) = -11/108$ with the self-polarizability being $(\pi_{1,1}) = +43/108$. Suppose we wish to calculate the charges in pyridine, then putting $(\alpha_N - \alpha_C) = 0{\cdot}5\beta$ we get the following result.

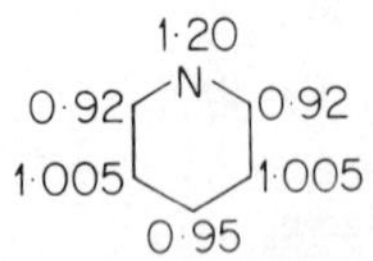

π-Electron distribution

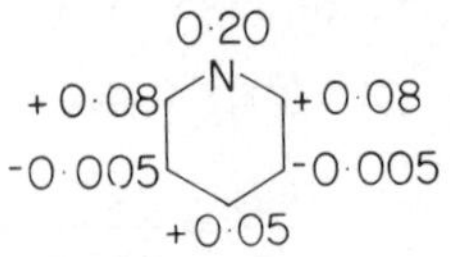

Net atom charges

† Coulson and Longuet-Higgins, *Proc. Roy. Soc.*, **A191**, 34 (1947).

In this treatment we have supposed that the heterosubstituent only alters one coulomb integral. Changes in the coulomb terms of the atoms next to the heteroatom, and the resonance integrals of the bonds connected to it, have a much smaller effect on the electron density.

15.8 A critique of Hückel theory

In chapter 12 we looked at the approximations inherent in MO theory. On top of these, Hückel theory has a further set of approximations which seem to be of rather doubtful validity. To what extent can they be justified?

The first justification is that Hückel theory works, and it works because it does not set its sights too high. In Hückel theory one never really calculates anything in isolation, but correlates one experimental fact with another. The calculated energies, electron densities, and so on, in Hückel theory are given in terms of the energy parameters α and β. These are formally defined by integrals involving some Hamiltonian **H** but the relationship of **H** to the complete Hamiltonian of the molecule is not specified.

Providing there are no heteroatoms in the molecule, relative energies in Hückel theory are all given in terms of one parameter, β. This means that there should be a linear relationship between the experimental energies and the Hückel energies, and the slope of the line determines the value of β. This is illustrated in figure 15.7, where we plot the frequency of the first strong

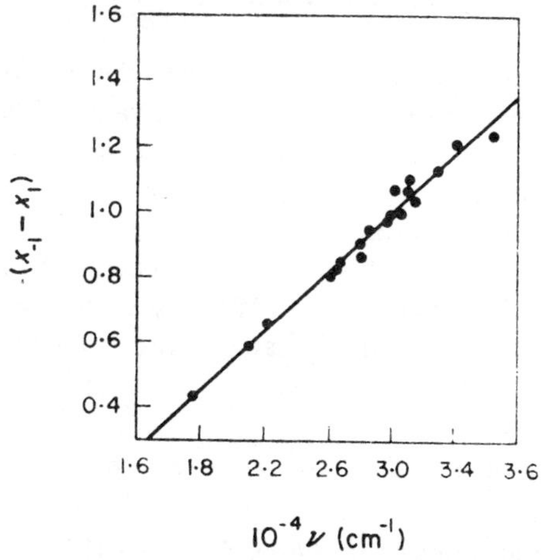

FIGURE 15.7 The correlation between the frequency of the first strong electronic absorption band and the first Hückel excitation energy

electronic absorption band of the aromatic hydrocarbons built up from fused benzene rings (e.g. naphthalene, pyrene) against the energy necessary to excite an electron from the highest occupied orbital ψ_1 (the highest bonding) to the lowest vacant (antibonding) orbital ψ_{-1}, in units of β. The regression line, that is, the best straight line through the points, satisfies the equation†

$$\nu(\text{cm}^{-1}) = 8200 + 21{,}900\,(x_{-1} - x_1). \qquad (15.42)$$

The regression line does not go through the origin, indicating that there is more to this excitation energy than would appear from Hückel theory. But if we have a compound of this type whose spectrum is unknown, then we could predict the position of this absorption band to within ± 1500 cm^{-1}, to a confidence limit of 95%. The value of β which best fits the experimental facts is $-21{,}900\ \text{cm}^{-1}$, which corresponds to an energy of $-2{\cdot}71$ ev.

As a second example we take the polarographic half-wave reduction potentials ($\varepsilon_{1/2}$) of fifty aromatic hydrocarbons (a wider class than for figure 15.7 including, for example, azulene and biphenyl). This can be related to the energy given up when an electron is added to the hydrocarbon and so should be correlated with the Hückel energy of the lowest vacant orbital ψ_{-1}. The regression line in this case is (figure 15.8)‡

$$\varepsilon_{\frac{1}{2}}(\text{aq dioxane}) = \frac{2{\cdot}37}{\mathscr{F}}\,x_{-1} - 0{\cdot}92\ (\text{volts}), \qquad (15.43)$$

where $\mathscr{F}$ is the Faraday constant. The slope of the line gives the value of β in electron volts, $\beta = -2{\cdot}37$ ev.

Lastly, we examine the correlation between the Hückel delocalization energy E_D (that is the total Hückel energy less a value of 2β for each formal double bond) and the observed delocalization or resonance energies. The latter are usually obtained by taking the difference between the heat of combustion of the hydrocarbon and the heat of combustion that is predicted on the basis of an additivity of bond energies§ (e.g. 54·0 kcal for each C—H bond, 49·3 for each C—C, 121·2 for each C=C, etc.). The correlation for a set of compounds similar to those in figure 15.7 is shown in figure 15.9 and the regression line is§

$$E_D(\text{obs}) = -16\,\frac{E_D(\text{Hückel})}{\beta}\ \text{kcal mole}^{-1}. \qquad (15.44)$$

† Heilbronner and Murrell, *J. Chem. Soc.*, 2611 (1962).
‡ Streitweiser, *Molecular Orbital Theory for Organic Chemists*, Wiley, 1961.
§ Klages, *Ber.*, **82**, 358 (1949).

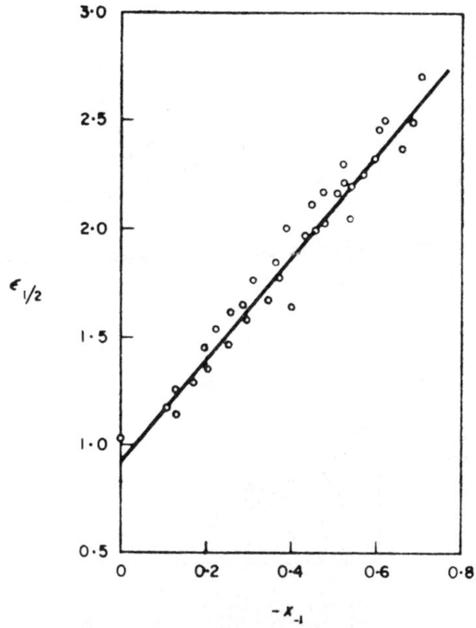

FIGURE 15.8. The correlation between the half-wave reduction potential and the Hückel energy of the lowest vacant orbital

The observed results are then obtained using a value of $\beta = -16$ kcal mole^{-1} or $-0{\cdot}69$ ev.

The three examples given above show that there is a very good linear relationship between the Hückel energies and the properties we have studied. The values of β deduced in the three cases are, however, rather different from one another; in particular, the best β for delocalization energies is much less than the best β's for reduction potential or excitation energies. The reason for this is that energies should properly be evaluated using the complete Hamiltonian for the molecule, and not the pseudo one-electron Hamiltonian of Hückel theory. If this is done then it is found that such things as excitation energies, electron affinities and dissociation energies are expressed not simply in terms of one-electron integrals like α and β, but they depend also on the integrals which are derived from the electron-interaction terms in the Hamiltonian (see section 10.7). By taking β as an empirical parameter it is found that we can compensate, in some cases (but not always), for having ignored these terms. But β now represents a different set of integrals for each type of property of the molecule that is being studied, so that the value of β which fits the experimental

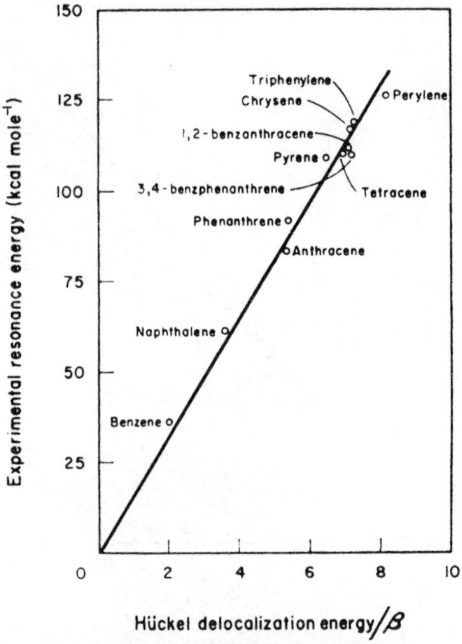

FIGURE 15.9 The correlation between the observed resonance energy and the Hückel delocalization energy

results in one case may be different from the value which fits the results in another.

This is the reason that there are no unique values for the parameters α and β of heteroatomic molecules. The value of α_N which best fits the basicity of the nitrogen heterocyclic molecules is not necessarily the best for predicting the unpaired electron density in nitrogen-containing radicals, because α_N stands for a different combination of one- and two-electron integrals in the calculated values of these two properties.

There are some cases, even amongst conjugated hydrocarbons, where Hückel theory in its simplest form gives rather poor results. For example, the excitation frequency of butadiene calculated from (15.42) is 35,300 cm^{-1} but the observed value is 46,100 cm^{-1}. The reason for the discrepancy in this case is that the bonds in butadiene are not all of the same length, so that it is a poor approximation to take β as being the same for each bond. Even acknowledging that β is not a well-defined quantity it is clear that it must be related roughly to the amount of overlap between the two atomic orbitals involved in the bond, and this varies roughly as the inverse exponential of the bond length.

In molecules where there is strong bond alternation, like the linear conjugated polyenes, it is important that the dependence of β on bond length should be allowed for. If, for example, one takes for the short (double) bonds $\beta = -4{\cdot}00$ ev and for the long (single) bonds $\beta = -2{\cdot}88$ ev, very good agreement is obtained between the observed and calculated position of the first absorption band of these compounds, as can be seen from figure 15.10.

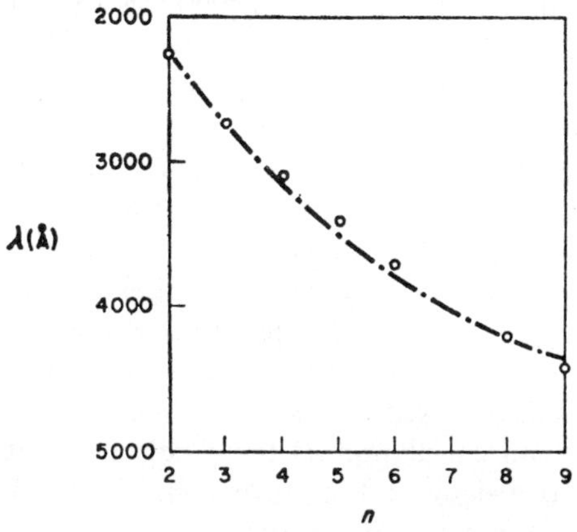

FIGURE 15.10 The correlation between the first electronic absorption band of the linear polyenes having n double bonds and the first Hückel excitation energy with allowance for bond-length alternation

Another case where a variable β must be considered is when the conjugated carbon atoms do not all lie in one plane. For example, experiment shows that biphenyl (in solution or the gas phase) is not a planar molecule because of steric interference between the hydrogens *ortho* to the central bond. If ϑ is the angle between the planes of the two benzene rings then one expects β for the central bond to vary as $\cos \vartheta$; when the two rings are perpendicular to one another then $\beta = 0$ and in the Hückel method we have two independent conjugated systems.

The results of Hückel theory are usually appreciably better for alternant hydrocarbons than for non-alternants. For example, the dipole moment of azulene calculated from the Hückel electron densities is 7 D, whereas the observed value is about 1 D. The reason for this discrepancy is that although α_μ in Hückel theory is a measure of the energy of an electron in orbital ϕ_μ, this takes no account of the interaction between this electron and the other π electrons in the molecule. It is clear, however, that if there is a high electron density on atom μ this must make α_μ less negative.

The simplest method of correcting for the effect of electron density on α_μ is through the Wheland–Mann formula;

$$\alpha_\mu = \alpha_0 + (1 - q_\mu)\omega\beta. \tag{15.45}$$

In this q_μ is the π-electron charge calculated from expression (15.8). When this is 1, as for a neutral alternant, then α_μ is taken to have a standard value α_0. If this is different from 1 then α_μ is corrected by an amount $\omega\beta$ times this difference, where ω is a constant. A value $\omega = 1{\cdot}4$ has been proposed to give the best agreement with experiment†. Because β is a negative quantity, α_μ is less negative than α_0 when q_μ is greater than one, and more negative when q_μ is less than one.

The Wheland–Mann type of calculation (sometimes called the ω-technique) is, as we shall see, a first approximation to a SCF calculation, and the method of obtaining the Wheland–Mann orbitals follows the usual SCF iterative procedure. Starting, say, with Hückel orbitals one calculates q_μ and hence α_μ. This value of α_μ is then used in a new Hückel-type calculation and from the resulting orbitals a new value of q_μ is derived. The calculation proceeds until self-consistency is reached.

A Wheland–Mann calculation of the dipole moment of azulene with $\omega = 1{\cdot}4$ gives 4·7 D, a significant improvement over the Hückel calculation, although still much larger than experiment.

Lastly we turn to what appears to be the grossest of the Hückel approximations; that of zero-overlap even between neighbouring atomic orbitals. For a bond length of 1·4 Å the overlap integrals between neighbouring $2p\pi$ atomic orbitals is 0·25, which is certainly not small. But the justification of the approximation really lies again in the fact that the parameters α and β are not rigorously defined. Instead of thinking of these parameters as integrals involving carbon $2p\pi$ atomic orbitals we can equally well consider them as integrals between orthogonalized atomic orbitals of the type defined by expressions (12.7) and (12.10).

Orbitals which are defined in terms of constants α and β but without making the zero-overlap assumption are called Wheland orbitals. If, in the Hückel approximation, an orbital has energy

$$E_r = \alpha - x_r\beta,$$

then in the Wheland approximation its energy is

$$E_r = \frac{\alpha - x_r\beta}{1 - x_r S}, \tag{15.46}$$

† STREITWEISER, *loc. cit.*

where S is the overlap integral between neighbouring orbitals. Also, if the Hückel orbital has coefficients c_{rv} then the Wheland orbital has coefficients $(1 - x_r S)^{-\frac{1}{2}} c_{rv}$.

There is no case to our knowledge where the correlation between an experimental observation and a calculation using the Wheland approximations is significantly better than the correlation with a Hückel calculation. For this reason little work is now done with Wheland orbitals.

To summarize some of our conclusions: Hückel theory provides a method of correlating experimental results for a class of related compounds. It is most useful when the class is large (say, all aromatic hydrocarbons), and the parameters involved are few. When we come to deal with heteroatomic systems we find that the number of parameters that have to be determined from experiment increases rapidly, and at the same time the number of molecules of each class gets smaller. For this reason the parameters appropriate for heteroatoms are even now not firmly established. The parameters of Hückel theory stand in place of the complication of integrals that would arise in any non-empirical theory. For this reason the best values of the parameters depend on what experimental fact is under study.

In the next two chapters we shall look at the application of Hückel theory to problems of chemical equilibrium and reactivity, but first we shall look briefly at a more sophisticated π-electron theory which deals explicitly with the electron repulsion terms.

15.9 Introduction of electron-interaction terms into π-electron theory

Hückel theory makes no specific reference to electron interaction, although it is assumed that the Hückel Hamiltonian does contain some average of the electron-interaction terms. The Wheland–Mann theory takes account of electron interaction in a very crude way by recognizing that the coulomb integral will be sensitive to the electron density on the atom in question. A more satisfactory treatment (which we refer to as the P-method) was developed by Pariser and Parr, and by Pople†. It is essentially a treatment of electron interaction within the framework of Hückel's zero-overlap approximation.

† For a fuller description of the P-method, with references, see MURRELL *The Theory of the Electronic Spectra of Organic Molecules*, Methuen and Wiley 1964.

If we evaluate the energy of an LCAO MO wave function using the complete Hamiltonian we shall come across electron repulsion integrals of the following type

$$\iint \phi_\mu(i)\phi_\nu(j)\left(\frac{1}{r_{ij}}\right)\phi_\rho(i)\phi_\sigma(j)\,dv_i\,dv_j. \tag{15.47}$$

The physical picture of this integral is that it represents the interaction of the electron densities $\phi_\mu\phi_\rho(i)$ and $\phi_\nu\phi_\sigma(j)$. In Hückel theory we neglect the overlap of two orbitals which are on different atoms, so that if we work within such a zero-overlap approximation it is consistent to neglect all the two-electron integrals which depend on an overlap density. That is, we only retain integrals of the type

$$\iint \phi_\mu(i)\phi_\nu(j)\left(\frac{1}{r_{ij}}\right)\phi_\mu(i)\phi_\nu(j)\,dv_i\,dv_j \equiv \gamma_{\mu\nu}. \tag{15.48}$$

This simplification does in fact follow from an earlier approximation suggested by Mulliken

$$\begin{aligned}\iint \phi_\mu(i)\phi_\nu(j)\left(\frac{1}{r_{ij}}\right)\phi_\rho(i)\phi_\sigma(j)\,dv_i\,dv_j \\ = \frac{S_{\mu\rho}S_{\nu\sigma}}{4}(\gamma_{\mu\nu}+\gamma_{\rho\sigma}+\gamma_{\mu\sigma}+\gamma_{\rho\nu}) \\ = \delta_{\mu\rho}\delta_{\nu\sigma}\gamma_{\mu\nu} \text{ in the zero-overlap approximation.}\end{aligned} \tag{15.49}$$

Hückel theory is built around the two parameters α and β which are integrals involving some undefined one-electron Hamiltonian. Similar parameters come into the P-method, but they are now well defined: they are called the core integrals (10.58)

$$\begin{aligned}\mathbf{H}^c_{\mu\mu} &= \int \phi_\mu \mathbf{H}^c \phi_\mu\,dv \\ \mathbf{H}^c_{\mu\nu} &= \int \phi_\mu \mathbf{H}^c \phi_\nu\,dv.\end{aligned} \tag{15.50}$$

$\mathbf{H}^c$ consists of the terms in the Hamiltonian which are a function of the position of electron 1: the kinetic energy operator and the potential field of the hydrocarbon which has been stripped of all its π electrons. This potential field can be written as a sum of contributions from each carbon atom V_ρ, that is

$$\mathbf{H}^c = -\tfrac{1}{2}\nabla_1^{\,2} + \sum_\rho V_\rho. \tag{15.51}$$

Now $\mathbf{H}^c$ differs from molecule to molecule, and the potential differs from place to place in a molecule. It is clear therefore that $\mathbf{H}^c_{\mu\mu}$ and $\mathbf{H}^c_{\mu\nu}$ cannot be taken, as in Hückel theory, to be constants for all carbon atoms and C—C bonds. If we substitute (15.51) into (15.50) we have

$$\mathbf{H}^c_{\mu\mu} = \int \phi_\mu[-\tfrac{1}{2}\nabla^2 + V_\mu]\phi_\mu \, dv + \sum_{\rho \neq \mu} \int \phi_\mu[V_\rho]\phi_\mu \, dv, \tag{15.52}$$

$$\mathbf{H}^c_{\mu\nu} = \int \phi_\mu[-\tfrac{1}{2}\nabla^2 + V_\mu + V_\nu]\phi_\nu \, dv + \sum_{\rho \neq \mu,\nu} \int \phi_\mu[V_\rho]\phi_\nu \, dv. \tag{15.53}$$

The first term in $\mathbf{H}^c_{\mu\mu}$ can be taken as a constant for a given type of atom. It is roughly equal to the ionization potential I_μ appropriate to removing an electron from a $2p$ orbital in the appropriate sp^2 valence state. Likewise, the first term in $\mathbf{H}^c_{\mu\nu}$ can be taken as a constant for the same atoms and the same bond lengths. The terms involving V_ρ are not, however, negligible. If we make the approximation that the energy of a density $\phi_\mu{}^2$ in the field of a distant core of net charge Z_ρ is equal but opposite in sign to the energy of $\phi_\mu{}^2$ in the field of a distant charge $Z_\rho\phi_\rho{}^2$, then we can say that

$$\int \phi_\mu[V_\rho]\phi_\mu \, dv = -Z_\rho \gamma_{\rho\mu}. \tag{15.54}$$

On the other hand,

$$\int \phi_\mu[V_\rho]\phi_\nu \, dv = -Z_\rho \iint \phi_\mu(i)\phi_\rho(j)\left(\frac{1}{r_{ij}}\right)\phi_\nu(i)\phi_\rho(j) \, dv_i \, dv_j \tag{15.55}$$

and these two-electron integrals are neglected in the zero-overlap approximation (15.48). It follows then that in the P-method $\mathbf{H}^c_{\mu\nu}$ is a parameter which can be treated as a constant in the same way as the Hückel β, but $\mathbf{H}^c_{\mu\mu}$ differs from the Hückel α_μ since it varies from atom to atom according to the expression

$$\mathbf{H}^c_{\mu\mu} = I_\mu - \sum_{\rho \neq \mu} Z_\rho \gamma_{\mu\rho}. \tag{15.56}$$

To calculate the energy of a many-electron wave function built up from orthogonal spin orbitals we must use expressions like (10.63) and (10.79), and use an LCAO expansion of the molecular orbitals. In general this leads to very complicated equations, but within the approximations of the P-method the problem is not too unwieldy. This can be seen by examining the form taken by the Roothaan equation (10.83).

$$\mathbf{F}_{\mu\nu} = \mathbf{H}^c_{\mu\nu} + \sum_{s=a}^{k} \sum_\rho \sum_\sigma c_{s\rho} c_{s\sigma} \Big[2 \iint \phi_\mu(i)\phi_\rho(j)\left(\frac{1}{r_{ij}}\right)\phi_\nu(i)\phi_\sigma(j) \, dv_i \, dv_j$$
$$- \iint \phi_\mu(i)\phi_\rho(j)\left(\frac{1}{r_{ij}}\right)\phi_\sigma(i)\phi_\nu(j) \, dv_i \, dv_j \Big]. \tag{15.57}$$

In the case of $\mu = \nu$ we have after introducing the zero-overlap approximation

$$\mathbf{F}_{\mu\mu} = \mathbf{H}_{\mu\mu}^{c} + \sum_{s=a}^{k} \left[\sum_{\rho} 2c_{s\rho}{}^{2}\gamma_{\mu\rho} - c_{s\mu}{}^{2}\gamma_{\mu\mu} \right]. \quad (15.58)$$

This can be simplified by introducing the definition of the atom charges (15.8) together with (15.56). The result is

$$\mathbf{F}_{\mu\mu} = I_{\mu} + \tfrac{1}{2}q_{\mu}\gamma_{\mu\mu} + \sum_{\rho \neq \mu} (q_{\rho} - Z_{\rho})\gamma_{\mu\rho}. \quad (15.59)$$

In the case of $\mu \neq \nu$ we have

$$\mathbf{F}_{\mu\nu} = \mathbf{H}_{\mu\nu}^{c} - \sum_{s=a}^{k} c_{s\mu}c_{s\nu}\gamma_{\mu\nu} \quad (15.60)$$

and using the definition of the bond order (15.9) we have

$$\mathbf{F}_{\mu\nu} = \mathbf{H}_{\mu\nu}^{c} - \tfrac{1}{2}p_{\mu\nu}\gamma_{\mu\nu}. \quad (15.61)$$

With these definitions of the **F** integrals it is a relatively easy matter to evaluate the energy of any wave function or to evaluate SCF orbitals through equations (10.84).

For a neutral alternant hydrocarbon it can be shown that the pairing properties of the molecular orbitals, which is an important feature of Hückel theory, is retained when one calculates SCF orbitals using the expressions given here for the **F** integrals. Likewise, the property of uniform charge density is also retained, which means that in these cases (15.59) reduces to

$$\mathbf{F}_{\mu\mu} = I_{\mu} + \tfrac{1}{2}\gamma_{\mu\mu}, \quad (15.62)$$

which can be treated as a constant analogous to the Hückel α. The Wheland-Mann correction for molecules without uniform charge densities is a crude approximation to expression (15.59) in which the effect of distant charges is neglected ($\gamma_{\mu\nu} = 0$, $\mu \neq \nu$) and $\omega\beta$ is equivalent to $-\tfrac{1}{2}\gamma_{\mu\nu}$.

As it is used now the P-method is a semi-empirical approach in which parameters like $\mathbf{H}_{\mu\nu}^{c}$ and $\gamma_{\mu\nu}$ are taken to fit experiment. Since there are more parameters than in Hückel theory it would be surprising if agreement with experiment were not better. However, there are some real advantages in the P-method. Firstly a parameter like $\mathbf{H}_{\mu\nu}^{c}$ is not something which differs from one property to another as does the Hückel β. Secondly there are some observations which can never be explained in Hückel theory since they are a direct result of electron interaction. Prominent in this class are many features of electronic spectra. For example, the difference between the singlet and triplet excited states of a molecule depend on the electron-interaction terms (given through the γ integrals in the P-method), as we have seen in the case of the helium atom (section 9.5). Thus the P-method is a very useful extension of the empirical theories of unsaturated hydrocarbons.

15.10 The free-electron model of π-electron molecules

The dominant characteristic of π electrons is their delocalization. The simplest model which brings out this property is the free-electron model. In essence one ignores any of the fine details of the potential arising from the individual electrons and nuclei in the molecule and simply assumes that the electrons are contained in a constant potential volume whose dimensions are determined by the size of the molecule. We shall illustrate this point by the simple example of a linear polyene.

Suppose we have a linear polyene of $2n$ carbon atoms. The π electrons are delocalized along the chain so that the essential features of the problem are brought out by considering a one-dimensional potential. We take the length of the potential box to be extended one bond length past the terminal carbon atoms (sometimes only half a bond length is taken but this detail is not important). The molecular orbitals and their energies are therefore given by introducing the length of the box $a = (2n + 1)d$, where d is the bond length, into the particle-in-a-box solutions (2.4)–(2.7)

$$\psi_r = \left(\frac{2}{(2n+1)d}\right)^{\frac{1}{2}} \sin \frac{\pi r x}{(2n+1)d}, \qquad E_r = \frac{h^2 r^2}{8md^2(2n+1)^2}. \quad (15.63)$$

Figure 15.11 shows the form of the first four free-electron molecular orbitals for butadiene. There is a strong resemblance between these and

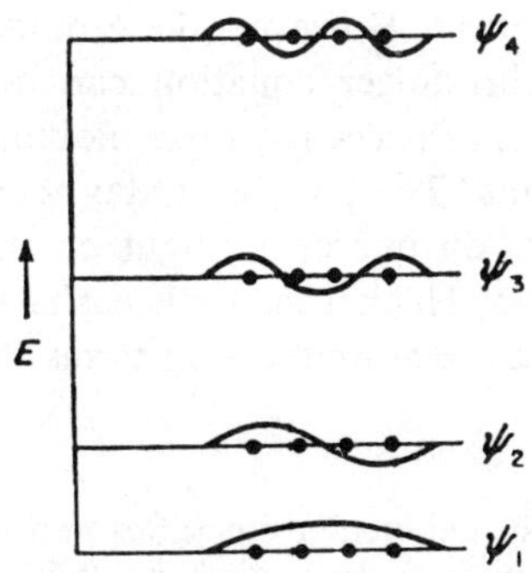

FIGURE 15.11 The free-electron molecular orbitals for butadiene

the Hückel orbitals given on page 273 ψ_1 has no nodes; ψ_2 has a node through the central bond; ψ_3 has two nodes and ψ_4 has nodes through each bond. This parallelism between the free-electron and Hückel orbitals extends to branched conjugated chains and to aromatic molecules: the amplitudes of the free-electron orbitals at the atoms are similar to the corresponding Hückel coefficients.

The free-electron theory has been used quite extensively to interpret the electronic spectra of unsaturated hydrocarbons. The attractive feature of the model is that, unlike Hückel theory, no empirical parameters appear in the calculated energies. For example, the ground state of butadiene has the electron configuration $\psi_1^2\psi_2^2$, the first excited state has the configuration $\psi_1^2\psi_2\psi_3$. The energy of this transition is therefore

$$E_3 - E_2 = \frac{h^2}{8md^25^2}(3^2 - 2^2) = \frac{h^2}{40md^2}.$$

Taking a C—C bond length of 1·4 Å this gives a predicted wavelength for the first absorption band of butadiene as

$$\lambda = \frac{hc}{E_3 - E_2} = 3200 \text{ Å}.$$

The observed value is 2200 Å.

When one considers the overall crudity of the free-electron model it is perhaps surprising that one gets as close as this to experiment. The linear polyenes are not in fact a very good case to test the theory since the presence of alternating long and short bonds means that the electrons are not freely delocalized along the molecule. We noted earlier that Hückel theory also gives poor results for these molecules unless one allows for different values of β for the long and short bonds.

Modifications of the free-electron model have been investigated, in which the constant potential is modulated by a fluctuating term to allow for bond-length alternation. However, these modifications have the disadvantage that the Schrödinger equation can only now be solved by numerical methods, and, of necessity, these modulating potentials involve some unknown constants. The position today seems to be that the free-electron model presents an interesting light on the nature of π-electron molecules, but in practice Hückel theory is easier to handle and easier to develop to include the electron-interaction terms.

Problems 15

15.1 Write down the Hückel secular equations and solve them for the hypothetical molecule trimethylene methane $(CH_2)_3C$. Evaluate the bond orders (cf. expression 15.13).

15.2 Determine the most stable configurations (linear or triangular) for the molecules H_3^+, H_3 and H_3^- using Hückel approximations.

15.3 Determine the Hückel orbitals of fulvene and hence determine the electron charges q_μ and the dipole moment.

15.4 Prove expression (15.40).

15.5 Prove the geometrical construction for the orbital energies of the cyclic polyenes described on page 283.

15.6 Examine the possibility that a resonance integral β' between π orbitals on different double bonds will increase the stability of the following molecule.

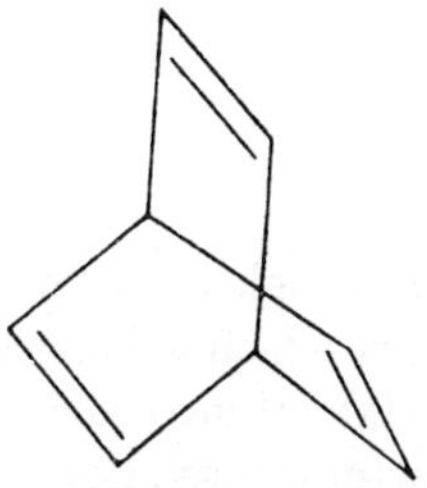

15.7 Calculate the mutual atom polarizabilities for butadiene and hence determine the electron densities in acrolein CH_2=CH—CH=O.

15.8 Write down the expression for the ground state π-electron energy of ethylene given by the P-method.

15.9 Show that the free-electron model predicts a relationship $\lambda \propto n$ for the first absorption band of a linear polyene having $2n$ carbon atoms.

Chapter 16

The electronic theory of organic chemistry

16.1 Introduction

In previous chapters of this book we have been mainly concerned with those aspects of valence theory which tell us something about the structure and stability of the ground state of molecules. If this theory is satisfactory we might hope to obtain useful information about the structure and stabilities of excited and reacting molecules. Typical questions asked by the organic chemist are: why is monofluoroacetic acid ($K_a = 2{\cdot}2 \times 10^{-3}$) stronger than acetic acid ($K_a = 1{\cdot}8 \times 10^{-5}$); and why does toluene nitrate more rapidly than benzene and give mainly 2- and 4-nitrotoluene while nitrobenzene nitrates very much more slowly than benzene and yields almost exclusively *m*-dinitrobenzene? The difficulty in treating these problems theoretically is very great indeed, for not only are the energy differences between two reaction paths very small but the factors which control the choice of path can be very numerous. On the other hand, there are more accurate kinetic and equilibrium data in physical organic chemistry than in any other field of chemistry. On the basis of these data organic chemists have developed empirical theories which have, by and large, proved extraordinarily successful. We shall look at these theories and see to what extent they can be accommodated in the concepts and methods of quantum mechanics.

The theories of organic chemistry have grown out of the pictorial ideas of Kekulé and van't Hoff and thus have their origins dating back long before any satisfactory theory of valency was available. The first real attempt to put these ideas on a physical basis was made by Robinson when he showed that the early ideas such as the 'partial valence' of Thiele and the 'alternating polarity' of Lapworth could be much more adequately accounted for in terms of the Lewis theory of valency involving electron pairs. Robinson's ideas were developed by Ingold and his colleagues and their work coincided with the development of the Heitler–London theory. As a result, organic chemists have used a pictorial representation of some of the concepts of vb theory, like 'structure' and 'resonance', without taking over the mathematical complications of the theory. This approach is generally called resonance theory. It is also possible, though not so widely recognized, to build the ideas of these early workers into the framework of mo theory. As we have seen, mo theory is mathematically easier to handle than vb theory, so that, if one is after a semi-quantitative theory of organic chemistry, mo theory is probably a more fruitful approach.

16.2 Resonance Theory and its Relationship to Hückel Theory.

The qualitative resonance theory used with great effect by organic chemists is based on the idea that the true electronic state of a molecule can often not be correctly depicted by any single classical structure, but is intermediate between those represented by two or more valence-bond structures. The relationship between this pictorial theory and valence-bond theory on which it is based is immediately apparent.

The acetate anion is regarded as being a 'resonant hybrid' of two 'canonical forms'. The double headed arrow is used specifically

$$CH_3-C\overset{O}{\underset{O^-}{\big\langle}} \longleftrightarrow CH_3-C\overset{O^-}{\underset{O}{\big\langle}}$$

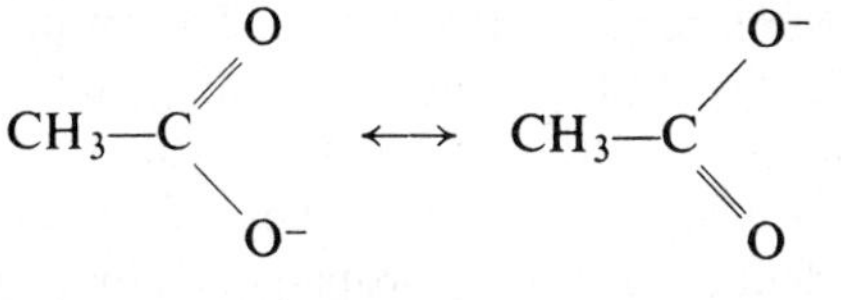

to indicate that 'resonance' occurs between two structures, i.e. that the true electronic state is intermediate between them.

Let us look at the acetate ion in terms of Hückel Theory. If we attempt a straightforward Hückel calculation on the acetate ion we have to introduce values for the coulomb and resonance integrals of the oxygen atoms. This complication can be avoided by considering instead the iso-electronic anion $(\overline{CH}_2-CH\!=\!CH_2)$.

The Hückel secular equations for the allyl system are

$$xc_1 + c_2 \qquad\quad = 0$$

$$c_1 + xc_2 + c_3 \ = 0$$

$$c_2 + xc_3 = 0.$$

Solving these equations we obtain the following three Hückel orbitals

$$\psi_3 = \tfrac{1}{2}(\phi_1 - \sqrt{2}\phi_2 + \phi_3) \qquad E_3 = \alpha - \sqrt{2}\beta$$

$$\psi_2 = \sqrt{\tfrac{1}{2}}(\phi_1 - \phi_3) \qquad E_2 = \alpha$$

$$\psi_1 = \tfrac{1}{2}(\phi_1 + \sqrt{2}\phi_2 + \phi_3) \qquad E_1 = \alpha + \sqrt{2}\beta$$

In the allyl cation there are two π-electrons in ψ_1 and in the anion there are two in ψ_1 and two in ψ_2. The calculated charge densities are as follows

the cation $\qquad \overset{+\frac{1}{2}}{C_1}-C_2-\overset{+\frac{1}{2}}{C_3}$

the anion $\qquad \overset{-\frac{1}{2}}{C_1}-C_2-\overset{-\frac{1}{2}}{C_3}$

These are in agreement with the VB picture in which, for example, the anion is an equal mixture of the two following structures.

$$\overset{+}{C}-C=C \leftrightarrow C=C-\overset{+}{C}$$

We can arrive at the above result without having to go through a complete Hückel calculation because we can see that these charge densities are just what we obtain by taking a neutral alternant (the allyl radical), which has a uniform charge density (page 276) and adding an electron to or removing an electron from ψ_2. Because allyl is an odd alternant its orbital must satisfy the general pairing theorem that we proved on page 277 and there must be a non-bonding orbital with energy $E = \alpha$ ($x=0$). We have shown on page 278 that the coefficients of such an orbital can be obtained very easily, and for a non-branching odd alternant having $2k$-1 atoms the coefficients of the non-bonding orbital of the starred set (the more numerous set) are $c = \pm(k)^{-1/2}$, and of the unstarred set are zero. This result was deduced from the property of a non-bonding orbital that the sum of the coefficients around any atom is zero. From this alone we can determine the coefficients of the non-bonding orbitals of starred atoms even if the molecule is branched. We take benzyl as an example.

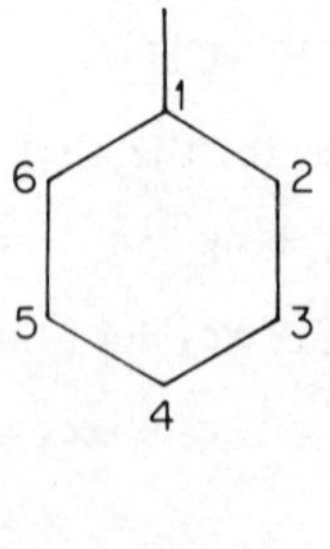

By considering the summation of coefficients about atom 7 we deduce $c_1 = 0$. Summing around atom 2 we have $c_1 + c_3 = 0$, hence $c_3 = 0$. Summing around atom 4 gives $c_5 = 0$. Thus atoms 1, 3 and 5, are the unstarred set and have zero coefficients. Let us now take $c_4 = a$. Summing the coefficients around atom 3 gives $c_4 + c_2 = 0$, hence $c_2 = -a$. Likewise $c_6 = a$. Finally summing around atom 1 gives $c_2 + c_6 + c_7 = 0$, hence $c_7 = 2a$. The normalization condition is $7a^2 = 1$, hence $a = (7)^{-1/2}$.

The benzyl radical will have one electron in this orbital, and it will be distributed around the starred atoms, the net π-electron charge at each atom is however, equal to one. In the anion there is one more electron in the non-bonding orbital; in the cation one less, the net charges are therefore as follows

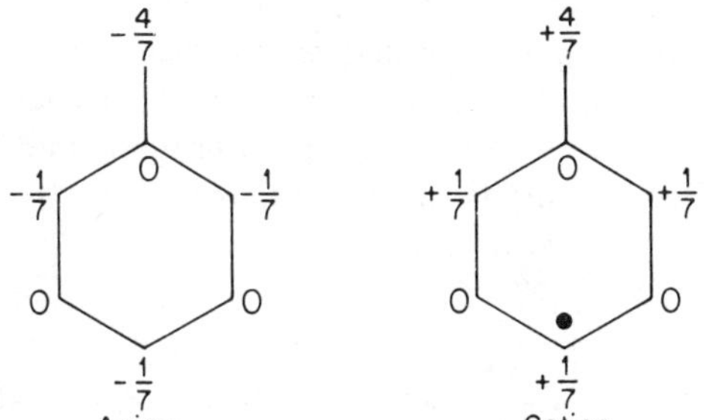

Anion Cation

This procedure, originally developed by Longuet-Higgins, provides us with a very simple method of estimating the distribution of charge in a conjugated anion or cation. For example, in terms of resonance theory the phenoxide ion and the diazonium ion would be represented as follows

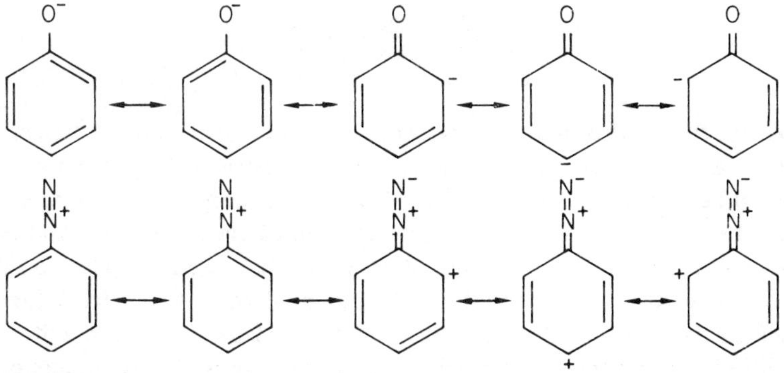

According to resonance theory, therefore, the negative charge on the oxygen atom of the phenoxide ion, and the positive charge on the nitrogen atoms of the diazonium salt are delocalized through the adjacent benzene ring at the *ortho-* and *para-*positions.

If we take the benzyl anion as a model for the phenoxide anion and the benzyl cation as a model for the diazonium cation, then we see that Hückel theory leads to the same result. Hückel and resonance theories are therefore in agreement in this example, if it is assumed that the three ionic structures in resonance theory have equal weight. This agreement has been proved to be generally true for molecules containing six-membered rings, but for the alternants which have four-membered rings the two theories do not always agree and, as we shall show in the next chapter, Hückel theory is in better agreement with experiment.

The greatest success of the early electronic theory of organic chemistry was the way in which it gave a mechanism for the transfer of electron pairs from one group to another, through a conjugated chain. For example, in resonance theory we would describe the ground state of the molecule R_2N—C=C—C=O in terms of the structures

$$R_2N-C{=}C-C{=}O \longleftrightarrow R_2\overset{+}{N}{=}C-C{=}C-\overset{-}{O}$$

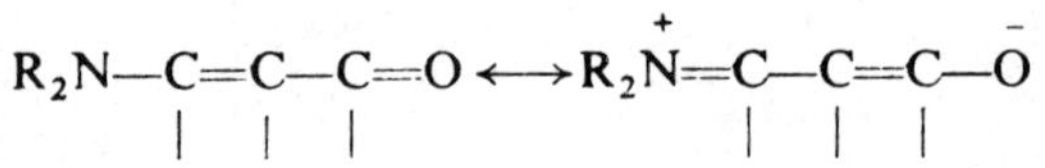

In this way the organic chemist indicates that the electron-donating R_2N group has an appreciable effect on the properties of the electron-accepting C=O group. Another method of representing this is

$$R_2\overset{..}{N}\!\!-\!C{=}C\!-\!C{=}\overset{.}{O}$$

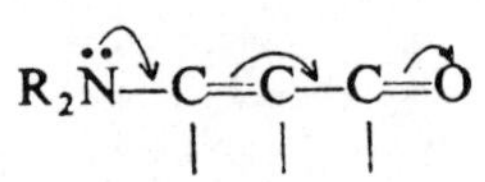

however, this is also used to represent complete redistribution of electron pairs in a reaction and it is probably better to restrict it to this usage.

Ingold carefully distinguished between polarization in the ground state, which he called the mesomeric effect (denoted by the symbol M) and possible polarizability in an excited state which he called the electromeric effect (denoted by the symbol E). The sum of these two effects Ingold called the tautomeric effect, but this is a most unsatisfactory term now that the difference between tautomerism and resonance is clearly understood.

We shall therefore use the term *resonance effect* (R) for either the M or E effects because although we consider it very important to distinguish between polarization in the ground state and possible polarization in a reacting molecule, this is a question of degree and not of kind.

In discussing the resonance effect we are concerned with the interaction of orbitals of π symmetry belonging to the substituent, with the π molecular orbitals of the remainder of the molecule. In styrene, for example, the interaction between the π orbitals of the vinyl group leads to a lower π-electron energy for the molecule. The total Hückel energy of styrene is less than that of the separate benzene and ethylene molecules by an amount $-0{\cdot}42\beta$.

Because styrene is an alternant hydrocarbon, it has a uniform charge density so that there is no net transfer of charge between the vinyl group and the ring. However, most common substituent groups which have π orbitals can either be classified as net donors ($+R$) or net acceptors ($-R$) of π electrons. For a substituent to be an electron donor it must have a filled orbital of π symmetry which has a relatively low ionization potential. For a substituent to be an electron acceptor it must have a vacant π orbital with a relatively high electron affinity.

All the $+R$ groups have lone pairs of electrons which have the symmetry of π orbitals (i.e. antisymmetric to reflection in the plane of the conjugated system). Table 16.1 gives the order of donating strength of the most common groups, as deduced from ionization potentials, and this seems to be generally in agreement with the order deduced from chemical evidence

TABLE 16.1 Donor ($+R$) groups

$+R$ strength of X	NH_2	>	SH	>	OH	>	Cl
Ionization potential of HX (ev)	10·15		10·46		12·59		12·74

All the common acceptors are unsaturated groups and have low lying antibonding π molecular orbitals which are not occupied by electrons Electron affinities are very difficult to obtain experimentally so it is not possible to give a table corresponding to 16.1. From spectroscopic evidence we can give a rough indication of $-R$ strength as follows.

TABLE 16.2 Acceptor ($-R$) groups

$-R$ strength	NO_2, SO_2R > RCO, SO_3R, CO_2R > CN

A substituent like SH can have acceptor properties through its vacant $3d$ orbitals, although overall it would behave like a donor. Similarly, the carbonyl group can donate electrons from its bonding molecular orbital but overall it will behave like an acceptor.

We can regard a methylene cation as an extreme example of an electron acceptor and a methylene anion as an extreme electron donor. We can therefore use the non-bonding orbital method described in the previous section to give us a qualitative picture of the resonance effect of any group. In fact, of course, different groups have very different electron withdrawing properties. In order to treat this problem in a slightly more sophisticated way we shall consider three classes of resonating substituents, the benzyl anion-type, the styrene-type and 2-phenylpropenyl anion-type.

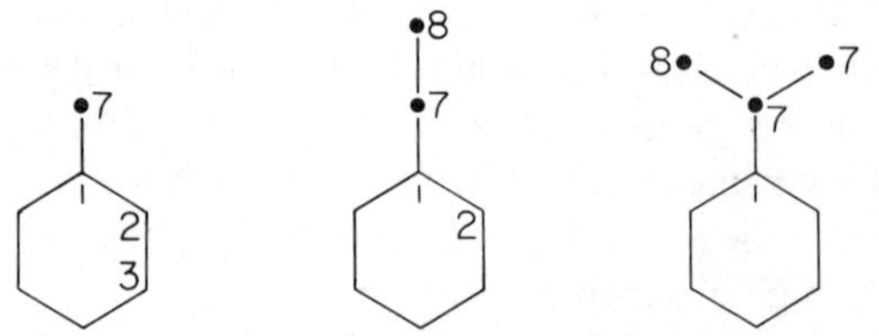

Benzyl anion Styrene 2-Phenylpropenyl anion

The benzyl anion is the isoelectronic conjugated system for phenol, anisole, aniline, etc. To carry out a Hückel calculation on these molecules we need to choose a value for the coulomb integral of the heteroatom (α_7) and for the resonance integral (β_{1-7}). Also, if the 1–7 bond is sufficiently polar this will affect the coulomb integral α_1, of the substituted carbon atom. This we shall deal with in the next section when we describe the I_π effect.

It is important to realize that the donor strength of these substituents depends on the values of both α_7 and β_{1-7}. If α_7 is very large and negative, then the pair of electrons contributed by the substituent will remain on the substituent. Also if β_{1-7} is very small this will inhibit the transfer of electrons to the ring. This effect is clearly seen experimentally from the fact that in *N,N*-dimethyl *o*-toluidine the *ortho-para* positions are much less active than those in *N,N*-dimethylaniline.

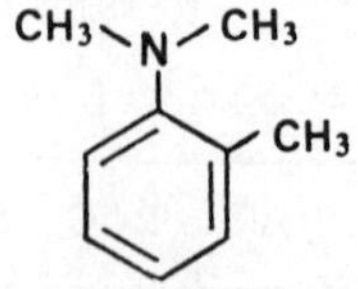

If the *N,N*-dimethyl group is twisted by an angle ϑ out of the plane of the benzene ring then β_{1-7} will be proportional to cos ϑ: the steric effect of the methyl group produces such a twist. Figure 16.1 shows the effect of a variation in α_7 and β_{1-7} on the Hückel charge densities in a molecule of this type.

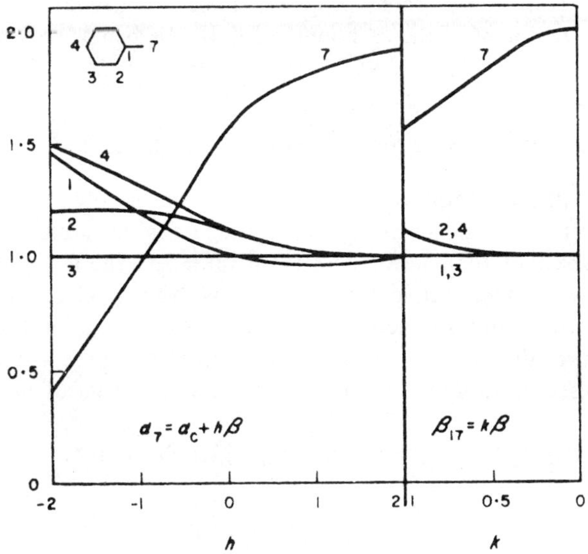

FIGURE 16.1. The variation of charge densities in the benzyl anion as a function of the Hückel parameters

The styrene and 2-phenyl anion systems are both models for electron-accepting groups e.g. RC=O, NO_2. The number of heteroparameters is increased but the general considerations about the relative acceptor strengths of groups follow the same arguments as we gave above for the donor substituents. However, all acceptor groups contain atoms of high electronegativity and this is the justification for their treatment by the benzyl cation model that we have given earlier. For example, the Hückel orbitals of the nitro groups will be found by solving the secular determinant

$$\begin{vmatrix} \alpha_O - E & \beta_{ON} & 0 \\ \beta_{ON} & \alpha_N - E & \beta_{ON} \\ 0 & \beta_{ON} & \alpha_O - E \end{vmatrix}$$

If α_O is very much less than α_N then there will be two molecular orbitals mainly localized on the oxygen atoms which will contain the four π electrons of the group, and one vacant orbital mainly localized on the nitrogen atom. Only the latter will have an energy close to that of the molecular orbitals of the benzene ring, so that the interaction between the nitro group and the ring is better represented by a perturbed benzyl cation model than by a perturbed 2-phenylpropenyl model. A similar argument can justify the treatment of the carbonyl group by the benzyl cation model.

16.3 The inductive effect in saturated systems I_σ

Superficially the simplest concept of the Robinson–Ingold theory is that of an inductive effect. For example, the reason that monofluoroacetic acid is a stronger acid than acetic acid is attributed to the fact that fluorine, being more electronegative than carbon, withdraws electrons from the carboxyl group and so facilitates the separation of a proton. But the mechanism of this withdrawal is by no means clear and it will be easier in the first instance to consider the dissociation constants of dibasic acids.

$$HOOC(CH_2)_nCOOH \underset{k_{-1}}{\overset{k_1}{\rightleftharpoons}} HOOC(CH_2)_nCOO^- + H^+ \qquad K_1 = \frac{k_1}{k_{-1}}$$

$$HOOC(CH_2)_nCOO^- \underset{k_{-2}}{\overset{k_2}{\rightleftharpoons}} {}^-OOC(CH_2)_nCOO^- + H^+ \qquad K_2 = \frac{k_2}{k_{-2}}.$$

There are two mobile protons in the undissociated acid and two sites for their return in the completely dissociated acid, hence, if the carboxyl groups were completely insulated from one another, we would have $k_1 = 2k_2$ and $k_{-2} = 2k_{-1}$, hence $K_1 = 4K_2$. In practice, the ratio K_1/K_2 is always greater than 4 but approaches 4 for very large values of n. If we regard this additional free energy ($\Delta G^\circ = -RT\log K_1 + RT\log 4K_2$) as entirely due to the energy necessary to remove the second proton against the attraction of the negative charge of the COO^- group, we get the expression due to Bjerrum

$$-\Delta G^\circ = RT\log\frac{K_1}{4K_2} = \frac{e^2}{Dr}, \qquad (16.1)$$

where D is the dielectric constant and r is the distance between the carboxyl groups. However, in order to fit experiment, this simple expression needs values for r which are much too small to be comparable to the molecular geometry.

Kirkwood and Westheimer were able to improve the Bjerrum equation by treating dibasic acids as ellipsoidal cavities of low dielectric constant surrounded by the solvent of high dielectric constant. Kirkwood and Westheimer's mathematically difficult calculations are entirely classical and will not be reproduced here. The important point is that this very simple electrical-field theory is able to account in large measure for the 'inductive effect' of the COO^- group.

We now return to the change in acidity in going from acetic acid to monofluoroacetic acid; in place of the negative charge on the carboxylate anion we must consider the electric field produced by the dipole of the C—F bond. We thus replace the Bjerrum expression by one proposed by Waters appropriate to the interaction between a dipole and a charge,

$$\Delta G^\circ = -RT \log \frac{K_H}{K_F} = \frac{e(\mu \cos \vartheta)}{Dr^2}, \tag{16.2}$$

where $(\mu \cos \vartheta)$ represents the difference between the dipole moments of the C—F and the C—H bonds in the direction of the ionizable proton. Like the Bjerrum equation this simple expression requires values of r which are much too small, but again reasonable values can be obtained by modifying the equation in terms of the Kirkwood and Westheimer hypothesis.

We can conclude, therefore, that the energy changes due to a long range inductive effect in aliphatic systems can best be understood in terms of a 'direct field' electrostatic phenomenon. But in addition there may be a small short range polarization of the electron density propagated through the carbon–carbon chain. For example, in the system X—C_1—C_2—C_3 if X is more electronegative than carbon then the electron pair making up the X—C_1 bond will be drawn towards X. This will make C_1 more electronegative than C_2 so that the C_1—C_2 electron pair will be drawn towards C_1. This effect is represented by the diagram

$$X \longleftarrow C_1 \longleftarrow \quad C_2 \leftarrow \quad C_3$$

In order to discuss this effect quantitatively we shall look at the wave functions for saturated compounds.

If we simply take a linear combination of hydrogen $1s$ and carbon sp^3 hybrid atomic orbitals and employ the same approximations that are used in the Hückel theory for π electrons, we then obtain completely localized bonds; this follows because orbitals only overlap in pairs. In order to obtain an inductive effect propagated through a chain, some orbital delocalization is needed. A very simple theory has been proposed by

Sandorfy, the so-called 'C' approximation, which introduces limited delocalization†.

Let us take propane as an example and consider only the C—C chain and the sp^3 hybrids involved in this (see figure 16.2). That is, we disregard not

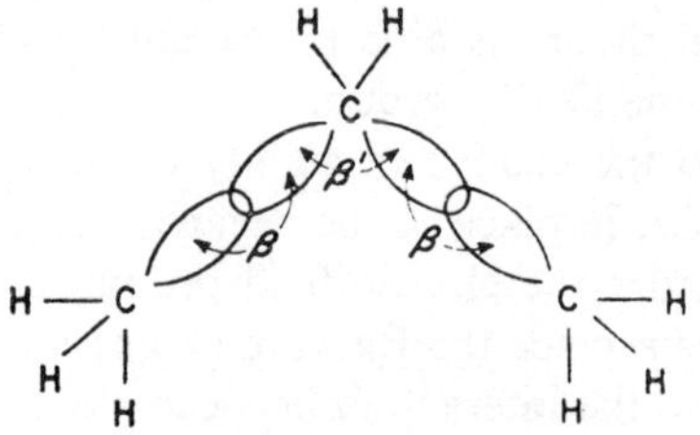

FIGURE 16.2. Resonance integrals in a saturated C–C chain: the Sandorfy C approximation

only the hydrogen atoms but the sp^3 carbon orbitals forming the C—H bonds. We shall put all the coulombic terms equal and, as in Hückel π-electron theory, neglect resonance integrals between non-adjacent carbon atoms and all overlap integrals. Two hybrid orbitals on the same carbon atom are orthogonal but we shall assume there is some interaction and introduce a resonance integral between them, β', and define a parameter k by $\beta' = k\beta$, where β is the resonance integral between two overlapping sp^3 hybrids in neighbouring carbon atoms. Using the abbreviation introduced earlier $x = (\alpha - E/\beta)$ the secular determinant for n-propane becomes

$$\begin{vmatrix} x & 1 & 0 & 0 \\ 1 & x & k & 0 \\ 0 & k & x & 1 \\ 0 & 0 & 1 & x \end{vmatrix} = 0.$$

It is of interest to note that this secular determinant would be the same as that for the π orbitals of butadiene in Hückel theory if $\beta' = \beta$, and if $k = 0$ we have two independent C—C bonds as mentioned before.

In order to investigate the inductive effect we shall replace a terminal CH_3— group by a heteroatom X, which we shall assume to be more electronegative than carbon $[\beta_{X-C} = \beta_{C-C}; \alpha_X = \alpha_C + h\beta]$. We shall investigate the case $h = 2$ and $k = 0{\cdot}25$, the calculation of the coefficients being entirely analogous to that for π-electron systems. If we define an *orbital electron charge*

$$q_\mu = 2 \sum_{r\,\mathrm{occ}} c_{r\mu}^2 \,, \qquad (16.3)$$

† SANDORFY, *Can. J. Chem.*, 33, 1337 (1955).

which is analogous to the π-electron charge of Hückel theory (15.8), and the net electron density on an atom by $\sum_{\mu} q_{\mu}$, where the summation is over all orbitals emanating from atom μ, then we obtain the following values:

$$
\begin{array}{lccccccc}
q_{\mu} & \mathrm{X} & \xrightarrow{\;1{\cdot}713\,|\,0{\cdot}314\;} & \mathrm{C} & \xrightarrow{\;1{\cdot}005\,|\,0{\cdot}971\;} & \mathrm{C} & \xrightarrow{\;1{\cdot}001\,|\,0{\cdot}998\;} & \mathrm{C} \\
\sum_{\mu} q_{\mu} & 1{\cdot}713 & & 1{\cdot}319 & & 1{\cdot}972 & & 0{\cdot}998
\end{array}
$$

The change in charge density resulting from the interaction between X and the carbon chain is then as follows.

$$
\begin{array}{cccc}
-0{\cdot}713 & +0{\cdot}681 & +0{\cdot}028 & +0{\cdot}002 \\
\mathrm{X}\text{———} & \mathrm{C}\text{———} & \mathrm{C}\text{———} & \mathrm{C}
\end{array}
$$

Net atom charges

The important feature of this calculation is that although the orbital charges oscillate down the chain, as in a π-electron system, the atom charges do not. Almost all the excess charge taken on by atom X is obtained at the expense of the first carbon atom.

We therefore conclude that the difference in the energy required to remove a proton from fluoroacetic and acetic acids arises mainly from the electrostatic effect of the C—F bond dipole and to a smaller extent from the polarization of the σ electrons propagated down the chain. In many cases there is no need to distinguish between these two effects and we can retain the symbol I for the combined effect, where $-I$ means electron attracting and $+I$ means electron repelling. However, we shall see later that the inductive effect on π electrons is quite different from that on σ electrons. It is therefore important to distinguish the inductive effect in these two systems and we use the symbols I_{σ} and I_{π}. The following table is a qualitative classification of the I_{σ} effect of some common groups.

TABLE 16.3. Inductive effect I_{σ} of groups

	$-I_{\sigma}$ groups		$+I_{\sigma}$ groups	
—NR_3	—F	—OR	—CH_3	—O^-
—NO_2	—Cl	—SR	—CH_2R	—S^-
—C≡N	—Br	—C=CR_2	—CHR_2	—COO^-
—C≡CR	—C=O		—CR_3	

There have been more sophisticated treatments of the bonding in saturated hydrocarbons. The most satisfactory are probably those of Pople and

Santry and of Hoffmann†. They base a Hückel-type theory on a linear combination of hydrogen and carbon hybrid orbitals, and take the resonance integrals to be proportional to the corresponding overlap integrals. Their parameters for two adjacent tetrahedral carbon atoms are summarized in figure 16.3 (φ is the azimuthal angle difference between the two tetrahedral orbitals involved in β''_{CC}). Sandorfy's C method amounts to ignoring all the primed terms except β'_{C}.

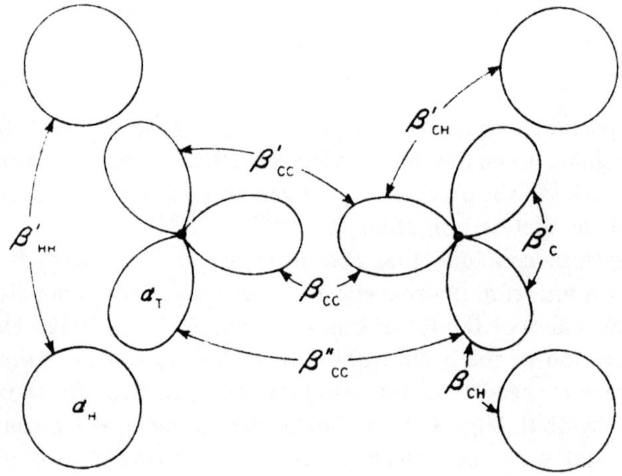

$\alpha_H = -13{\cdot}60$, $\beta_{CC} = -6{\cdot}47$, $\beta'_C = -1{\cdot}20$, $\beta'_{CH} = -1{\cdot}50$,
$\beta''_{CC} = -0{\cdot}07 - 1{\cdot}28 \cos\varphi$, $\alpha_T = -12{\cdot}40$, $\beta_{CH} = -6{\cdot}88$,
$\beta'_{CC} = -1{\cdot}08$, $\beta'_{HH} = -1{\cdot}90$ all in ev units.

FIGURE 16.3 The interaction integrals in the Pople–Santry treatment of σ-electron systems

† POPLE and SANTRY, *Mol. Phys.*, 7, 269 (1964); HOFFMANN, *J. Chem. Phys.*, **39**, 1397 (1963).

16.4 The inductive effect in unsaturated systems I_π

Probably the least satisfactory aspect of resonance theory is the lack of any clear picture of the effect an inductive group will have on a π bond. The methyl group in toluene is *ortho-para* directing and if we assume that the orientation of electrophilic substitution depends on electron densities, the only way of accommodating this result in resonance theory is to draw canonical forms involving hyperconjugations.

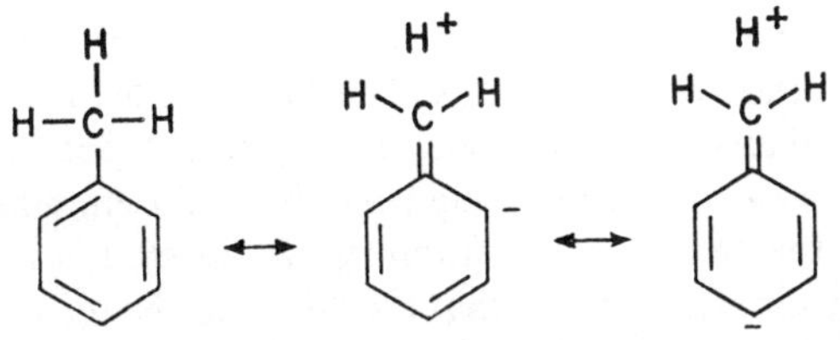

Similarly, the trifluoromethyl group and the quaternary ammonium group are *meta* directing (the latter less exclusively than was originally believed) and the canonical forms necessary to explain this must bring in vacant $3p$ orbitals of the substituents.

However, these directing properties can be explained without the introduction of a resonance effect since the inductive effect on π electrons produces a polarization of charge. We have seen that in a saturated system a heteroatom produces quite a substantial effect on the electron density at the adjacent atom, but very little effect further along the saturated chain. Thus an electronegative heteroatom or substituent attached to a conjugated chain can be regarded as altering the coulomb integral (α) of the carbon atom to which it is attached, while leaving the coulomb integral of more remote atoms unaffected, at least within the Hückel approximation of considering α to be independent of π-electron density.

To calculate the inductive effect on π electrons in Hückel theory we have to introduce a suitable value of α for the substituted carbon atom. If we just want to know the effect on the net electron charges then this can be found by using the mutual atom polarizabilities of the unperturbed hydrocarbon and the expression (15.41)

$$q_\mu = 1 + \pi_{\mu 1}(\alpha_1 - \alpha_C),$$

where α_1 is the selected value for the coulomb integral of the substituted carbon atom. We can illustrate this with a substituted butadiene. The mutual atom polarizabilities were calculated in problem 15.7, and if we assume that the substituent changes the coulomb integral of the substituted atom by an amount β, then the following charges are obtained

X—	—C_1—	—C_2—	—C_3—	—C_4
$\pi_{\mu 1}(\beta^{-1})$	0·626	−0·402	0·045	−0·268
π-electron charges q_μ	1·626	0·598	1·045	0·732
total charges	0·626	−0·402	0·045	−0·268

The effect of an inductive substituent on benzene follows the same type of calculation as was given for the charge distribution in pyridine (page 298). We note from these results the important fact that the inductive effect on π electrons produces a polarization of charge between the starred and unstarred atoms of an alternant hydrocarbon. To emphasize this result we give below the charge polarization produced by a $+I_\pi$ substituent in naphthalene; it should be noted that the effect on the unsubstituted ring is rather small.

We see that both an I_π and an R substituent produce a long range effect on a conjugated system. The difference between the two is that the resonance effect produces a change in the electron densities only at every

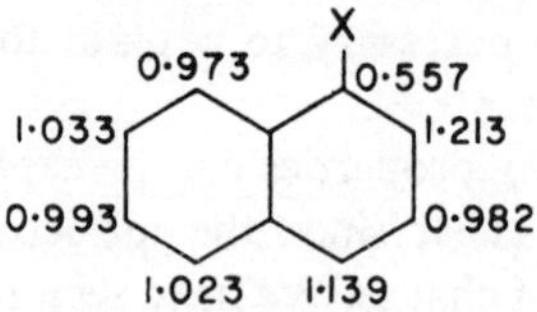

$$q_\mu = 1 + \pi_{\mu 1}(\alpha_1 - \alpha_C)$$

$$\alpha_1 - \alpha_C = -\beta$$

alternate atom (this is only strictly true if the substituent is treated by the equivalent hydrocarbon model), whilst the inductive effect builds up charge on one set of atoms but decreases it on the other. We can illustrate this as follows, where δ represents a small amount of charge.

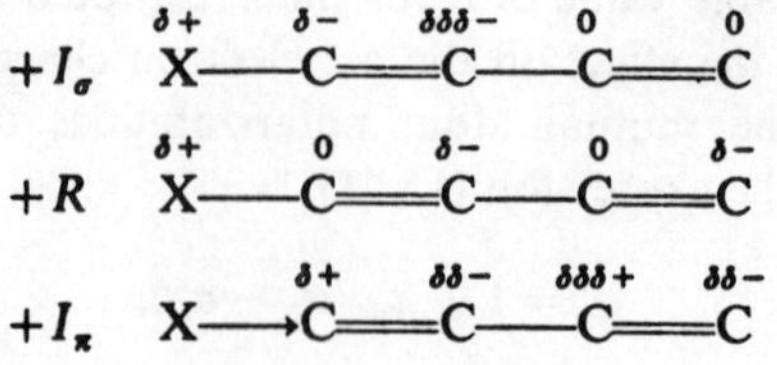

In reactions that depend on electron densities, a substituent which produces alternating reactivities along a conjugated chain, may be exerting an R effect or an I_π effect or a combination of the two.

We now turn to the vexed question of the electronic effects of halogens. There is no doubt that the inductive effect $-I_\sigma$ of the halogens is in the order F > Cl > Br > I, i.e. in the order of their relative electronegativity. The halogens, when attached to an organic molecule, have electrons in non-bonded orbitals of π symmetry and they would therefore be expected to have a $+R$ effect on a conjugated system. Data from physical organic chemistry have provided a wide variety of evidence to support the belief that the $+R$ strengths are in the order F > Cl > Br > I, which is the reverse of what would be expected on the grounds of their electronegativity or ionization potentials (see page 317). Spectroscopic data, on the other hand, indicate a relative order of resonant release I > Br > Cl > F which is in accord with electronegativity. The very well established physical organic data apparently require that fluorine should release its electrons more readily than chlorine, while spectroscopic data and theoretical considerations predict the reverse.

Various explanations have been offered to account for the unexpected order obtained from physical organic data. These mostly depend on the idea that the orbitals of the larger halogens will overlap less effectively with the π orbitals of the conjugated carbon system. But overlap will probably play a smaller part than electronegativity. The probable solution of this paradox comes from considering the I_π effect of the halogens. We have already agreed that the halogens attract σ electrons ($-I_\sigma$) by virtue of their greater electronegativity than carbon. However, π electrons lying on an adjacent carbon atom will be affected not only by the dipole created in the σ bond but by the repulsion due to the halogen atom lone-pair electrons of π symmetry as illustrated in figure 16.4. This repulsion may be attributed

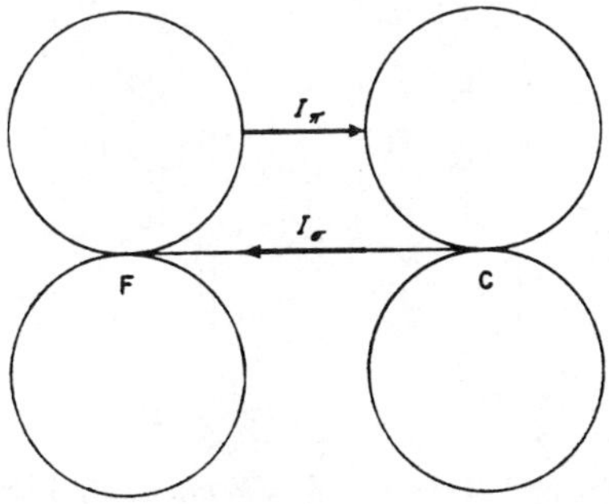

FIGURE 16.4. Repulsion and attraction of electrons in a C—F bond

to the Pauli exclusion principle rather than to an electrostatic effect. If an orbital containing two electrons overlaps with one containing a single electron, this will lead to a repulsion, since two electrons with the same spin are trying to occupy the same space. But we have shown that in many respects the electron density changes produced by a $+R$ and a $+I_\pi$ effect are similar, so that if one is looking at some property of the ground state of a molecule which depends mainly on electron density, it is difficult to separate the two. It is our contention that what the physical organic chemists have been measuring is the sum of the $+R$ and $+I_\pi$ effects†.

There is an additional factor to be considered; the heavier halogen atoms have vacant orbitals into which they can accept electrons and so exert a $-R$ effect. In chlorine this implies the occupation of $3d$ orbitals (or less likely $4p$). In fluorine such a process would be most unfavourable, but with bromine and iodine it becomes increasingly probable. The $+R$ and $-R$ effects of these substituents will produce electron density changes which cancel one another but will produce energy changes which add to one another, both being stabilizing. This means that the order of the R effect deduced from spectroscopic evidence (which is mainly from energy changes) is apparently different from the order deduced from ground state properties (which depend mainly on electron density). This emphasizes a point which has been sadly glossed over in the literature: when one talks about an R effect or an I effect one should state clearly whether this is an effect on electron density or on the energy of the system.

16.5 Hyperconjugation

Hyperconjugation is the term given to the process by which a group such as a methyl group interacts with an unsaturated system as though it possessed electrons of π symmetry. In resonance theory this is brought in through structures such as I and II but a proper VB treatment has not been carried out.

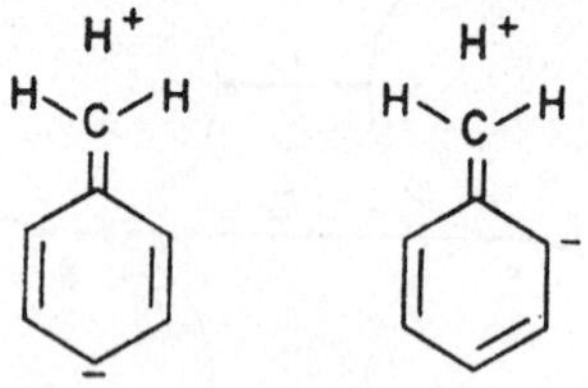

I II

† CLARK, MURRELL and TEDDER, *J. Chem. Soc.*, 1250 (1963); CRAIG and DOGGETT, *Mol. Phys.*, **8**, 485 (1965).

In a MO treatment we have to break down the orbitals of the CH_3 group in such a way that we obtain the components of π symmetry. A method due to Coulson includes the carbon $2p$ atomic orbitals and group orbitals of the three hydrogen atoms as part of a conjugated system. If we take an orientation of the methyl group as shown in figure 16.5, then the group orbitals, which are depicted in figure 16.6, have the following form (see page 106)

$$\psi_\sigma = \frac{1}{\sqrt{3}}(\phi_a + \phi_b + \phi_c)$$

$$\psi_{\pi_1} = \frac{1}{\sqrt{6}}(2\phi_a - \phi_b - \phi_c)$$

$$\psi_{\pi_2} = \frac{1}{\sqrt{2}}(\phi_b - \phi_c).$$

ψ_σ is symmetrical about the C—C bond axis, and can only interact with the σ electrons of the molecule. ψ_{π_1}, on the other hand, has a nodal plane roughly in line with the $2p\pi$ orbital of the adjacent carbon atom. It is therefore similar to a normal π-type orbital and delocalization is possible

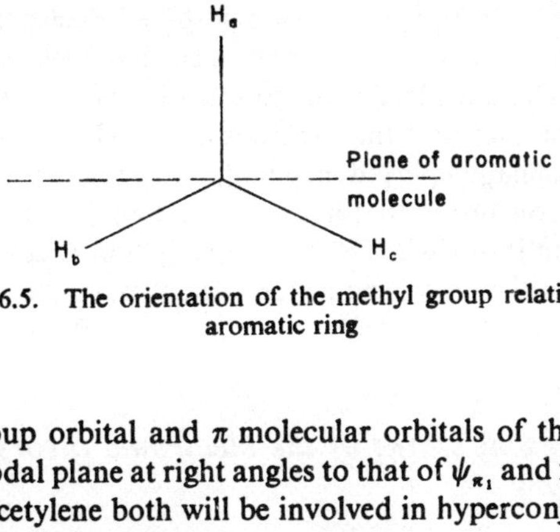

FIGURE 16.5. The orientation of the methyl group relative to the aromatic ring

between the group orbital and π molecular orbitals of the benzene ring. ψ_{π_2} also has a nodal plane at right angles to that of ψ_{π_1} and for a compound such as methylacetylene both will be involved in hyperconjugation.

$$H_3{\equiv}C{-}C{\equiv}C{-}H$$

We thus treat the methyl group in the same way as one does a halogen substituent. Although we have assumed a particular orientation of the CH_3 group in this discussion it can be shown that the extent of electron delocalization between the methyl group and the ring is independent of the orientation of the group.

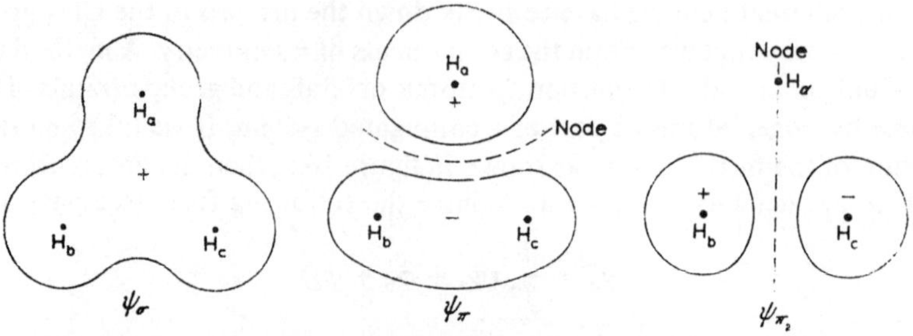

FIGURE 16.6. The group orbitals of the hydrogen atoms in a methyl substituent

Hyperconjugation is not a concept that is confined to a methyl group. Any group which has orbitals with components of π symmetry can extend the conjugation of an unsaturated system. For example the C—C bond orbitals of a *t*-butyl group could be taken in symmetry combinations in just the same way as were the hydrogen orbitals of the methyl group.

Early literature on hyperconjugation is much concerned with observations by Baker and Nathan concerning the relative rates of hydrolysis of a series of alkyl substituted *p*-benzyl halides. In order to explain their results, Baker and Nathan invoked the idea that a methyl group could release electrons as though it possessed unshared electrons of π symmetry, i.e. by hyperconjugation. Since then evidence has been presented to show that the Baker–Nathan effect is due to solvation effects rather than hyperconjugation. The fact that the experimental results for which hyperconjugation was originally invoked may no longer require the concept, should not lead to the conclusion that hyperconjugation does not occur. What is really at question is not whether hyperconjugation occurs, but its importance in any particular phenomenon. Spectroscopic evidence is clearly in favour of a methyl group behaving as a weak $+R$-type substituent.

16.6 The present status of the electronic theory of organic chemistry

The Robinson–Ingold electronic theory of organic chemistry is nearly forty years old and it is reasonable to ask why we still have to make use of it, if we can obtain the same data from simple quantum mechanics.

The theory is concerned with the effect various substituents have on the electron distribution in an organic molecule and we have seen that the original ideas based on the Lewis theory of valency are capable of re-interpretation in terms of simple MO theory. We have assumed that this electron distribution is related to chemical reactivity and equilibria, but we have been very careful to be rather vague about this.

In the next chapter we shall consider what specific data we can obtain from the simplified forms of the quantum-mechanical theories of valency in order to compare and predict reaction rates and positions of equilibria. We shall find that even for the simplest reaction it is impossible to find a set of theoretical parameters which can be used to calculate reaction rates accurately. The organic chemist is often working with complex molecules for which at the moment one cannot expect reliable MO calculations. It is therefore still extremely valuable to have a pictorial theory which can be used to make intelligent guesses about the effect of changing substituents.

Since its conception, the electronic theory has undergone many modifications but it is still a vital part of organic chemistry. What is important, and what we have tried to show, is that 'arrow pushing' and MO calculations are not in contradiction to each other, but are complementary.

Chapter 17

Reactions and relative reactivity in organic chemistry

17.1 Introduction

This book is concerned with theories of valency and not with reaction kinetics. It is nonetheless pertinent for us to ask whether valency theory can throw any light on the problem of relative reaction rates.

Let us consider the very simplest chemical reaction $H + H_2 \rightarrow H_2 + H$. As the hydrogen atom approaches the hydrogen molecule there will be an increase in the electronic potential energy of the system (H_3 is not a stable molecule). Calculations show that the most stable configuration for the H_3 system is a linear one, which means that the easiest path for the reaction is one in which the three atoms remain in a straight line. The potential energy

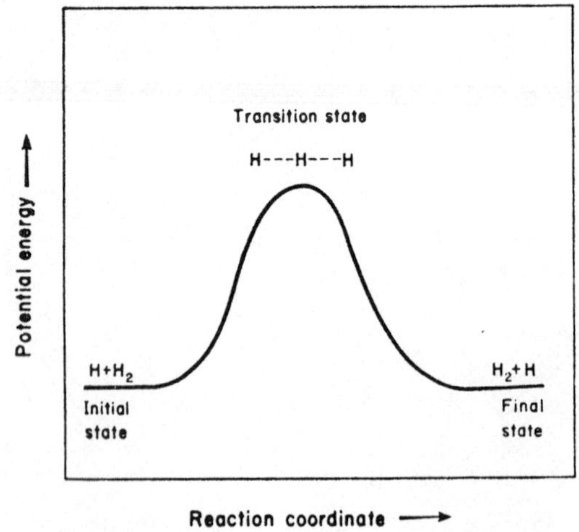

FIGURE 17.1.
The potential energy diagram for the $H + H_2$ reaction

along the reaction path will be qualitatively as shown in figure 17.1. At the position of maximum potential energy the molecules are said to be in the *transition state*, and the difference between the energies of the initial and transition states is called the *activation energy*.

Even for this most simple of all chemical reactions the results of a non-empirical calculation of the potential energy surface are only in moderate agreement with experiment†, so that for more complicated reactions we must certainly turn to empirical theories. However, the rate of a chemical reaction depends on the difference in free energy between the reactants and the transition state and not just on the activation energy. This is because experiments are not done on isolated molecules but on an ensemble of molecules with different vibrational, rotational and translational energies. Thus, in order to calculate the rate of a chemical reaction we need to employ the methods of statistical mechanics, the details of which are outside the scope of this book. The results of such a statistical theory of reaction rates can be summarized as follows. The rate constant can be expressed in terms of two parameters $\Delta S^{\ddagger}$ (the entropy of activation) and $\Delta H^{\ddagger}$ (the enthalpy of activation) according to the equation

$$\text{Rate constant} = \frac{kT}{h}\, e^{\Delta S^{\ddagger}/R}\, e^{-\Delta H^{\ddagger}/RT}. \qquad (17.1)$$

$\Delta H^{\ddagger}$, the enthalpy of activation, is directly related to the difference in potential energy between the initial and transition states, and it is this quantity which we may hope to estimate by valence theory.

In solution the solvation of the reactants and activated complex makes an important contribution to $\Delta H^{\ddagger}$ and $\Delta S^{\ddagger}$ so that, at present, a complete calculation of these quantities is quite impractical. For most organic reactions we are thus driven to comparing rates of reaction, that is, free-energy data, with calculations on isolated molecules or complexes in free space, that is, idealized potential energy data. The only justification for this procedure is that, if we compare a particular reaction for a related series of compounds (e.g. the nitration of aromatic hydrocarbons) under identical conditions of concentration and temperature, there is good reason to believe that the entropy and solvation factors will remain constant within the series, or at least vary in a regular manner.

The field in which this approach has been most employed is that of aromatic substitution reactions. Hückel theory provides a very simple approach on which to base our calculations, and in the next section we shall examine some of the ways in which this has been done.

† Boys *et al.*, *Nature*, **178**, 1207 (1956).

17.2 *Electrophilic aromatic substitution*

Aromatic substitution reactions are not simple one stage processes. For example, it is generally agreed that an electrophilic substitution can be depicted as follows.

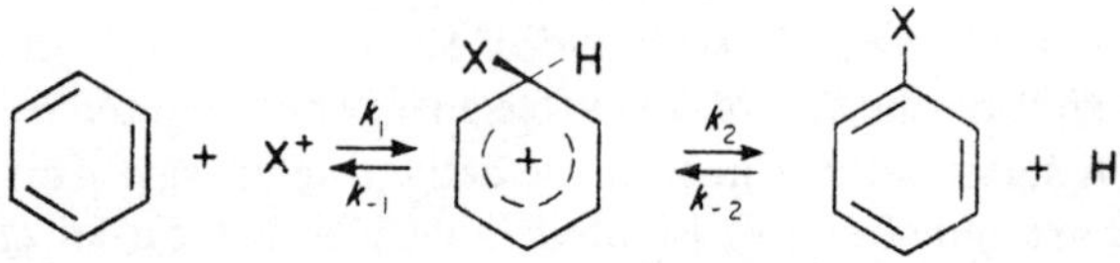

In the reaction intermediate the carbon atom at which the substitution is taking place has an approximately tetrahedral configuration, the attacking group X and the leaving hydrogen atom being on opposite sides of the ring. This intermediate is called a σ *complex* or 'Wheland intermediate'.

We shall only consider reactions in which k_1 is rate determining, which includes the important cases of nitration and halogenation. The potential energy diagram appropriate to this case has the qualitative form shown in figure 17.2

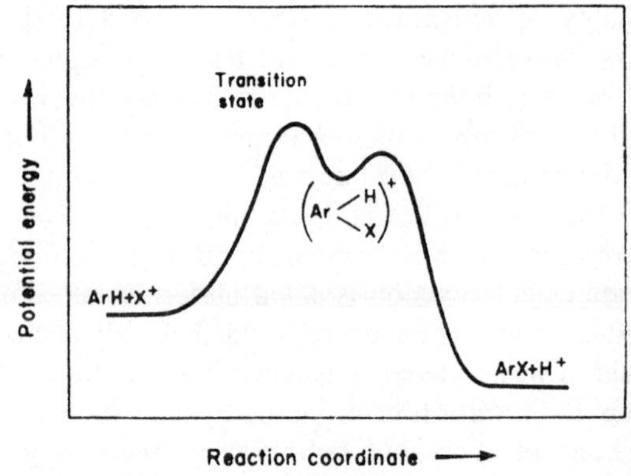

FIGURE 17.2. The potential energy diagram for an electrophilic substitution

The σ complex is not the transition state, and, since a number of these complexes have actually been isolated at low temperatures, it is probably quite stable. We can only guess about the structure of the transition state and are therefore reduced to predicting the shape of the potential energy diagram from the structures of the reactants and products; *the products in this case being the σ complex*, not the final product of the reaction.

We would anticipate that the smaller the initial repulsion energy between the reactants the smaller will be the activation energy; similarly, the more stable the σ complex relative to the reactants the smaller will be the activation energy. Ideally we should always consider both these factors, but in practice this is rarely done. Hückel theory is used in two ways; either we examine the properties of the molecular orbitals of the initial aromatic compound or we examine the molecular orbitals of the σ complex. The first approach is called the *isolated-molecule approximation* and the second is usually referred to as *localization theory* as it basically attempts to assess the energy required to localize the π electrons needed to form the σ bonds of the complex.

17.3 The isolated-molecule approximation

There is good evidence that in aromatic nitration in concentrated sulphuric acid the electrophile is the positively charged nitronium ion NO_2^+. We would therefore expect a correlation between the relative rates of nitration at different sites in a molecule and the electron densities at those sites. This is what is found when large differences are expected between the electron densities at different positions. Unfortunately these large differences only occur when the reacting hydrocarbon already contains heteroatoms so that the calculated electron densities depend on the choice of Hückel parameters for these atoms. Because there is still some uncertainty in the value of these parameters it is difficult to get a convincing test of the correlation between electron density and reactivity. Qualitatively, however, one finds that $+R$ groups, which donate electrons to an aromatic ring, lead to enhanced attack at the *ortho* and *para* positions, and $-R$ groups lead to attack at the *meta* position with a lower rate than for the unsubstituted molecule.

There may be cases of electrophilic substitutions in which the attacking species is not a positive ion, and for these we would not necessarily expect a correlation between reaction rate and electron density. In addition, electron density can give us no guide for the substitution of unsubstituted polycyclic aromatic hydrocarbons because, as we have previously shown, the π-electron charge is unity at every atom in an alternant hydrocarbon. Even for non-alternants the calculated differences in charges are small. For these molecules a more important factor is likely to be the polarization of charge produced by the attacking substituent.

When a charged species approaches very close to atom μ it will alter the electron density at that site. If we assume that the attacking group changes the coulomb integral α_μ then we can calculate the change in electron density through the expression for self-polarizability (15.37)

$$\delta q_\mu = \pi_{\mu,\mu}\delta\alpha_\mu. \tag{17.2}$$

$\delta\alpha_\mu$ is only likely to be large if the attacking group is highly polar (e.g. NO_2^+), so that it is only in these cases that we expect a correlation between the rate of a reaction and the index $\pi_{\mu,\mu}$. As we shall see later (table 17.1) the correlation is quite good. Because the induced charges δq_μ will generally be small $\pi_{\mu,\mu}$ is expected to be an unreliable reactivity index when permanent charges are present. For this reason it will not play an important role in the theory of the reactivity of heterosubstituted molecules. Fukui† suggested that only the density of the electrons in the highest bonding molecular orbital is important in determining the position of electrophilic attack, and only the density in the lowest antibonding orbital is important for nucleophilic attack. These orbitals he called the *frontier* orbitals. To take a simple example, an electron in the highest occupied orbital of butadiene, ψ_2 is distributed 36% on each of atoms 1 and 4, and 14% on each of atoms 2 and 3 (see page 274). According to frontier-orbital theory we therefore expect attack by an electrophile at the terminal positions. An electron in the lowest antibonding orbital is distributed in the same way so we also expect nucleophilic attack at the terminal positions.

There is no direct theoretical justification for Fukui's suggestion, but one can be found indirectly through the relationship between frontier-electron density ($\mathfrak{F}$) and a superdelocalizability factor also introduced by Fukui‡.

An electrophile has a low lying vacant orbital. If this orbital overlaps with the π orbitals of the hydrocarbon then there will be a stabilization of the transition state through electron delocalization. We can calculate the delocalization energy using second-order perturbation theory. Let E_s be the energy of the orbital ψ_s of the electrophile and E_r be the energy of the bonding orbitals ψ_r of the hydrocarbon, and let us consider only the case $E_s > E_r$. If the resonance integral between the substituent orbital and the atomic orbital of the attacked carbon atom is β', then the interaction between ψ_s and ψ_r will lower the energy of orbital ψ_r by an amount (6.54)

$$\frac{\mathbf{H}_{sr}^2}{E_s - E_r} = \frac{c_{r\mu}^2\beta'^2}{E_s - E_r}. \tag{17.3}$$

† Fukui *et al.*, *J. Chem. Phys.*, **20**, 722 (1952).
‡ Fukui *et al.*, *J. Chem. Phys.*, **27**, 1247 (1957).

The total stabilization is obtained by summing over all the bonding orbitals and including a factor 2 because there are two electrons in each orbital

$$2 \sum_{r\,\mathrm{occ}} \frac{c_{r\mu}^2 \beta'^2}{E_s - E_r}. \tag{17.4}$$

This expression has an obvious similarity to Fukui's superdelocalizability factor which is defined by

$$S_\mu = 2 \sum_{r\,\mathrm{occ}} \frac{c_{r\mu}^2}{x_r}, \tag{17.5}$$

where x_r is the dimensionless Hückel energy parameter defined before (15.4). The larger S_μ is, the larger the delocalization energy will be and the lower the activation energy.

The most important term in the summation in (17.5) will, in general, come from the frontier orbital, ψ_1, since this is the one which has the smallest value of x_r. An approximation to the superdelocalizability (not always a good one) is therefore

$$S_\mu' = \frac{2c_{1\mu}^2}{x_1}. \tag{17.6}$$

But $c_{1\mu}^2$ is the frontier-electron density so if one finds a correlation between the rate of reaction and S_μ' one should also find it between the rate and $c_{1\mu}^2$. This provides some justification of the frontier density idea.

Several indices for correlating reaction rates have been based on a charge-transfer model. In these models it is assumed that in the transition state an electron has been transferred from the hydrocarbon to the electrophile. The less the energy required for this transfer the more facile the reaction. We shall describe briefly Brown's treatment of this model and show that it also leads to a justification for the frontier-electron approach.

We can consider two different states for the hydrocarbon plus electrophile; these differ according to which species has the positive charge. We can draw potential energy diagrams for each of the two states, as in figure 17.3. Because of the interaction between the two states, and the non-crossing rule (section 10.5), the potential energy will really follow the dotted line shown†.

The reactants approach one another with a potential energy represented by the upper curve, but when they are close together there is the possibility of jumping to the lower curve with an electron being transferred from the hydrocarbon to the electrophile.

† This type of interaction is discussed more fully in section 18.2.

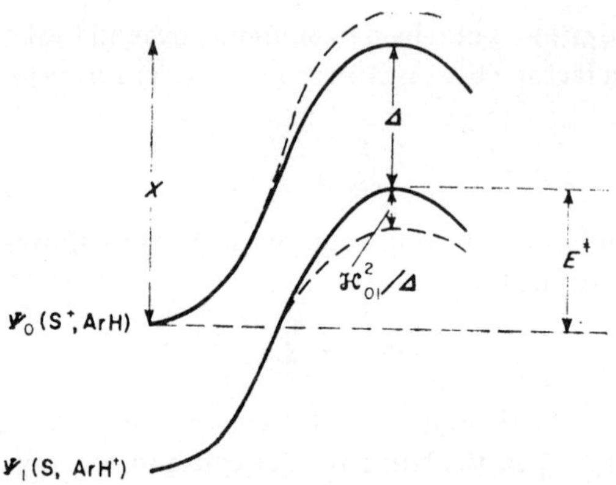

FIGURE 17.3. The potential energy diagram for a charge-transfer model of electrophilic substitution

Let Ψ_0 and Ψ_1 be the wave functions of the states as defined in the figure, and Δ be the difference in their energies at the transition state before their interaction. Let X be the energy of the state Ψ_0 relative to that of the reactants, and $\mathscr{H}_{01}$ the interaction integral between Ψ_0 and Ψ_1 at the transition state. The activation energy is then given from (6.54) by

$$E^{\ddagger} = X - \Delta - \frac{\mathscr{H}_{01}^2}{\Delta}. \tag{17.7}$$

Δ is given by the difference between the electron affinity of the electrophile and the ionization potential of the hydrocarbon. The latter is, in Hückel theory, equal to βx_1, hence we write

$$\Delta = -\beta(x_s - x_1), \tag{17.8}$$

where x_s is a constant of the electrophile. The interaction integral $\mathscr{H}_{01}$ is proportional to the Hückel coefficient of the frontier orbital at the site of attack: we write this as follows

$$\mathscr{H}_{01} = \sqrt{2}c_{1\mu}g\beta. \tag{17.9}$$

The $\sqrt{2}$ comes in via a spin factor for the state Ψ_1 (9.24). The resonance integral between the accepting orbital of the electrophile and the π atomic orbital at the position of attack is $g\beta$. Substituting (17.8) and (17.9) into (17.7) we obtain

$$E^{\ddagger} = X + \beta\left(x_s - x_1 + \frac{2g^2c_{1\mu}^2}{x_s - x_1}\right). \tag{17.10}$$

There is bound to be some solvation energy contributing to $E^{\ddagger}$, however one may hope that both this and X are constant for similar reactions. Brown therefore defines an index of reaction Z_{μ}, by†

$$Z_{\mu} = x_s - x_1 + \frac{2g^2 c_{1\mu}^2}{x_s - x_1}. \qquad (17.11)$$

The parameter g is a measure of the extent of interaction in the transition state: both this and x_s are treated as constant parameters for a given type of reaction. For example, for nitration Brown has taken $x_s = 2$ and $g = 0{\cdot}6$ and finds a reasonable correlation with reaction rate. For different sites in a particular hydrocarbon Z_{μ} varies only through the frontier density $c_{1\mu}^{\ 2}$. This therefore provides another justification for the frontier model.

Finally we must mention the free valence F_{μ} which was defined in equation (15.13). Modern opinion is that this index is probably rather more suspect than the others we have described.

17.4 Localization theory

In localization theory one makes the assumption that the lower the energy of the σ complex the lower will be the energy of the transition state. Localization approximations involve an assessment of the change in π-electron energy between the initial hydrocarbon and the σ complex. If the Hückel energy of the hydrocarbon before attack by an electrophile is $a\alpha + b\beta$ and that of the σ complex is $(a - 2)\alpha + b_{\mu}^{+}\beta$ (there are two less π electrons in the σ complex which have been used to form the bond with the electrophile), then the localization energy is defined by

$$L_{\mu}^{+} = (b - b_{\mu}^{+})\beta \qquad (17.12)$$

This represents only that part of the total localization energy which will change from one position of attack to another.

Evaluating localization energies is simply a question of working out the total Hückel energy for the hydrocarbon and the total Hückel energies for the fragment that is obtained by removing one carbon atom and two π electrons from the conjugated system. For example, the Hückel energy of benzene is $6\alpha + 8{\cdot}00\beta$; localization of a pair of electrons at one atom leaves a pentadienyl cation whose Hückel energy is $4\alpha + 5{\cdot}46\beta$. Thus L_{μ}^{+} for benzene is $2{\cdot}54\beta$. If there are different types of carbon atom in the molecule, these will have different localization energies. The lower the localization

† BROWN. *J. Chem. Soc.*, 2224 (1959).

energy, the more facile the reaction at that position. For example, the localization energies for azulene are as follows.

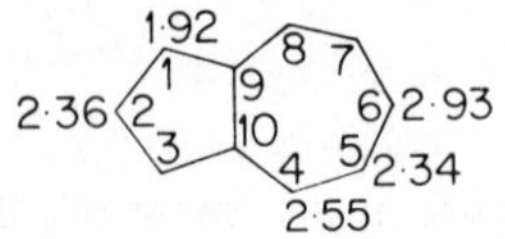

We therefore predict that electrophilic substitution should occur far more readily in the 1-position than elsewhere, and this is in complete accord with experiment.

We saw in section 15.5 how we could simplify the calculation of π-electron energy of a large molecule by breaking it down into two fragments R and S. The energy of the combined molecule RS was given approximately by

$$E_{\mathrm{RS}} \approx E_{\mathrm{R}} + E_{\mathrm{S}} + 2\beta \left(\sum_{\nu \to \mu} c^{\mathrm{R}}_{\mathrm{o}\mu}\, c^{\mathrm{S}}_{\mathrm{o}\nu'}\right) \tag{15.32}$$

where $c^{\mathrm{R}}_{\mathrm{o}\mu}$ and $c^{\mathrm{S}}_{\mathrm{o}\nu}$ were the coefficients of the non-bonding orbitals of R and S respectively for atoms μ (of R) and ν (of S) which are joined together.

We can use this method to calculate approximate localization energies. To do this we take the particular case in which the fragment R is the single atom where substitution occurs, so that $c^{\mathrm{R}}_{\mathrm{o}\mu} = 1$ and the two atoms of S joined to μ are ν and ν' (15.32) then becomes

$$E_{\mathrm{RS}} - E_{\mathrm{R}} - E_{\mathrm{S}} \approx 2\beta(c_{\mathrm{o}\nu} + c_{\mathrm{o}\nu'}) \tag{17.13}$$

where c_{ov} and c_{ov}' are the coefficients of the non-bonding orbital adjacent to the site of S. The sum $2\,(c_{ov}+c_{ov}')$ is called the *Dewar Number* N_μ. The smaller the Dewar number the smaller will be the interaction energy between R and S, that is, the easier it will be to isolate the carbon atom from the π-electron system.

For example, attack on naphthalene can proceed through two σ complexes shown below. We evaluate the coefficients of the non-bonding orbitals of the two π-electron fragments in the manner described earlier (page 314).

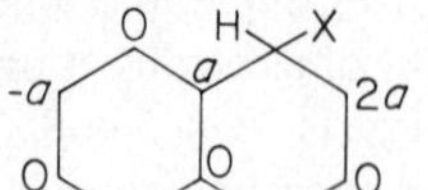

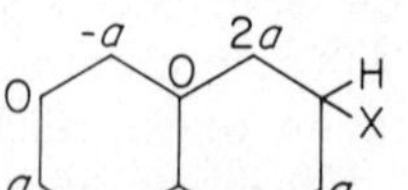

$a = \sqrt{11};\ N_1 = 6 \times \sqrt{(11)} = 1{\cdot}81 \qquad a = \sqrt{8};\ N_2 = 6 \times \sqrt{8} = 2{\cdot}12$

N is smaller in the 1-position than in the 2, and substitution at the 1-position should be easier, in agreement with experiment.

Dewar's method cannot be used for non-alternants because their radical fragments do not generally have non-bonding orbitals.

17.5 *Nucleophilic and radical attack*

Nucleophilic substitution is easily discussed using similar indices to those we have already considered for electrophilic substitution. The attacking reagent usually has a negative charge (e.g. OH^-, CN^-) and an occupied orbital of relatively high energy. We therefore expect a correlation between electron density at the position of attack and the rate of reaction, and this is qualitatively what is found.

The mechanism of nucleophilic substitution in aromatic compounds is similar to that of electrophilic substitutions and again σ complexes have been isolated as in the following case.

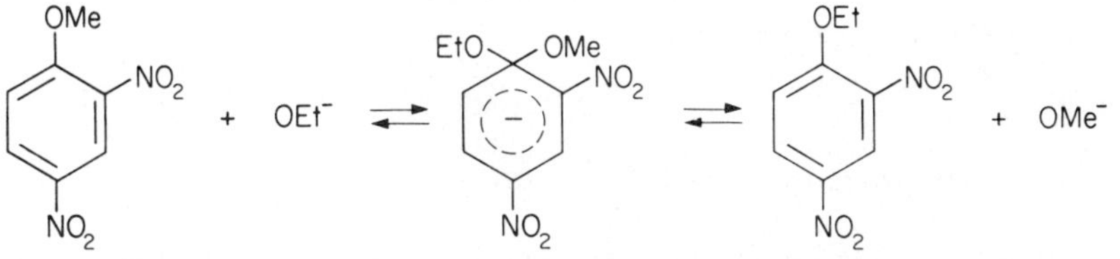

We have seen that electron density is not a useful reaction index for unsubstituted alternant hydrocarbons, but this is not a serious failure for nucleophilic substitution, because unsubstituted alternant hydrocarbons are not, in general, attacked by nucleophilic reagents. The self-polarizability index predicts the same site of attack for electrophilic and nucleophilic reagents, because this index simply measures the ease with which the electron density at a particular site is changed. The frontier orbital for nucleophilic attack is the lowest antibonding orbital. For alternant hydrocarbons this gives the same frontier-electron density as the highest bonding orbital.

Localization theory also predicts identical positions for electrophilic and nucleophilic attack in alternant hydrocarbons, because the σ complexes differ only in the number of electrons occupying the non-bonding orbital of the hydrocarbon fragment. For non-alternants this is not generally the case. For example, the energy required to localize a negative charge (two π electrons) for azulene is least in the 1-position, but the energy required to localize a positive charge is least in the 4-position. This result is in agreement with experiment: azulene is attacked by electrophilic reagents in the 1-position but by nucleophilic reagents in the 4-position.

Free-radical addition is mechanistically much more difficult to discuss. Such reactions undoubtedly involve a 'σ complex' but the subsequent steps are by no means certain and k_1 may not be rate determining. However, calculations are usually made assuming that it is.

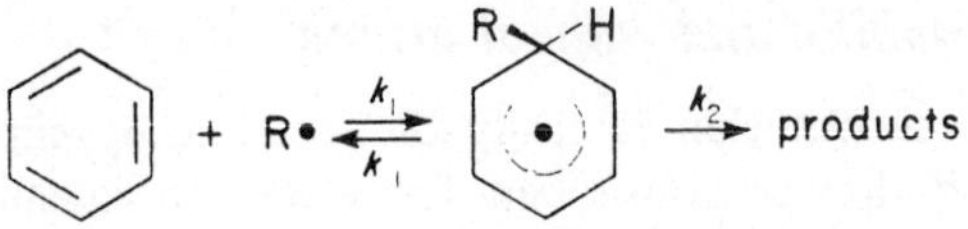

Since the attacking radical is not usually a charged species we do not expect any correlation between electron density or self-polarizability and the rate of radical attack. Free valence has been used quite a lot and is fairly successful. The most satisfactory approach, though, is based on localization energy. For alternant hydrocarbons this index is the same as for electrophilic and nucleophilic attack (we calculate the energy required to isolate a single π electron at a carbon atom), but for non-alternants we again get a different parameter.

17.6 *Comparison of reactivity indices*

Let us now summarize the position. For aromatic substitution, particularly electrophilic substitution, a large number of indices have been proposed. We have briefly mentioned them all because at present the whole subject is in a state of flux and the student may meet any of them. Because they are all derived from the Hückel coefficients they are not independent quantities and it is perhaps not too surprising that they all lead to similar predictions for the position of attack. This is particularly so for benzenoid hydrocarbons, in which case resonance theory also agrees with the MO indices. Table 17.1 compares some experimental and theoretical results.

Non-alternants and non-benzenoid hydrocarbons provide a better test of the theories but unfortunately there is much less experimental data for these molecules. We shall illustrate the differences between the various indices for two cases.

Biphenylene

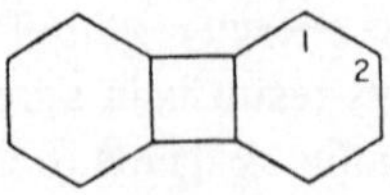

	q	$\pi_{\mu,\mu}$	F_μ	L_μ	N_μ	$\mathscr{F}$
1	1	0·419	0·428	2·40	2·00	0·027
2	1	0·443	0·420	2·35	1·73	0·088

TABLE 17.1. A comparison of reactivity indices and experimental data

Hydrocarbon		Experimental data: log K_b[a]	Experimental data: Partial rate factor for nitration[b]	Experimental data: Partial rate factor for chlorination[b]	Experimental data: Relative rate of trichloromethyl radical attack	Experimental data: Methyl affinities[c]	Isolated molecule indices: F_μ	Isolated molecule indices: $\pi_{\mu,\mu}$	Isolated molecule indices: S_μ	Isolated molecule indices: Z_μ	Localization energies: L_μ	Localization energies: N_μ
Benzene		−10·2	1	1	1	1	**0·40**	**0·40**	**0·83**	**1·21**	**2·54**	**2·31**
Naphthalene	1		470				**0·45**	**0·44**	**0·99**	**1·47**	**2·30**	**1·81**
	2	−4·6	50	6·6 × 10^4	10^2	33	0·40	0·41	0·87	1·42	2·48	2·12
Phenanthrene	1		360				**0·45**	**0·44**	0·98	1·46	2·32	1·86
	2		92				0·40	0·40	0·86	1·40	2·50	2·18
	3	−4·1	300	3 × 10^5	40	40·5	0·41	0·41	0.89	1·45	2·45	2·40
	4		79				0·44	0·43	0·94	1·42	2·37	1·96
	9		490				**0·45**	**0·44**	**1·00**	**1·48**	**2·30**	**1·80**
Triphenylene	1		600				**0·44**	**0·43**	0·89	**1·38**	**2·38**	**2·00**
	2	−5·4	600	2·3 × 10^4	70	—	0·41	0·41	**0·93**	**1·38**	2·48	2·12
Biphenyl	2		30				**0·44**	**0·42**	**0·91**	1·35	**2·40**	**2·07**
	3	−6·3	—	—	7	7·5	0·40	0·40	—	—	2·54	2·31
	4		18				0·41	0·41	0·89	**1·38**	2·45	**2·07**

[a] K_b is the basicity constant for the equilibrium $A + H^+ \leftrightarrows AH^+$ corrected for the number of positions having the same basicity (i.e. log K for benzene = log K_{exp} − log 6). Data taken from MACKOR, HOFSTA and VAN DER WAALS, *Trans. Faraday Soc.*, **54**, 66 (1958).

[b] Partial rate factors are the relative rate of attack at a particular site compared with the rate of attack at a single site in an unsubstituted benzene molecule. Data taken from DE LA MARE and RIDD, *Aromatic Substitution, Nitration and Halogenation*, Butterworths, 1959.

[c] Methyl radical affinities are the relative rate of attack per available reactive positions. Data from KOOYMAN and FARENHOST, *Trans. Faraday Soc.*, **49**, 58 (1959); SZWARC and BINKS, *Chem. Soc. (London), Spec. Publ.* (1959).

F_μ = free valence, $\pi_{\mu,\mu}$ = self-polarizability, S_μ = superdelocalizability, Z_μ = Brown's Z factor, L_μ = localization energy, N_μ = Dewar number, bold numbers indicate the predicted site of attack.

The 2-position is predicted to be the more reactive by all the indices except the free valence F (and the charge density q which is not relevant in this case). Experiment has shown that both electrophiles and radicals do attack at the 2-position. Resonance theory gives the wrong result. For the σ complex of electrophilic attack at the 1-position one can draw 11 structures of the type (a), and for the 2-position one can draw 10 structures of the type (b). According to resonance theory therefore the 1-position should be slightly the more readily attacked.

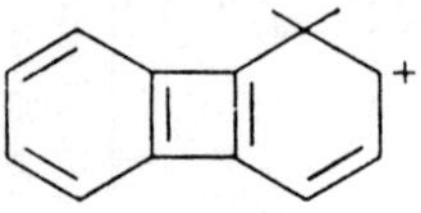

(a)

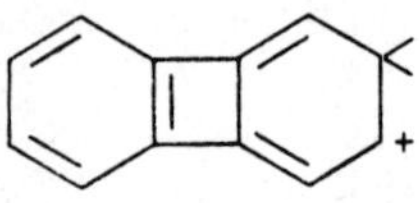

(b)

Fluoranthene†

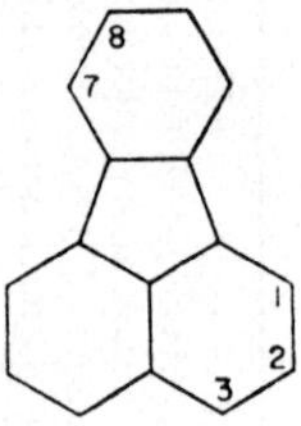

	q	$\pi_{\mu,\mu}$	F_μ	L^+	S_μ
1	0·947	0·440	0·453	2·47	0·828
2	1·005	0·400	0·398	2·50	0·860
3	0·959	0·462	0·470	2·34	0·930
7	0·997	0·427	0·438	2·37	0·936
8	1·008	0·510	0·409	2·44	0·872

None of the indices for electrophilic attack quoted in the table for fluoranthene are in complete agreement with the observed order of reactivities for nitration: $3 > 8 > 7 > 1 > 2$. The localization energy is clearly the best, disagreeing only in the order of 7 and 8. The charge and free valence give particularly poor correlations. This result suggests that the reason for the success of some of these reactivity indices for benzenoid hydrocarbons is just that they are related to the localization energies.

† Streitweiser, Brauman and Bush, *Tetrahedron*, **19**, *Supl. 2*, 379 (1963).

17.7 Effect of substituents on reactivity

In chapter 16 we have already given a general account of the treatment of substituent effects within the Hückel model and compared this with the traditional approach. We now wish to return briefly to the effect of substituents on reactivities in the light of the topics we have discussed in this chapter.

It has been usual to discuss substituent effects mainly in terms of their effect on electron density. However, we have seen, in the case of non-alternant hydrocarbons, that small differences in electron density are obscured by large differences in localization energy. Even when large differences in electron density may occur, the rate of reaction may still be dominated by the localization energy. In many cases, however, both electron density and localization energy predict the same position for attack. For example, a Hückel calculation on aniline† using the parameters $h_N = 1{\cdot}5$, $k_{C-N} = 0{\cdot}8$, $h_1 = 0{\cdot}1$ gives the following results. Both

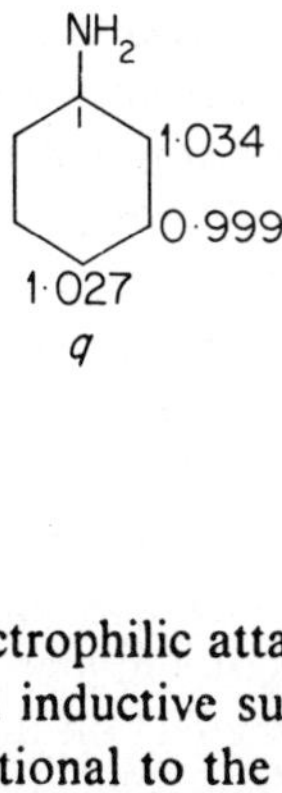

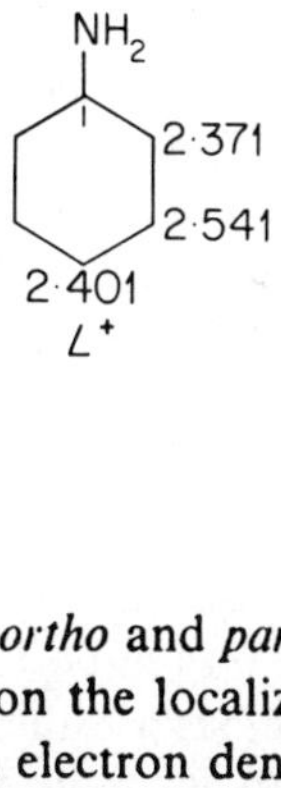

indices predict electrophilic attack at the *ortho* and *para* positions.

The effect of an inductive substituent on the localization energy is, to first order, proportional to the change in electron density at the position of substitution in going from the hydrocarbon to the σ complex. For an alternant hydrocarbon which has a uniform charge density this change is given by the electron density in the σ complex minus one, and for electrophilic and nucleophilic substitution the charges in the σ complex are obtained by squaring the non-bonding-orbital coefficients.

† STREITWEISER, *Molecular Orbital Theory for Organic Chemists*, Wiley, 1961.

For example, electrophilic attack at the 9-position of anthracene gives a σ complex whose charge densities are as follows.

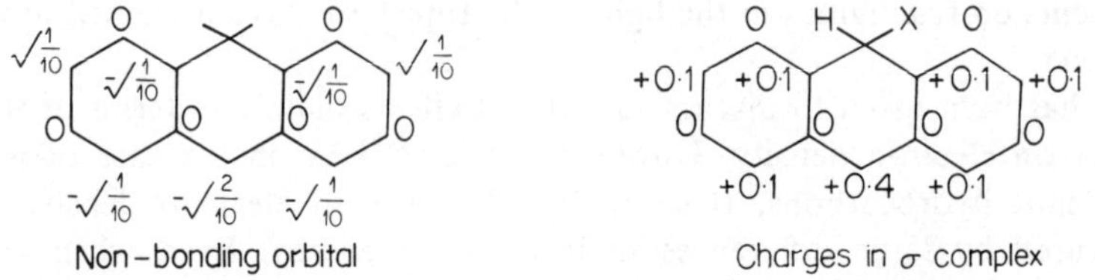

A substituent which repels π electrons $(+I_\pi)$ such as a methyl group, will lower the energy of this σ complex if it is in the 2-, 4- or 10-positions but not if it is in the 1- or 3-positions. Its effect in the 10-position should be four times as large as its effect in the 2- or 4-positions. In accord with this anhydrous HF protonates anthracene at the 9(10)-position. The basicity constant of 2-methylanthracene is ten times larger than that of anthracene, and that of 9-methylanthracene is another factor of ten larger.

The conjugating effect of a group on the σ complex may be much larger than it is on the parent hydrocarbon. For example, a $+R$ group which donates electrons to an aromatic ring will be expected to have a larger effect on the positively charged σ complex given by an electrophile but a smaller influence on the negatively charged σ complex given by a nucleophile. Dewar has given cogent arguments against the importance of the hyperconjugation of a methyl group, as far as ground state properties of aromatic hydrocarbons are concerned, whilst recognizing that hyperconjugation will be very important for carbonium ions. Similarly, we have argued that fluorine has a rather weak $+R$ effect on the ground state of a hydrocarbon due to its high ionization potential, but its affect on an electrophilic reaction intermediate may be appreciable.

Our aim in this chapter has been to see what light valency theories could throw on reaction rates. We were very careful to stress that these theories by themselves can never be used to calculate absolute reaction rates; at best they can only give information about the potential energy diagrams. Nonetheless we now see that although no precise predictions can be made, approximate methods like Hückel theory form a very useful background against which the experimental results may be discussed.

17.8 Pericyclic reactions†

Pericyclic reactions include the Diels-Alder reaction and similar cycloaddition reactions. They also include electrocyclic intramolecular cyclization of polyolefins

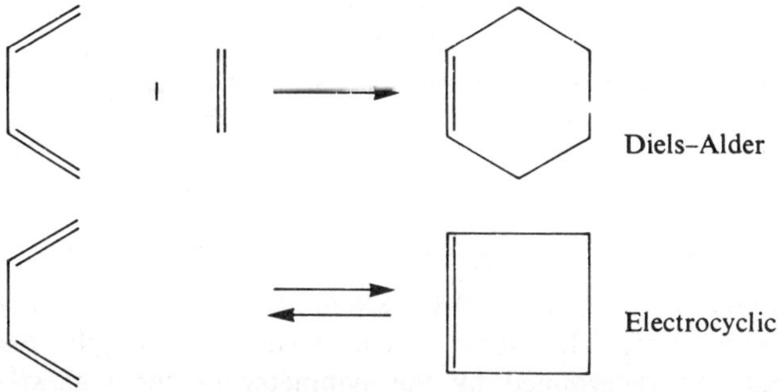

Some of these reactions proceed thermally, others require light, but they usually proceed with a defined stereochemistry. If a compound undergoes both thermally and photochemically induced reactions, the products often have a different stereochemistry. For example, hexatrienes cyclize to cyclohexadienes. Thus *trans,cis,trans*-1,6-dimethylhexa-1,3,5-triene on heating cyclizes to *cis*-1,2-dimethylcyclohexa-3,5-diene; whereas the same compound cyclizes on irradiation to yield the *trans*-1,2-dimethyl-cyclohexane-3,5-diene. The former process is called *disrotatory* and the latter *conrotatory*, as shown in figures 17.4 and 17.5.

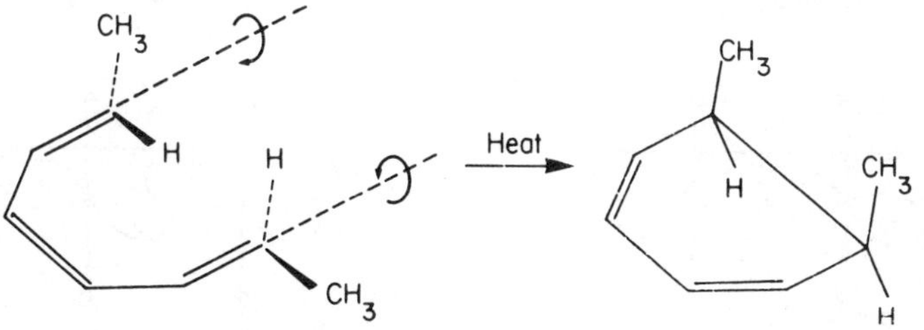

FIGURE 17.4. Disrotatory—Thermal cyclization

†R. B. WOODWARD and R. HOFFMANN, *Angewante Chemie*, (*Intern. Ed.*), **8**, 781–853 (1969).

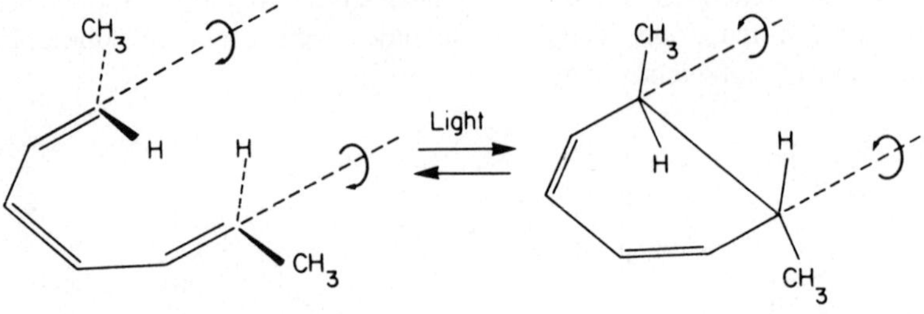

FIGURE 17.5. Conrotatory—Photochemical cyclization

The simplest hypothesis used to rationalize these reactions is a *frontier electron* hypothesis. The steric course of such electrocyclic reactions is assumed to be determined by the symmetry of the highest occupied molecular orbital of the open chain form. Attention is paid to the nature of the overlap between the orbitals which combine to form a σ-bond during the cyclization process. If we have an orbital pattern shown in figure 17.6 a disrotatory process leads to a bonding interaction, whilst a conrotatory process leads to an anti-bonding interaction.

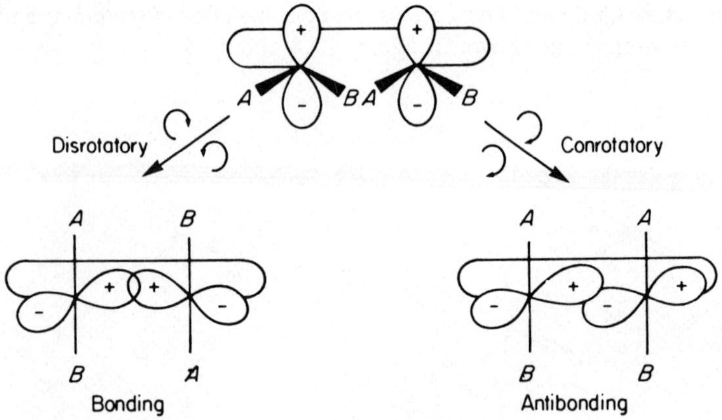

FIGURE 17.6.

However if the bonding pattern is as shown in figure 17.7 it is now a conrotatory process which leads to a bonding interaction

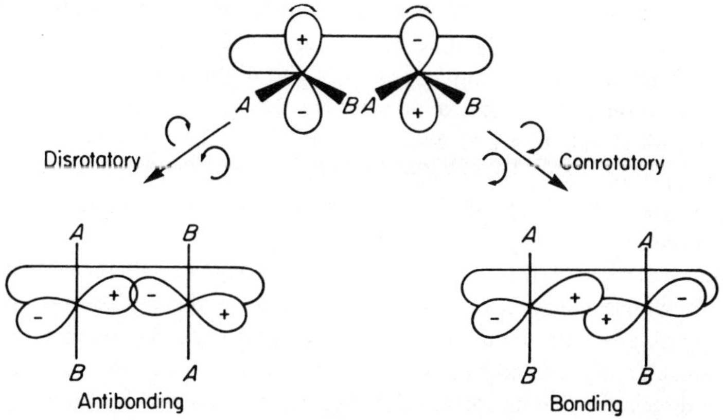

FIGURE 17.7.

Let us consider as a specific example the symmetries of the highest bonding and lowest anti-bonding π-orbitals of hexa-1,3,5-triene. These may be obtained by the Hückel method described in chapter 15 and are depicted below. The highest occupied orbital in the ground state is ψ_3 but in a photochemical process an electron is promoted from ψ_3 to ψ_4 which now becomes the highest occupied orbital, as shown in figure 17.8.

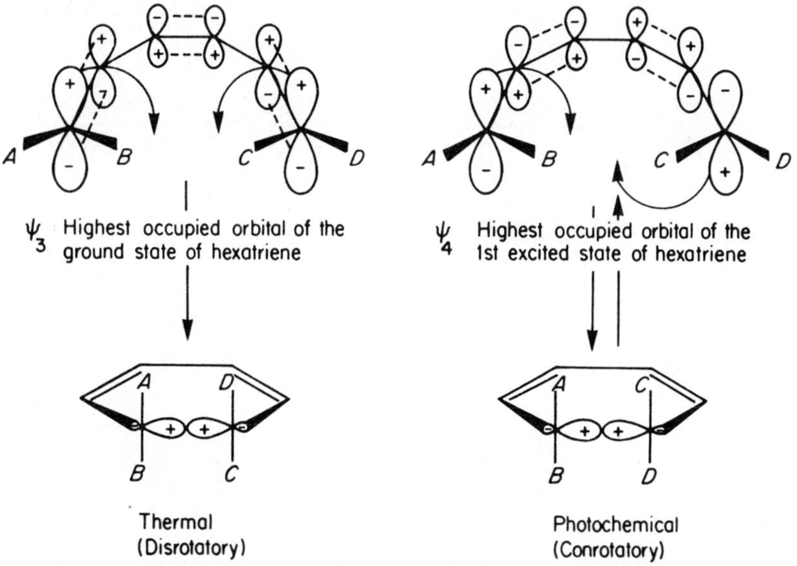

FIGURE 17.8. Cyclizations involving the 'Frontier Orbitals' of hexatriene

From our knowledge of the pairing properties of Hückel orbitals we know that the number of nodes in the highest bonding orbital of a linear polyene will be $\frac{1}{2}n-1$ where n is the number of carbon atoms in the chain (n must of necessity be even). Thus if $n/2$ is odd ($n = 6, 10, \ldots$) the coefficients of the terminal atoms of the highest bonding orbital will have the same sign so that a disrotatory thermal process will lead to a bonding σ-orbital. If $n/2$ is even ($n = 4, 8 \ldots$) the terminal coefficients have opposite signs hence the thermal process will be conrotatory. Similarly, a photochemical electrocyclic process will be disrotatory if $n/2$ is even and conrotatory if $n/2$ is odd.

The weakness of this treatment is that it neglects all other occupied orbitals whose energies certainly change during the cyclization. Rather than considering the behaviour of each orbital occupied by electrons we can consider the overall behaviour of the ground state and excited state of the molecule. The important features can be deduced by the use of the symmetry operations discussed in chapter 9. The first feature shown in figure 17.9, is that in a conrotatory process a C_2 axis is preserved throughout the reaction while a disrotatory process is characterized by the preservation of a mirror plane.

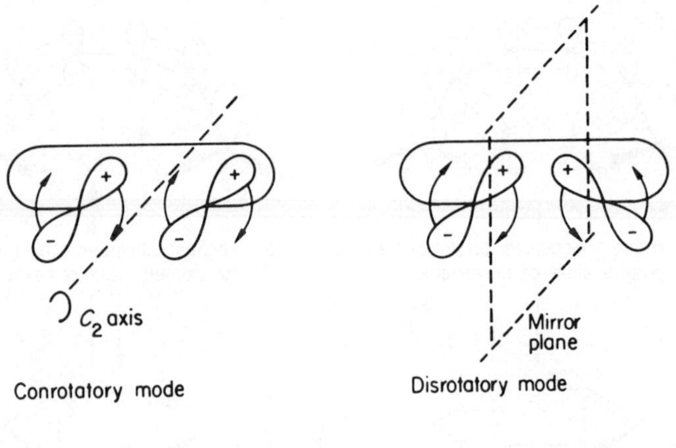

FIGURE 17.9. The symmetry of the conrotatory and disrotatory processes

Figure 17.10 shows the characteristics of the relevant orbitals of cyclobutene and of *cis*-butadiene under C_s and C_2 symmetries.

C_s	I	σ
A'	1	1
A''	1	-1

C_2	I	C_2
A	1	1
B	1	-1

Character Tables

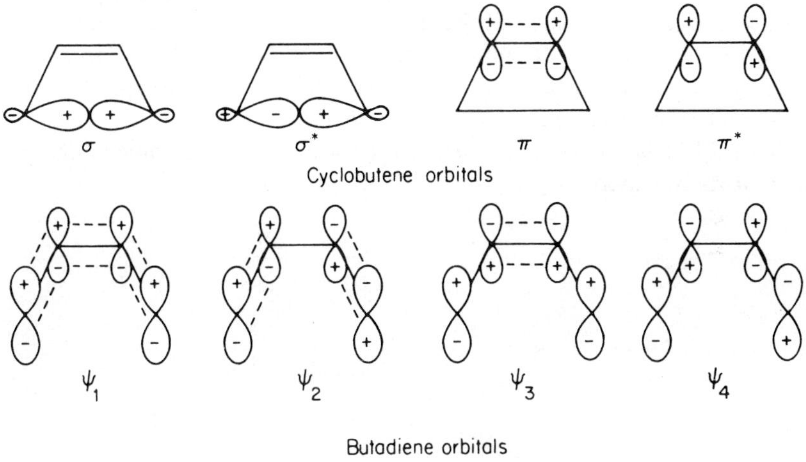

FIGURE 17.10.

In table 17.2 we have indicated the orbital correlation between butadiene and cyclobutene orbitals.

TABLE 17.2. Symmetry properties of butadiene and cyclobutene orbitals under conrotatory and disrotatory modes

		C_2 (Conrotatory mode)	C_s (Disrotatory mode)
Cyclobutene	σ (Bonding)	a (1)	a' (1)
	σ^* (Anti-bonding)	b (2)	a'' (2)
	π (Bonding)	b (1),	a' (2)
	π^* (Anti-bonding)	a (2)	a'' (1)
Butadiene	ψ_1 (Bonding)	b (1)	a' (1)
	ψ_2 (Bonding)	a (1)	a'' (1)
	ψ_3 (Anti-bonding)	b (2)	a' (2)
	ψ_4 (Anti-bonding)	a (1)	a'' (2)

Thus under a conrotatory mode the σ-orbital of cyclobutene correlates with ψ_2 of butadiene and the π-orbital with ψ_1, while in the disrotatory mode σ correlates with ψ_1 and π with ψ_3. The ground state configuration of cyclobutene is $\sigma^2\pi^2$ (as in previous chapters the superscript being the extent of their occupancy). Hence the symmetry configuration in the conrotatory mode is $a(1)^2\ b(1)^2$ which is totally symmetric, A. This will correlate with the ground state of butadiene which is also totally symmetric. The first excited state of cyclobutene is $\sigma^2\pi\pi^*$ [$\equiv a(1)^2\ b(1)a(2) = B$] for a conrotatory mode and it correlates with the excited $\psi_1\ \psi_2^2\ \psi_4$ [$\equiv b(1)a(1)^2a(2) = B$] state of butadiene. Notice that this is not the lowest excited state of butadiene which would be $\psi_1^2\ \psi_2\ \psi_3$ [$\equiv b(1)^2 a(1)b(2)$] which correlates with an upper excited state of cyclobutene ($\sigma\ \pi^2\ \sigma^{*2}$) [$\equiv a(1)b(1)^2b(2) = B$].

Figure 17.11 is a correlation diagram for cyclobutene-butadiene under the conrotatory mode.

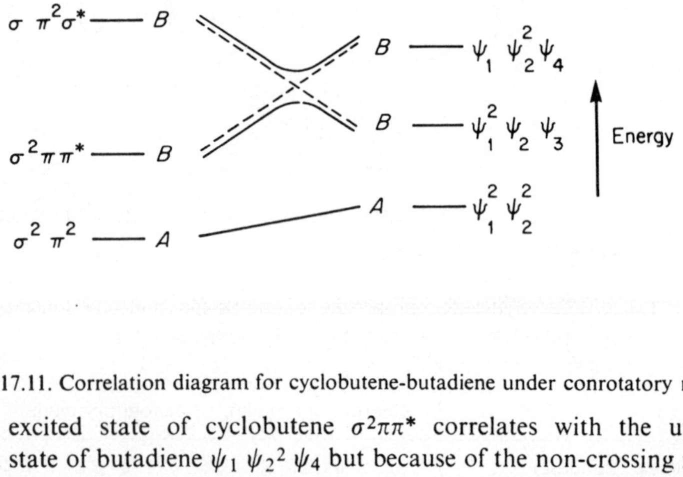

FIGURE 17.11. Correlation diagram for cyclobutene-butadiene under conrotatory mode

The excited state of cyclobutene $\sigma^2\pi\pi^*$ correlates with the upper excited state of butadiene $\psi_1\ \psi_2^2\ \psi_4$ but because of the non-crossing rule, (see section 10.2), the actual correlations are given by the full curved lines.

We can construct an exactly analogous diagram for the disrotatory mode, as in Figure 17.12. Here we have the ground state $\sigma^2\pi^2$ [$\equiv a'(1)^2 a'(2)^2 = A'$ for a disrotatory mode] of the cyclobutene correlating with $\psi_1^2\ \psi_3^2$ [$\equiv a'(1)^2a'(2)^2A'$] and the first excited state $\sigma^2\pi\pi^*$ [$a'(1)^2a'(1) a''(1) = A''$] correlating with $\psi_1^2\ \psi_2\ \psi_3$ [$a'(1)^2a''(1)a'(1) = A''$] and again the non-crossing rule applies

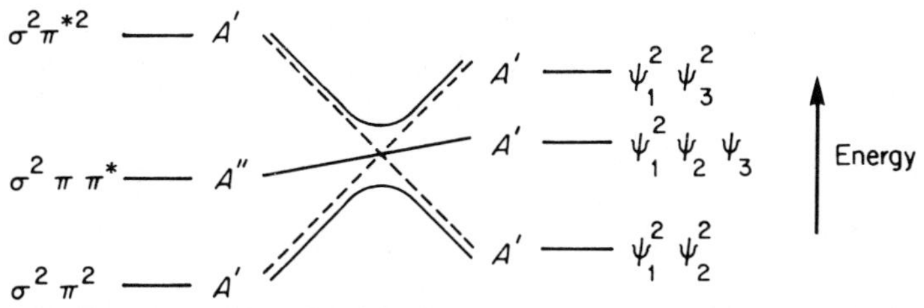

FIGURE 17.12. Correlation diagram for cyclobutene butadiene under disrotatory mode

We see from the correlation diagrams that there is no energy barrier for the transformation of the ground state of cyclobutene into the ground state of butadiene in a conrotatory mode and in the same way there is no energy barrier between the first excited states in the disrotatory mode. This is the same conclusion we would have obtained from our frontier theory but this time we have taken the other occupied orbitals into consideration. For the general case we find that if none of the bonding orbitals correlate with anti-bonding orbitals then a thermal process is permitted, but if bonding orbitals correlate with anti-bonding orbitals then a photochemical process is likely to be preferred.

We must now consider pericyclic reactions involving two separate molecules like the Diels–Alder reaction. The simplest case would be ethylene dimerization, and we will assume that the two ethylene molecules approach each other in parallel planes, one above the other (figure 17.13). There are two mirror planes, σ_v and σ_v' for this transition state which is of D_{2h} symmetry. It is sufficient, however to use C_{2v} symmetry. The actual reaction involves four π-orbitals of the ethylenes and four

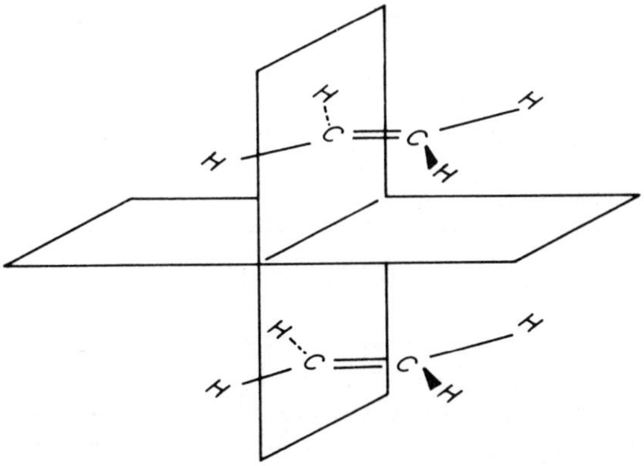

FIGURE 17.13.

σ-orbitals of the product cyclobutane. The C_{2v} character tabie is given on page 95 (Table 8.4) and we can depict the relevant orbitals as in figure 17.14.

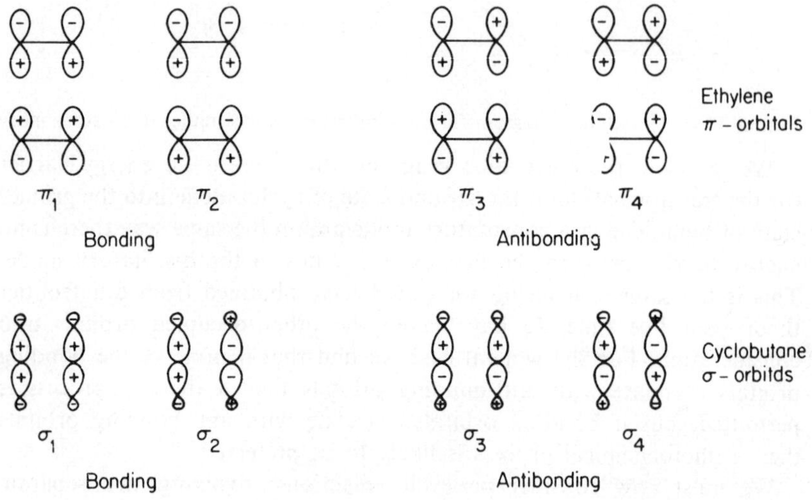

FIGURE 17.14.

In table 17.3 we designate the symmetries of these orbitals under the operations of the C_{2v} point group.

TABLE 17.3. The symmetries of the orbitals involved in the dimerization of ethylene

Ethylene π-orbital	Symmetry'symbol	Cyclobutane σ-orbital
π_1	a_1	σ_1
π_3	b_1	σ_2
π_2	b_2	σ_3
π_4	a_2	σ_4

The ground state of the ethylenes $(\pi_1)^2\ (\pi_2)^2 = a_1{}^2b_2{}^2$ correlates with $(\sigma_1)^2\ (\sigma_3)^2$, a doubly excited state of the cyclobutane, but an excited state of the ethylenes $(\pi_1)^2\ (\pi_2)\ (\pi_3) = a_1{}^2b_2b_1$ correlates with a corresponding excited state of cyclobutane $(\sigma_1)^2 (\sigma_2) (\sigma_3)$ and we can draw a correlation diagram figure 17.15 as before

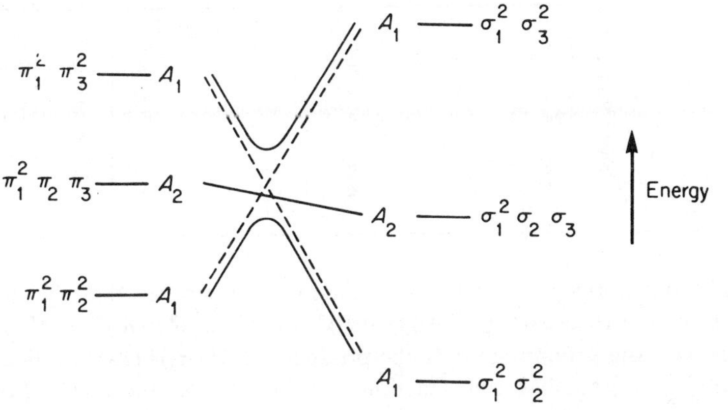

FIGURE 17.15. Correlation diagram for ethylene dimerization

From the diagram and the above discussion it would appear that ethylene dimerization (or dissociation of cyclobutane) was favoured for a photochemical process.

We can now consider the reaction between butadiene and ethylene as an example of the general case in which we have a σ-plane (see figure 17.16 and table 17.4). The symmetry is C_s, for which the character table has been given above.

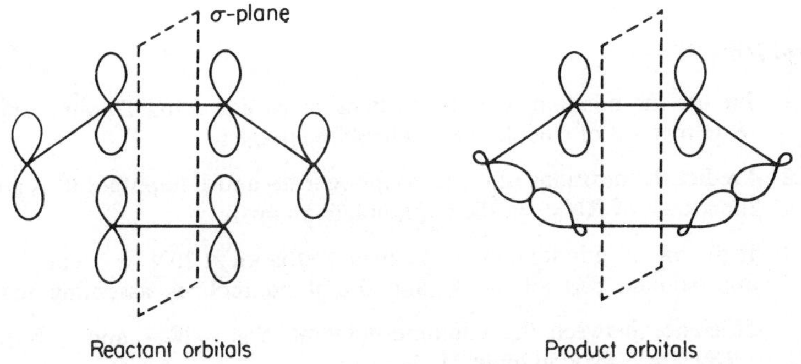

FIGURE 17.16. The cycloaddition of ethylene and butadiene

TABLE 17.4. The symmetries of the orbitals involved in the cycloaddition of ethylene and butadiene

	Reactant orbitals	Symmetry symbol	Product orbitals	
Butadiene	ψ_1	a'	σ_1	Cyclohexene
	ψ_2	a''	σ_2	
	ψ_3	a'	σ_3^*	
	ψ_4	a''	σ_4^*	
Ethylene	π_e	a'	π_c	
	π^*_e	a''	π_c^*	

Without drawing a correlation diagram we can see that the ground states of the reactants $(\psi_1)^2\,(\psi_2)^2\,(\pi_e)^2\;[(a')^2\times(a'')^2\times(a')^2 = A']$ correlates with the ground state of the products $(\sigma_1)^2\,(\sigma_2)^2\,(\pi_c)^2\;[(a')^2\times(a'')^2\times(a')^2 = A']$ so that this reaction should proceed thermally. For the general case, as with the intramolecular cyclizations, we take note of the number of carbon atoms in the chain (n is necessarily even). If $n/2$ is odd the reaction will proceed thermally, but if $n/2$ is even the reaction will only proceed photochemically through excited states.

Suggestions for further reading

STREITWEISER, *Molecular Orbital Theory for Organic Chemists*, chapter 11, Wiley, 1961.

BROWN, '*Molecular Orbital Theory of Electrophilic Substitution*,' *Tetrahedron* **19**, *Supl. 2*, pp. 337–349, (1963).

Problems 17

17.1 Predict the position of radical attack in pyrene using Dewar's approximate method of calculating localization energies.

17.2 Predict the positions of diazo-coupling in α- and β-naphthol (this goes by the attack of RN_2^+ on the naphtholate anions).

17.3 If the p*K* of quinoline in 50% aqueous ethanol at 20°C is 5, calculate the approximate p*K*s of 3-, 4- and 5-aminoquinolines assuming that the difference between the coulomb integrals for $-\text{N}{=}$ and $-\overset{+}{\text{N}}\text{H}{=}$ is $1{\cdot}2\beta$ ($\beta \approx -20$ kcal mole^{-1}).

17.4 Considering only the changes in π-electron energy calculate the difference between the p*K*s of cyclopentadiene and toluene.

Chapter 18

Weak chemical bonds

18.1 Van der Waals' forces

In this book we have been mainly concerned with the theory of the covalent bond in which the binding energy comes from electron delocalization. There are also forces between atoms (or molecules) which do not involve electron delocalization: the ionic bond is one example which we have already encountered. Another example is provided by the weak forces responsible for holding non-polar molecules together in liquids and solids. There is more than one contribution to these but collectively they are called Van der Waals' forces.

We shall not give a detailed discussion of these forces as they are rather on the fringe of valence theory and in order to discuss them properly one needs to go into the details of the polarization of molecules by an external electric field. We shall restrict ourselves to a brief statement of the classical explanation of these forces.

If a molecule has a charge or a permanent moment then this will induce a displacement of charge in a neighbouring molecule and there will be an attraction between the permanent and the induced moments. The energy of this interaction is called the *induction energy.*

A charged molecule will induce a dipole moment in a neighbouring molecule which is proportional to $1/R^2$ where R is the separation of the two molecules. The energy of attraction between a charge and a dipole varies as $1/R^2$ so that the charge-induced dipole of the induction energy varies as $1/R^4$. A dipolar molecule will induce a dipole in a neighbouring molecule which will vary as $1/R^3$, and the dipole-induced dipole energy will then vary as $1/R^6$.

If a neutral molecule has a spherically-symmetrical charge distribution it has no permanent moments and hence there will be no inductive contributions to the Van der Waals' forces. Only neutral atoms really fulfil this condition but a molecule like methane comes close to it. However, classically one might still expect that such a molecule (or atom) could possess an instantaneous dipole moment, the time average of which is zero. This instantaneous dipole will induce a moment in a neighbouring molecule (proportional to $1/R^3$) and the interaction between the instantaneous and the induced dipoles will give an energy varying as $1/R^6$. This is called the *London dispersion energy*.

Both the induction and dispersion forces depend on the polarizability of the molecule: this is defined by the dipole moment induced by a uniform unit electric field $\boldsymbol{\mu} = \boldsymbol{\alpha}\boldsymbol{\varepsilon}$ ($\boldsymbol{\alpha}$ has, in general, six components e.g. $\mu_x = \alpha_{xy}\varepsilon_y$, and is called a second-order tensor). An induction energy is proportional to the polarizability of the polarized molecule; the dispersion energy varies as the product of the polarizabilities of the two interacting molecules. Except possibly for the case when one molecule has a permanent charge, the dispersion energy is the larger.

An approximate expression for the dispersion energy between two molecules A and B has been given by London as follows

$$E_D = -\frac{3}{2}\frac{I_A I_B}{I_A + I_B}\frac{\alpha_A \alpha_B}{R^6}, \tag{18.1}$$

where I is the ionization potential and α the average polarizability $\frac{1}{3}(\alpha_{xx} + \alpha_{yy} + \alpha_{zz})$. For example, two methane molecules at 3·8 Å (their average separation in the liquid) have an attraction energy of about 0·02 ev (0·5 kcal mole^{-1}).

Dispersion and induction forces can occur not only between different molecules, but also between different atoms or groups within the same molecule. For example, dispersion energies between non-bonded C—H groups have been proposed to account for differences in heats of formation amongst paraffin hydrocarbon isomers†.

Dispersion forces alone are probably never responsible for the formation of a complex between two different molecules. Because the product of the polarizabilities of two different molecules, $\alpha_A\alpha_B$, is always less than $\frac{1}{2}(\alpha_A{}^2 + \alpha_B{}^2)$, one expects the dispersion energy of two dimers A_2 and B_2 to be greater than that of 2AB. However, complexes between unlike molecules are in fact more common than dimers, and we shall now turn to an examination of the forces involved in these cases.

18.2 Donor–acceptor complexes

The term molecular complex is used to describe a group of two or more molecules which are bound together. Typical examples are $BF_3 \leftarrow NH_3$,

† Pitzer and Catelano, *J. Am. Chem. Soc.*, **78**, 4844 (1956).

and quinone–hydroquinone (quinhydrone). These can both be prepared as crystalline compounds with a well defined 1 : 1 stoichiometry. There are, in addition, many examples where it is known that association of two molecules occurs in solution but no crystalline compounds can be prepared: iodine–benzene is one such case.

These examples all have one thing in common. This is that one of the molecules is an electron donor and the other an electron acceptor.

Donor	Acceptor
NH_3	BF_3
hydroquinone	quinone
benzene	I_2

An electron donor is a molecule which has weakly bound electrons (e.g. the non-bonding pair in NH_3), and a corresponding low ionization potential. In association with this property, the molecule may be a reducing agent or a Lewis base. The acceptor, on the other hand, has a low lying vacant orbital (e.g. the vacant $2p$ orbital in BF_3). Acceptors have high electron affinities and may be oxidizing agents or Lewis acids. However, donor and acceptor strength must really be considered relative to the other partner in the complex. Just as the phenyl group accepts electrons from the substituent in aniline but donates electrons to the substituent in nitrobenzene, so benzene behaves like a donor with respect to iodine, but behaves like an acceptor with respect to a strong donor like an alkali metal.

In many cases a donor–acceptor complex is highly coloured. For example, hydroquinone is colourless and quinone is yellow but quinhydrone crystals have a very characteristic green metallic appearance. Similarly, benzene/iodine solutions show a UV absorption band at 3000 Å which has no counterpart in the spectra of either iodine or benzene.

The first satisfactory quantum-mechanical treatment of these molecular complexes was given by Mulliken. He showed that both their stability and their characteristic electronic absorption bands could be linked with their donor-acceptor properties.

Suppose the wave functions of the separate donor and acceptor are $\Psi(D)$ and $\Psi(A)$ respectively. The wave function of the pair, still separated, will be

$$\Psi_0(D,A) = \Psi(D)\Psi(A). \tag{18.2}$$

Likewise, if we have ions D^+ and A^- separated by a large distance, then the wave function for this system is

$$\Psi_\infty(D^+,A^-) = \Psi(D^+)\Psi(A^-). \tag{18.3}$$

When the two molecules come together such that their electron clouds overlap one another, then both these wave functions must be amended to allow for electron exchange, according to the antisymmetry principle. This is done formally by introducing the antisymmetrizing operator which simply converts the simple product into an antisymmetrized sum of products†

$$\Psi(D,A) = \mathscr{A}\Psi(D)\Psi(A); \quad \Psi(D^+,A^-) = \mathscr{A}\Psi(D^+)\Psi(A^-). \tag{18.4}$$

We now write the total wave function of the complex as a linear combination of these two functions

$$\Psi = a\Psi(D,A) + b\Psi(D^+,A^-), \tag{18.5}$$

and determine the coefficients by solving the relevant secular equations. Two solutions can be obtained whose wave functions we write as follows

$$\Psi_1 = a_1\Psi(D,A) + b_1\Psi(D^+,A^-),$$

$$\Psi_0 = a_0\Psi(D,A) + b_0\Psi(D^+,A^-). \tag{18.6}$$

The energy of the lower, Ψ_0, is less than the energy of either $\Psi(D,A)$ or $\Psi(D^+,-A^-)$, so this interaction provides a possible explanation of the stability of the complex. Also, the electronic transition $\Psi_0 \rightarrow \Psi_1$, which has no counterpart in either the separate donor or acceptor, can be associated with the characteristic absorption bands of the complex.

Let us examine first the weak complexes, which are typically formed by organic molecules. For these the state $\Psi(D,A)$, which Mulliken calls the no-bond state, has a much lower energy than the ionic or *charge-transfer* state $\Psi(D^+,-A^-)$. In addition the interaction integral between these states is usually small. The final ground state therefore has a wave function

$$\Psi_0 = \Psi(D,A) + b_0\Psi(D^+,-A^-), \tag{18.7}$$

where $b_0 \ll 1$, and the excited state has a wave function

$$\Psi_1 = \Psi(D^+,-A^-) + a_1\Psi(D,A), \tag{18.8}$$

where again $a_1 \ll 1$. The electronic transition $\Psi_0 \rightarrow \Psi_1$ is therefore accompanied by the transfer of almost a whole electron from the donor to the

† $\mathscr{A}\phi_a(1)\phi_b(2) = \sqrt{\tfrac{1}{2}}\{\phi_a(1)\phi_b(2) - \phi_a(2)\phi_b(1)\}$ is the simplest example, but D and A may not have wave functions which are single products of spin-orbitals, so $\mathscr{A}$ does not necessarily convert the wave function into determinantal form.

acceptor, and because of this the absorption band associated with this transition is called the *charge-transfer band.*

The strongest evidence in support of this interpretation of the absorption band comes from its frequency. When the interaction between the two components of the wave function is weak, the energy of the transition $\Psi_0 \rightarrow \Psi_1$, is approximately equal to the energy of the transition

$$\Psi(\mathrm{D,A}) \rightarrow \Psi(\mathrm{D^+,A^-}).$$

This is (cf. 11.34)

$$E(\mathrm{D^+,A^-}) - E(\mathrm{D,A}) = I(\mathrm{D}) - A(\mathrm{A}) + Q. \tag{18.9}$$

For a series of complexes with a common acceptor but varying donor, one finds that a relationship

$$h\nu = I - K, \tag{18.10}$$

where K is a constant, holds surprisingly well, as shown in figure 18.1. This means that even though Q, the coulomb energy between the two ions, is expected to be quite large, it does not vary a great deal with the nature of the donor.

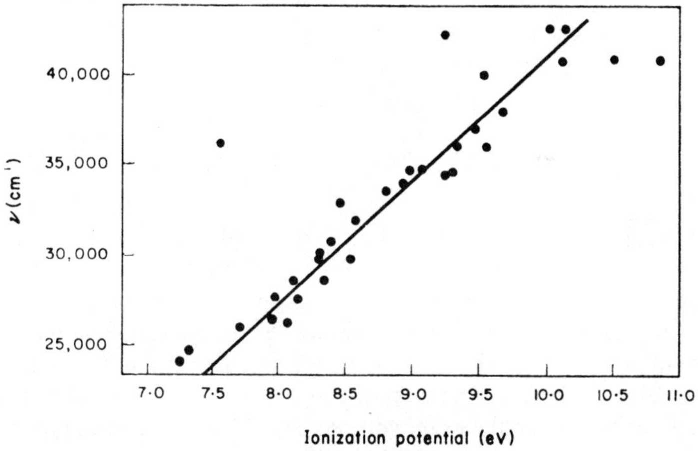

FIGURE 18.1. The correlation between the energy of the charge-transfer band and the ionization potential of the donor for some complexes between iodine and organic donors†

The interaction between the no-bond and charge-transfer states depends on the overlap between the orbital of D which loses the electron (the donor

† BRIEGLEB and CZEKALLA, *Angew. Chem.*, **72**, 401 (1960).

orbital) and that of A which accepts the electron (the acceptor orbital). This is easily seen if we take the simplest case, where we only have a one-electron system.

$$\Psi(\mathrm{D,A}) = \psi_{\mathrm{d}}, \quad \Psi(\mathrm{D^+,A^-}) = \psi_{\mathrm{a}}. \tag{18.11}$$

ψ_{d} is the donating orbital of D and ψ_{a} is the accepting orbital of A. Then

$$\int \Psi(\mathrm{D,A})\mathscr{H}\Psi(\mathrm{D^+,A^-})\,d\tau = \int \psi_{\mathrm{d}}\mathscr{H}\psi_{\mathrm{a}}\,dv \tag{18.12}$$

and this integral is zero unless the orbitals ψ_{d} and ψ_{a} overlap one another; if at any point either ψ_{d} or ψ_{a} is zero then the contribution to the integral from that region in space is zero. In general the interaction integral will be roughly proportional to the overlap integral S_{da}, and the stabilization will be proportional to S_{da}^2 (since it is a second-order interaction) (cf. equation 17.7).

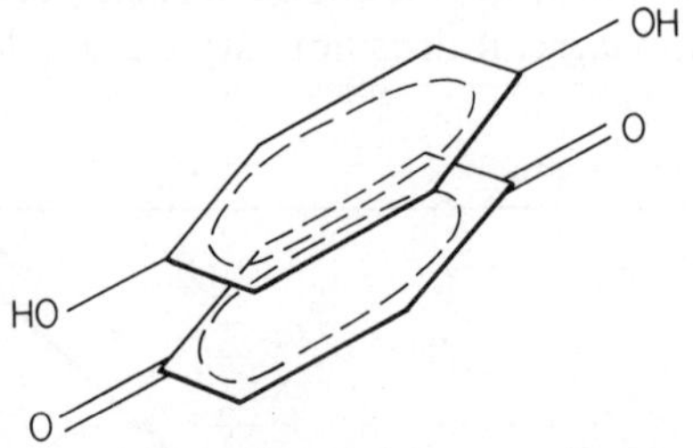

FIGURE 18.2. The structure of quinhydrone

In typical organic complexes like the quinhydrones the donor and the acceptor orbitals are delocalized π molecular orbitals and the region of overlap between these two orbitals is delocalized as shown in figure 18.2. If one wants to talk about a chemical bond between the two components of the complex then it has to be understood that this is a delocalized bond and not something that can be represented by drawing a line between two atoms. Complexes like the quinhydrones have binding energies typically in the range 0·1 to 0·5 ev (2–10 kcal mole^{-1}). This is not much larger than the energies one could expect from the London dispersion forces, so that although one has good reason for attributing complex formation in these cases to a charge-transfer interaction, a fairly large part of the binding energy probably comes from the dispersion terms.

The complexes like $BF_3 \leftarrow NH_3$ are quite a different case. The state $\Psi(\mathrm{D^+,A^-})$ is calculated to have about the same energy as $\Psi(\mathrm{D,A})$ and this alone will favour a large binding energy. But, more important, the donor and acceptor orbitals are not delocalized molecular orbitals but hybrid atomic orbitals which can overlap one another to a large extent. The interaction between the no-bond and charge-transfer states gives rise to a binding energy for $BF_3 \leftarrow NH_3$ of about 2 ev, which is much larger than that for complexes like the quinhydrones. This large stabilization of the ground state must be accompanied by a large de-stabilization of the excited state. In the case of $BF_3 \leftarrow NH_3$ the separation of the ground and excited state is too large for a charge-transfer band to be observed above 2300 Å, the wavelength at which NH_3 itself starts to absorb light.

The charge-transfer theory of molecular complexes that we have described has a strong similarity to the VB theory of a diatomic molecule. This is obvious if D and A are taken to be atoms rather than molecules. Now we have seen that a simple VB wave function is better than a simple MO function if the atoms are only weakly bound together, because the MO function has the wrong dissociation limit. The charge-transfer treatment of weak molecular complexes is therefore going to be better than a MO treatment in which electrons are allocated to orbitals which are delocalized over both the donor and the acceptor. On the other hand, the strong complexes, where the ground state may be an almost equal mixture of $\Psi(D,A)$ and $\Psi(D^+,A^-)$, may be as well, or better described in terms of delocalized molecular orbitals. MO theory makes no distinction between the B—N bond in $BF_3 \leftarrow NH_3$ and the bond in Li—H: both are electron-pair bonds with a large ionic character.

When we use the term charge-transfer to describe the bonding in a molecular complex we use it in the same sense that the terms 'structure' and 'resonance' are used. That is, it is a term linked to a particular quantum-mechanical model and not to something that is physically observable. On the other hand, when it is used to describe an electronic absorption band then it means that a large amount of electron density is transferred from the donor to the acceptor when the complex is excited: in this sense it has been raised above the level of a concept, and is much more useful.

18.3 The hydrogen bond

In many cases where there is an unusually strong association between molecules or between different groups in the same molecule, experiment shows that a hydrogen atom is involved. The association is then said to be through a hydrogen bond. There are four main sources of this evidence:

1. Physical properties such as dielectric constant and boiling point (e.g. water, which is hydrogen bonded, has a higher boiling point and dielectric constant than methane, which is not).
2. Infrared and Raman studies on A—H stretching vibrations show changes in the position, intensity and shape of the spectral band.
3. From nuclear magnetic resonance one finds that the chemical shifts of some protons are very sensitive to the nature of the solvent.
4. The position of hydrogen atoms in crystals (measured either directly by x-ray or neutron diffraction or indirectly by broad line nuclear magnetic resonance) shows in some cases that they are bonded to two atoms.

The following are typical examples:

(A) In crystals of KHF_2 the anion is $(F—H—F)^-$ which is linear and has the hydrogen atom midway between the two fluorines.

(B) Oxalic acid crystals have one crystalline form in which the following structure occurs

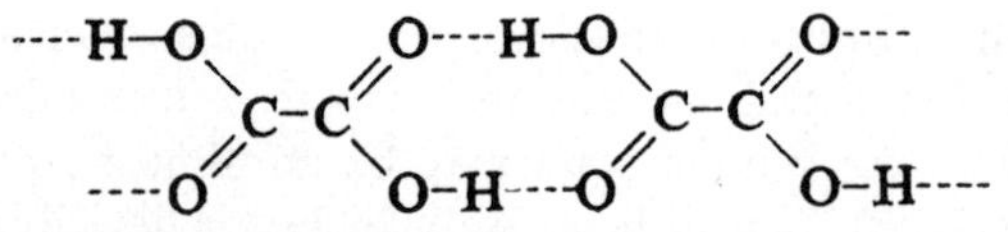

The O—O distance is unusually short (2·5 Å) but the hydrogen atom is not midway between the two oxygen atoms.

(C) In salicylaldehyde there is an internal hydrogen bond in which the phenolic hydrogen is weakly bonded to the aldehydic oxygen as in the following structure

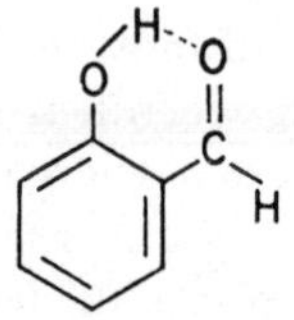

There are two points at which theoretical chemistry can help in an understanding of the hydrogen bond. Firstly, the vibrational properties of the bond must be analysed to give a potential energy surface, and secondly, valency theory must be used to explain this surface. The first problem does not concern us here except that we will just look at the types of surface that can be expected. Figure 18.3(a) shows a symmetrical single minimum potential (example A above), figure 18.3(b) is a symmetrical double minimum (example B), figure 18.3(c) an unsymmetrical double minimum (probably very uncommon) and figure 18.3(d) an unsymmetrical single minimum (example C).

There has been a lot of speculation about the nature of the hydrogen bond and several VB calculations have been published, but overall these still only allow one to arrive at a few qualitative conclusions.

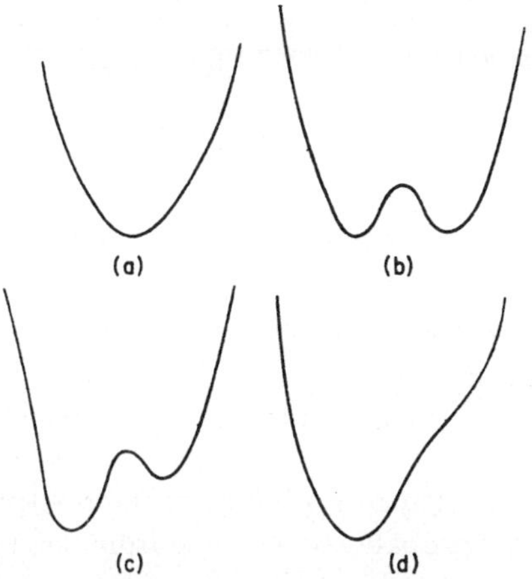

FIGURE 18.3. Potential energy curves for hydrogen bonds

Only polar bonds, for example, O—H, N—H, F—H, S—H, are involved in hydrogen bond formation. In all cases the hydrogen is at the positive end of the dipole. C—H bonds are not usually thought to be involved except perhaps for a molecule like $CHCl_3$ where the polarity of the C—H bond is increased by the electron-attracting chlorine atoms. This result suggests that the energy of the hydrogen bond is mainly electrostatic in origin and calculations on this basis give about 0·2–0·5 ev, the right order for hydrogen bond energies.

Let us consider the energy of an O—H ··· O bond of the type shown in example B. When the two oxygen atoms are far apart we have a normal two-electron O—H bond and the second oxygen has a lone pair of electrons pointing towards the hydrogen atom

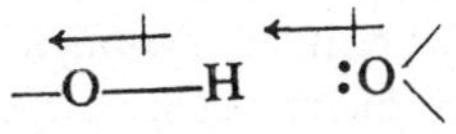

The O—H bond dipole and the lone-pair atomic dipole moments are clearly orientated in such a way that there is an electrostatic attraction between the two groups. However, there are many systems for which the electrostatic energy is attractive which do not form bonds (two helium

atoms for example). What usually happens in the case of interacting closed shells of electrons is that the exchange energy is repulsive and overrides the electrostatic attraction. The best explanation one can give for the formation of weak hydrogen bonds is probably that *their exchange energy is small.* It is small in the first place because the hydrogen atom is unique in having no inner shells of electrons, and in the second place because, in polar bonds like OH, the electron density around the proton is small. It is the absence of a strong exchange repulsion that enables the two atoms sharing the proton to come close together and so give a large electrostatic energy of attraction.

Although there has been no really convincing calculation on weak hydrogen bonds it does appear that 'long-bond' structures (charge-transfer states) such as $\overset{-}{O}\ H—\overset{+}{O}\langle$ make very little contribution to the total binding energy†. They may, however, have an important influence on the shape of the potential energy surface, and be responsible for the change in the OH bond length and force constant which occurs on hydrogen bonding.

The strong symmetrical hydrogen bonds (example A) present quite a different situation. They are probably best treated from the start as a delocalized electron system. For example, $(FHF)^-$ would seem to present the same type of problem as the bridge B—H—B bond in diborane. There does not seem to be much point in speculating on how much electrostatic forces contribute to the observed binding energy, which is about 50 kcal mole^{-1} relative to HF plus F^-‡. The F—F distance is only 2·26 Å which is much less than twice the ionic radius of F^-, so that a model in which a proton holds two F^- ions together does not seem to be very relevant. From a molecular-orbital viewpoint $(FHF)^-$ is somewhat similar to F_2: all the molecular orbitals which can be formed from 1*s*, 2*s* and 2*p* atomic orbitals are filled, except the most antibonding.

18.4 Sandwich compounds§

There are now many examples of so-called 'sandwich' molecules, in which a transition metal is bonded symmetrically to two unsaturated hydrocarbons. The structure of ferrocene, dicyclopentadienyl iron, which

† Tsubomura, *Bull. Chem. Soc. Japan*, **27**, 445 (1954).
‡ Waddington, *Trans. Faraday Soc.*, **54**, 25 (1958).
§ Although we do not wish to imply that all sandwich compounds have weak chemical bonds this seems to be the most convenient place to discuss their structure.

is perhaps the best known example, is shown in figure 18.4. It has also been found that just one such hydrocarbon may be attached to the metal provided that other ligands such as carbon monoxide or triphenylphosphine are also present. There seems to be a strong similarity between the bonding properties of cyclopentadienyl on the one hand and carbon monoxide or triphenylphosphine on the other.

FIGURE 18.4. The 'sandwich' structure of ferrocene

Carbon monoxide is thought to be bonded to a transition metal by both σ and π bonds. In the σ bonding there is a partial donation of electrons to the metal. In the π bonding there is a partial transfer of electrons from the metal to the lowest vacant orbital of the ligand. Overall, approximate neutrality is preserved. It is thought that the σ and π bonding enhance one another (this is frequently called a synergic process): donation of the σ electrons of the carbon monoxide makes the ligand a stronger π-electron acceptor, and vice versa.

Dewar†, for silver–ethylene, and Chatt and Duncanson‡, for platinum–ethylene complexes, extended this idea to hydrocarbon ligands. They proposed that the occupied π bonding orbital of the ethylene behaved as a donor and the vacant π antibonding orbital as an acceptor (figure 18.5).

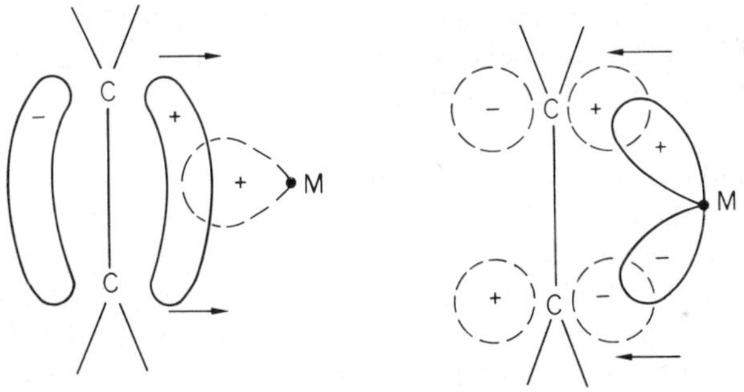

FIGURE 18.5. σ and π bonding in metal–olefin complexes

This picture has been generally accepted and extended to many other hydrocarbon ligands. However, it fails to explain why the stability of the

† DEWAR, *Bull. Soc. Chim. France*, **18**, C71 (1951).
‡ CHATT and DUNCANSON, *J. Chem. Soc.*, 2939 (1953).

complexes varies greatly with the nature of the olefin. Evidently more is involved than the simple presence of donating and accepting orbitals.

Ferrocene is of $\mathbf{D}_{5d}$ symmetry and a useful first step in discussing the bonding in this molecule is to classify the interacting orbitals according to their transformation properties in this group. We consider only the π orbitals of the cyclopentadiene rings since they have the highest energy and also the greatest overlap with the metal d orbitals. The C_5H_5 molecular orbitals have the symmetry A_2'', E_1'' and E_2'', in the $\mathbf{D}_{5h}$ group of cyclopentadienyl, and the two sets of molecular orbitals have the symmetries $A_{1g} + A_{2u} + E_{1g} + E_{1u} + E_{2g} + E_{2u}$ in the $\mathbf{D}_{5d}$ group of ferrocene. The atomic orbitals of the central atom have the following transformation properties.

$$A_{1g}: 4s,\ 3d_{z^2}$$

$$A_{2u}: 4p_z$$

$$E_{1g}: 3d_{xz},\ 3d_{yz}$$

$$E_{1u}: 4p_x,\ 4p_y$$

$$E_{2g}: 3d_{xy},\ 3d_{x^2-y^2}$$

By examining the overlap of ligand and metal orbitals belonging to the same symmetry species, we obtain the approximate MO energy level diagram given in figure 18.6.

Eighteen valence electrons are fed into these molecular orbitals so that all orbitals up to e_{2g} are filled. Addition of further electrons, as in the analogous cobalt and nickel compounds, leads to paramagnetism. However, their magnetic moments are only consistent with the unpaired electrons being in degenerate orbitals (probably e_{2u}^*), so the order of the antibonding orbitals appears to depend on the metal.

Semi-empirical SCF MO calculations have been carried out on ferrocene, but, unfortunately, the details of the bonding which emerge are strongly dependent on the orbitals used for the iron atom. Using Slater orbitals a net charge of $+0{\cdot}68$ is calculated on the iron atom†. Calculations using SCF atomic orbitals, on the other hand, give a charge of $-0{\cdot}69$‡. Although the sign of this charge is in doubt it seems clear that the charge itself is small, corresponding at most to the transfer of about two thirds of an electron between each ring and the iron atom. This supports the view that the Dewar, Chatt and Duncanson approach to ethylene complexes is of wider

† SHUSTOROVITCH and DYATKINA, *Dokl. Akad. Nauk SSSR*, **133**, 141 (1960).

‡ DAHL and BALLHAUSEN, *Kgl. Danske Videnskab. Selskab, Mat.-Fys. Medd.*, **5**, 33 (1961).

applicability, and that in ferrocene there is some cancellation of the residual charges resulting from the metal–cyclopentadiene and cyclopentadiene–metal delocalization processes.

A serious problem in the understanding of the ferrocene molecule is that it has proved theoretically very difficult to obtain what, experimentally, appears to be the correct ordering of the molecular orbital energy levels.

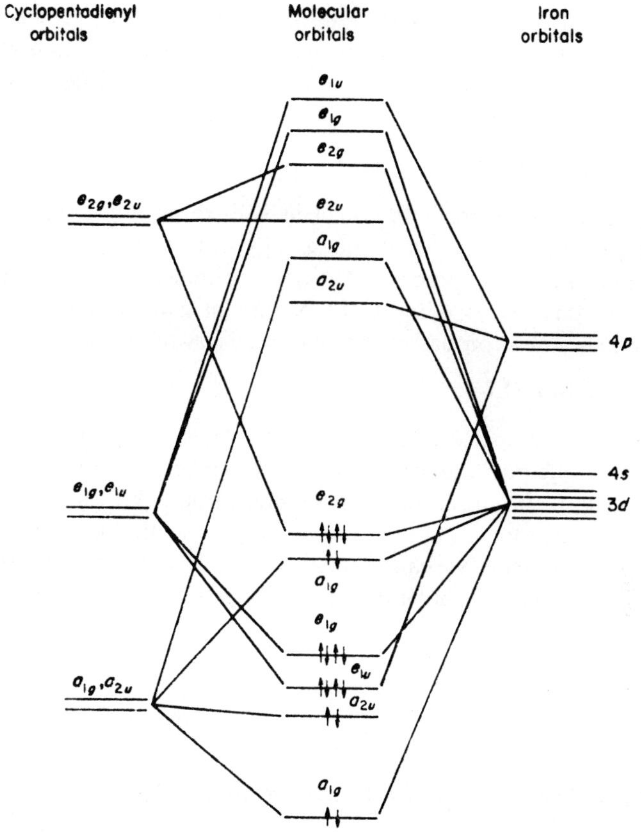

FIGURE 18.6. The molecular-orbital energy scheme for ferrocene

Hints to problems

Chapter 1

1.1 and **1.2** The Planck–Einstein relationship (1.2) relates the energy of a photon to its frequency or wavelength. When light is absorbed by or emitted from a molecule the molecular energy must change to conserve the total energy.

1.3 Use equation (1.1).

1.4 Only direct collisions with a nucleus will scatter a neutron, since it is a neutral particle (there is a small interaction between the magnetic moment of the neutron and any permanent magnetic moment of the atom but we shall ignore this here). What is the chance of a direct collision with a nucleus when a neutron passes through one layer of atoms?

1.5 Use equation (1.6).

1.6 If we confine an electron within a length 10^{-13} cm then this is the uncertainty in its position, and from the Heisenberg uncertainty principle one can deduce the uncertainty in the momentum. We can then assume that the average momentum is at least as large as the uncertainty in the momentum, and hence we can estimate a lower limit to the average kinetic energy. Is the coulomb potential energy large enough to counterbalance this kinetic energy?

Chapter 2

2.1 and **2.2** These both follow from expression (2.5).

2.3 There is only one variable in this problem which can be x (the distance around the circumference) or $\vartheta (x = a\vartheta)$. The one-dimensional Schrödinger equation in the distance coordinate x (2.1) can be converted to an equation in ϑ by $(\mathrm{d}^2\Psi/\mathrm{d}x^2) = (1/a^2)\ (\mathrm{d}^2\Psi/\mathrm{d}\vartheta^2)$. One can now substitute the two trial solutions Ψ_r and Ψ_r' into this equation and hence deduce an expression for E. The boundary condition is $\Psi(\vartheta) = \Psi(2\pi + \vartheta)$, and this imposes a condition on r. What happens when $r = 0$?

Chapter 3

3.1 The method of constructing a polar diagram is explained on page 25.

3.2 Treat the three component p orbitals as vectors pointing along the coordinate axes and the p orbital required as the (1, 1, 1) vector. Normalize the resulting function.

3.3 The 1s and 2s wave functions are obtained from table 3.3. Note that ρ depends on n. Keep the integration in terms of polar coordinates, using expression (3.3) for the volume element of integration.

3.4 To calculate the mean value of $1/r$ we use the same method as in classical mechanics: take the electron density (in quantum mechanics Ψ^2) multiply by $1/r$ and integrate. A more general treatment of mean values is given in chapter 6. The average potential energy is $-Ze^2(\overline{1/r})$. The average kinetic energy is obtained through the total energy (expression 3.21).

$$\int_0^\infty x^n e^{-ax}\,dx = n!/a^{n+1}; \quad n > -1, a > 0.$$

Chapter 5

5.1 Just evaluate e^2/r.

5.2 Find the distance r such that the coulomb attraction just balances $I - A$.

5.3 How does the Morse curve behave when $r \to 0$? Differentiate the Morse function to find k, and introduce this into the expression for ν.

5.4 Plot a graph of μ vs. the difference in electronegativity of the atoms.

5.5 There are three distinct ways in which d orbitals can overlap to give a large value of the overlap integral: $d_{z^2} - d_{z^2}$, $d_{xz} - d_{xz}$ (or $d_{yz} - d_{yz}$) and $d_{xy} - d_{xy}$ (or $d_{x^2-y^2} - d_{x^2-y^2}$).

5.6 Expand $\int \psi\psi\, dv$ in terms of atomic orbitals and find the condition that this be unity. Evaluate ψ^2 at a point half way between the two nuclei, and compare this with $1s_a{}^2 + 1s_b{}^2$.

5.7 What is the ratio of s to p character in an sp^3 hybrid? How much can p_x and p_y contribute to a hybrid pointing in the z direction?

5.8 There are three forms of hybrid used to describe the bonding in carbon compounds: sp, sp^2, sp^3. If an orbital overlaps equally with the orbitals of two (or more) neighbouring atoms then this is associated with a delocalized electron system.

5.9 The forms of the sp, sp^2 and sp^3 hybrids pointing along the z axis must first be established (see, for example, problem 5.7), then the overlap integrals will be of the type

$$\int (a2s + b2p)1s\, dv = a\int 2s1s\, dv + b\int 2p1s\, dv.$$

Chapter 6

6.1 Use expressions (6.18) and (6.33).

6.2 Start with (6.18) and multiply throughout by $\mathscr{B}$.

6.3 The eigenfunctions of d/dx are given in expression (6.6) If we want an eigenfunction which is everywhere finite so that it can be normalized then there is a further condition to be imposed on k.

6.4 The Hamiltonian is given by (6.13). What terms in it do not commute with x? What does $(\mathscr{H}x - x\mathscr{H})$ reduce to? Note that $x\mathscr{H}\Psi_j = xE_j\Psi_j$ and that $\mathscr{H}$ is a Hermitian operator so that

$$\int \Psi_i{}^*\mathscr{H}x\Psi_j\, dv = \int x\Psi_j\mathscr{H}^*\Psi_i{}^*\, dv.$$

6.5 Expand $\mathscr{B}\vartheta_b$ in terms of the complete set of functions ϑ_i and find the expansion coefficients by multiplying both sides by $\vartheta_j{}^*$ and integrating. This gives

$$\mathscr{B}\,\vartheta_b = \sum_i \left\{\int \vartheta_i{}^*\,\mathscr{B}\,\vartheta_b\, dz\right\} . \vartheta_i.$$

The theorem is then proved by multiplying throughout by $\vartheta_a{}^*\,\mathscr{A}$ and integrating.

Chapter 8

8.1 Determine the symmetry operations for each molecule and then use the rules given on page 93 to determine the groups to which they belong.

8.2 The C_{2h} group (e.g. *trans* CHCl=CHCl) has the following operations: I, C_2, σ_h, i. Determine the effect of successive operations of any two of these on the molecule and write the results in tabular form as table 8.2.

8.3 What sets of numbers satisfy the group multiplication table constructed in problem 8.2? Classify these sets according to the rules given on page 98.

8.4 What combination of the characters given in table 8.5 gives these sets of numbers? The vector decomposition rule given on page 103 can be used to facilitate the analysis.

8.5 How many of the functions are unchanged under the group operations (taking -1 for a change in sign)? These are the characters of the required representations. They can be resolved into component *I.R.*'s by the method described in problem 8.4.

8.6 What are the *I.R.*'s obtained from the representation based on these functions? To get the combinations which transform like these *I.R.*'s use the method given on page 104. Note that for CH_4 there is a complication through degeneracy, and one needs to get three orthogonal functions which together transform like the triply-degenerate *I.R.* (see page 105). The symmetry operations for the tetrahedron are shown in the following figure.

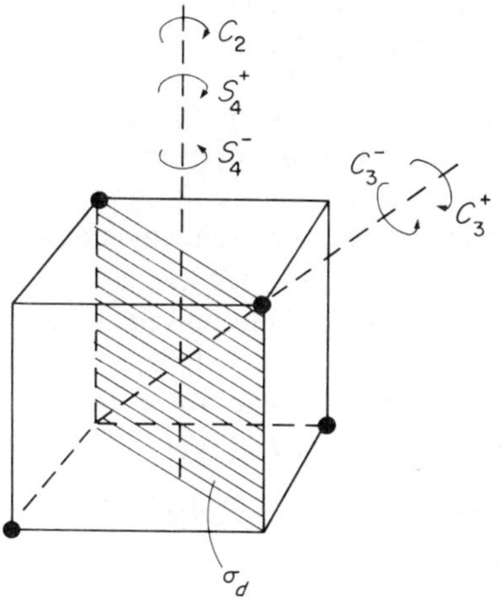

8.7 To get the direct products of two *I.R.*'s one just multiplies the corresponding characters of the *I.R.*'s together and the result, in the case of a group with no degeneracies, is the set of characters of another *I.R.*. The discussion following expression (8.32) answers the last part of the question.

Chapter 9

9.1 This only involves straightforward differentiation of products e.g.

$$\frac{\partial}{\partial z}\left(\frac{z\partial}{\partial x}\right) = \frac{\partial z}{\partial z}\frac{\partial}{\partial x} + \frac{z\partial^2}{\partial z \partial x} = \frac{\partial}{\partial x} + \frac{z\partial^2}{\partial z \partial x}.$$

9.2 Apply the rule given on page 116.

9.3 Combine the spins of two electrons and then combine their resulting spins with the spin of the third electron.

9.4 Simply carry out the same procedure as was used to obtain table 9.3 and then note the 1:1 correspondence between the M_L, M_S values of the two configurations.

9.5 The only configuration not yet analysed is p^3. This gives ^4S, ^2P and ^2D terms.

9.6 Work out the energy of the level with quantum numbers J, L, S and that with $J-1$, L, S and note the difference.

9.7 See page 119.

9.8 For $3p\ 3d$ one has an L–S coupling case, for $3p\ 6d$ a j–j coupling case. Confirm that there is a 1:1 correspondence in J values in the two cases.

9.9 Start with the wave function of the ^1D term having $L=2$, $M_L=2$. This has two electrons with opposite spins in the p orbital with $m=1$ (see table 9.3).
Write an antisymmetrized function which describes this situation. Operate on this with $\mathbf{L}_- = \mathbf{L}_-(1) + \mathbf{L}_-(2)$, the shift operation for both electrons 1 and 2. Use the formula (9.21) to find the result of this operation. $\mathbf{L}_-$ operating on the component with $L=2$, $M_L=2$ will convert it into the component with $L=2$, $M_L=1$. The form of this function is obtained by considering how $L_-(1)$ and $L_-(2)$ act on the p orbitals.

9.10 The selection rules for a Russell–Saunders (L–S) coupling case are given on page 124.

9.11 A small magnetic field splits a level of quantum number J into $2J+1$ equally spaced levels whose spacing is determined by expression (9.27).

9.12 Write the $1s$ orbital in the form $1s = 2\zeta^{3/2}e^{-\zeta r}Y_{00}(\vartheta, \varphi)$, and then integration over the angle coordinates gives unity. The integration over the coordinates of the first electron is done in two parts as described on page 129.

9.13 Write down the antisymmetrized form for the ^3S state having spin $M_S=1$. Square this and integrate over the spin to give $P(r_1, r_2)$. Then $P(r_1) = \int P(r_1, r_2)\,dv_2$ etc., from which the required expression can be derived.

Chapter 10

10.1 Start with expression (10.16) and introduce the explicit form of the Hamiltonian (10.17). The required relationship can then be established using expressions (10.18).

10.2 The following wave functions are required

$$1s(\mathrm{H}) = \pi^{-\frac{1}{2}}\mathrm{e}^{-r}$$

$$1s(\mathrm{He}^+) = \left(\frac{8}{\pi}\right)^{\frac{1}{2}}\mathrm{e}^{-2r}$$

$$2p\sigma(\mathrm{He}^+) = \pi^{-\frac{1}{2}}r\mathrm{e}^{-r}\cos\vartheta.$$

10.3 What is the electron configuration of C_2^+ (cf. table 10.2)? A configuration which needs k electrons to make a closed shell gives the same states as a configuration having k electrons in that shell.

10.4 The possible arrangements of electrons amongst orbitals for the configuration $(\pi_g 2p)^2$ are shown on page 149. Write the wave functions in terms of Slater determinants (page 83) using the correct form of the spin functions given by (9.23) and (9.24).

10.5 This will be of the form of figure 10.6, but one knows the exact form of the energy levels of H, He^+ and Li^{2+} (expression 3.19).

10.6 Solve the determinant

$$\begin{vmatrix} \mathscr{H}_{\mathrm{aa}} - ES_{\mathrm{aa}} & \mathscr{H}_{\mathrm{ab}} - ES_{\mathrm{ab}} \\ \mathscr{H}_{\mathrm{ab}} - ES_{\mathrm{ab}} & \mathscr{H}_{\mathrm{bb}} - ES_{\mathrm{bb}} \end{vmatrix} = 0$$

with the condition that a and b are equivalent orthogonal orbitals. The solutions should be the same as expressions (10.16).

10.7 and **10.8** These both require a reduction of the many-electron integrals to one- and two-electron integrals in the manner outlined on page 128.

Chapter 11

11.1 Integrate Ψ^2 and show that the result is unity, using expression (11.8) and the fact that the atomic orbitals are normalized.

11.2 Just substitute the numbers given into expressions (11.6) and (11.11) and hence plot expression (11.9).

11.3 What is the ground state symmetry of C_2 (table 10.2)? How must the *p* orbitals of carbon be paired together to give such a symmetry? The MO configuration given in table 10.2 will give a guide. Write a wave function analogous to (11.17) which represents two electron-pair bonds.

11.4 What is the normalizing constant? Expression (10.51) shows the form of the dipole moment operator. Expressions (11.30) and (11.31) are typical covalent and ionic functions; are these orthogonal? What about the homopolar dipole?

11.5 Solve the secular equations (6.67) using the data given. (cf. problem 10.6).

Chapter 12

12.1 How do the carbon 2*s* and 2*p* orbitals and the hydrogen 1*s* orbitals transform under the operations of the group? From this one can deduce the symmetries of the molecular orbitals (cf. the discussion of H_2O on page 190). The overall symmetry of a state is obtained from the direct product (page 107) of the orbitals which are not completely filled by electrons. What are the symmetries of the molecular orbitals in the $\mathbf{C}_{3v}$ group? How does the *s* and *p* character of the orbitals change on distorting to this symmetry?

12.2 Note that ψ_1, ψ_2 and ψ_3 have the same transformation properties as the nitrogen 2*s*, $2p_x$ and $2p_y$ orbitals respectively. From the form of sp^2 hybrid orbitals (5.15–5.17), which can be considered as equivalent orbitals of an atom having C_3 symmetry, one can deduce the form of the equivalent orbitals of the molecule.

12.3 Use formula (12.34). Consider only the energy of the structure I (figure 12.7); there are two pairs of orbitals which are forming bonds and four pairs not forming bonds.

12.4 We form *sp* hybrids of Be (5.4) and write down a VB function describing the two electron-pair bonds (cf. problem 11.3). The valence state function is then obtained by taking only that part of this function which involves the Be orbitals. What combination of atomic states is given by each determinant in this wave function?

Chapter 13

13.1 The complex contains Fe^{3+} which has five 3*d* electrons. The two possible ground states for a d^5 ion in an octahedral complex are $^6A_{1g}$

(weak field) and $^2T_{2g}$ (strong field). Where does F^- appear in the spectro-chemical series? What are the spin selection rules for allowed transitions?

13.2 The components are obtained in the same way as the components of a D state (page 222). The seven F functions depend on φ in the following way:

$$e^{3i\varphi}, \quad e^{2i\varphi}, \quad e^{i\varphi}, \quad 1, \quad e^{-i\varphi}, \quad e^{-2i\varphi}, \quad e^{-3i\varphi}.$$

13.3 No calculation is required. The atomic state eigenfunctions are chosen from those listed in tables 13.5, 13.6 (remembering the hole-electron analogy), and eigenfunctions in terms of d orbitals are the strong-field eigenfunctions (cf. page 231).

13.4 Use figure 13.9. $[Mn(H_2O)_6]^{2+}$ is a weak-field complex.

13.5 Remember that the crystal-field operator does not affect electron spins. Use the interaction integrals (13.7) and (13.8).

13.6 Construct a table similar to table 13.10. Note that the character of all triply-degenerate *I.R.*'s is zero for the C_3 and S_6 operations. The following figure shows a convenient method of orientating the p orbitals; a cross indicates the positive end of the orbital.

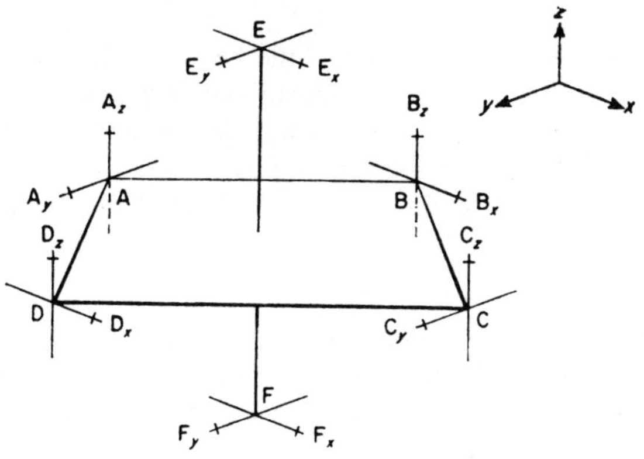

Chapter 14

14.1 Draw possible structures for the molecule and ask whether any structure satisfies the necessary condition for stability: that the bonding molecular orbitals are completely filled by electrons.

14.2 The form of the molecular orbitals will be obtained by solving the relevant secular equations (10.30). As the angle is increased the interaction $\mathbf{H}_{ac}$ will decrease relative to $\mathbf{H}_{ab}(=\mathbf{H}_{bc})$.

Chapter 15

15.1 This is quite straightforward, use exactly the same procedure as for butadiene, page 272. In a branched chain the order in which the carbon atoms are numbered is unimportant since, for example, if the centre carbon atom is numbered one instead of two this corresponds to interchanging both two columns and two rows of the secular determinant which leaves its numerical value unchanged. The π-electron bond order $p_{\mu\nu}$ is given by expression (15.9).

15.2 The secular equations for linear and triangular H_3 in Hückel theory are the same as the secular equation for allyl and cyclopropyl.

15.3 This problem is a straightforward Hückel calculation. The evaluation of the molecular-orbital coefficients can be simplified by the use of symmetry. First determine to which symmetry group fulvene belongs and then draw the appropriate diagrams as was done for naphthalene. The secular determinants follow directly. The electron charges q_μ are given by expression (15.8).

15.4 Write the molecular orbitals of the heterosubstituted hydrocarbon in terms of the orbitals of the unsubstituted hydrocarbon using (6.52). Expand the molecular orbitals in terms of the component atomic orbitals and retain only the most important terms. Hence obtain an expression for δq_ν, which, divided by $\delta\alpha_\mu$, gives $\pi_{\mu,\nu}$. Inspect the form of the final summations.

15.5 Draw a circle of radius $2|\beta|$, and insert radii to the bottom of the circle, along the horizontal diameter and to a point on the circumference representing an apex of the polygon. Draw a perpendicular from this point to the horizontal diameter. This perpendicular represents the energy of an orbital in units of β.

15.6 Consider the interactions of the bonding and antibonding π orbitals of three suitably orientated ethylene molecules.

15.7 The atom polarizabilities can be computed directly from (15.40), but in practice it is easier to simplify the expression, making use of the fact that butadiene is an even alternant, in which the orbitals occur in pairs with equal and opposite energies. Only the lowest $\frac{1}{2}n$ molecular orbitals

(n = number of C atoms) are occupied. The π-electron densities q are given by (15.41).

15.8 The energy is given by (10.79). Expand the molecular orbitals in terms of atomic orbitals and use (15.48) and (15.56) to simplify the expression for the energy.

15.9 Use (15.63) to calculate the difference in energy between the highest occupied and the lowest vacant orbitals, as a function of the number of atoms in the chain.

Chapter 17

17.1 This is a straightforward application of Dewar numbers. Remember that the sum of the coefficients of the non-bonding orbital on each side of an unstarred position is zero.

17.2 We assume that the RN_2^+ will attack the site of highest electron density, and that this is the same site which has the highest electron density in the equivalent hydrocarbon (that is, the hydrocarbon obtained by replacing O^- by CH_2^-). This electron density can be calculated from the non-bonding orbitals.

17.3 In this problem the π-electron energy calculated for the bases and for their conjugate acids must be related to the free-energy change associated with the protonation of the bases. Hammett has provided strong experimental evidence that the ratios of the activity coefficients of two similar bases in the same medium, at the same temperature are equal. Thus the free-energy change can be related directly to the potential energy change and then approximately equated to the difference in π-electron energy of a base and of its conjugate acid. Next, it is necessary to decide whether a proton will preferentially add to the ring nitrogen or to the exocyclic NH_2 group. Having decided this, the calculation is completed by considering the effect the other nitrogen atom will have on the π-electron energy. Remember that if the coulomb integral of atom μ is altered by a small amount $\delta\alpha_\mu$ then the change in π-electron energy is given by $E = q_\mu \delta\alpha_\mu$. Because the aminoquinolines are odd alternants q_μ can be calculated using the non-bonding orbitals.

17.4 In this question the number of conjugated centres is changed in going from the hydrocarbons to their anions. The same assumptions about the relationship of the free-energy change with the π-electron energy change must be made, as was required in problem 17.3. The π-electron energies of the cyclopentadienyl anion and of benzene (the obvious approximation for toluene) can be obtained by the graphical method (problem 15.5); but the π-electron energy of benzyl can only be obtained by solving the appropriate secular determinant (this calculation can be simplified by considering the symmetry of the molecule).

Answers to problems

Chapter 1

1.1 $E = h\nu = hc/\lambda$.
At 4000 Å $E = (6{\cdot}6 \times 10^{-27})\ (3{\cdot}0 \times 10^{10})/(4 \times 10^{-5})$ erg.
To convert to ev this must be divided by $1{\cdot}6 \times 10^{-12}$.
$E(4000\ \text{Å}) = 3{\cdot}09$ ev; $E(7500\ \text{Å}) = 1{\cdot}65$ ev. These are the limits to the change in energy of the electron for the wavelength of the emitted photon to be in the visible region.

1.2 Following the same procedure as in problem 1.1 one finds that 2·48 ev is equivalent to a wavelength of 5000 Å. Light of shorter wavelength will have enough energy to dissociate Cl_2. However, $\lambda < 5000$ Å is a necessary but not a sufficient condition for bringing about dissociation; a further necessary condition is that the light shall be absorbed by the molecule.

1.3 $$\tfrac{1}{2}mv^2 = (hc/\lambda) - W = \frac{(6{\cdot}6 \times 10^{-27})(3{\cdot}0 \times 10^{10})}{2537 \times 10^{-8}} - 2{\cdot}3 \times 1{\cdot}6 \times 10^{-12}$$
$$= 4{\cdot}1 \times 10^{-12} \text{ erg.}$$

1.4 Taking an atomic diameter as 10^{-8} cm and a nuclear diameter as 10^{-13} cm, then the fractional area of a single layer of atoms covered by the nuclei is 10^{-10}. In 1 cm there will be 10^8 layers hence the chance of a collision of a neutron with a nucleus is about one in a hundred.

1.5 *i* $\tfrac{1}{2}mv^2 = p^2/2m$. Hence, for a kinetic energy of $4{\cdot}1 \times 10^{-12}$ erg, we have $p = 8{\cdot}64 \times 10^{-20}$ g cm sec^{-1}. The wavelength associated with this momentum is then $\lambda = h/p = 7{\cdot}6$ Å.
ii Take a ball mass 100 g moving with a velocity 10^4 cm sec^{-1}, then $p = 10^6$ g cm sec^{-1} and $\lambda = 6{\cdot}6 \times 10^{-33}$ cm.

1.6 $\Delta x . \Delta p_x \simeq h/2\pi$. If $\Delta x = 10^{-13}$ cm, $\Delta p_x \simeq 10^{-14}$ g cm sec^{-1}. Similarly for Δp_y and Δp_z. The average kinetic energy is therefore expected to be at least
$$(1/2m)(\Delta p_x^{\ 2} + \Delta p_y^{\ 2} + \Delta p_z^{\ 2}) \simeq 0{\cdot}2 \text{ erg.}$$
The potential energy between a proton and an electron separated by 10^{-13} cm is
$$V = \frac{(4{\cdot}8 \times 10^{-10})^2}{10^{-13}} \simeq 2 \times 10^{-6} \text{ erg.}$$

The potential energy could not therefore counterbalance the kinetic energy associated with the uncertainty principle.

Chapter 2

2.1 As a tends to infinity in equation (2.5) the separation between levels of different r tends to zero. The energy is therefore not quantized under these conditions.

2.2 Equation (2.5) generalized to a three-dimensional box of sides 10^{-13} cm will give a kinetic energy

$$T = \frac{3h^2}{8m10^{-26}} = 2 \text{ erg}.$$

The algebra only differs from that of problem 1.6 by the omission of a π^2 in the denominator of the expression for the kinetic energy.

2.3 The Schrödinger equation in terms of the variable ϑ is, for a zero potential energy,

$$\frac{d^2\Psi}{d\vartheta^2} + \frac{8\pi^2 ma^2E}{h^2}\Psi = 0. \tag{a}$$

Try $\Psi_r = N \sin r\vartheta$ as a solution. Substituting this into (a) we get

$$N \sin r\vartheta\left(-r^2 + \frac{8\pi^2 ma^2E}{h^2}\right) = 0, \tag{b}$$

which is true if $E = h^2r^2/8\pi^2ma^2$. An equally good solution is $\Psi_r' = N \cos r\vartheta$ which leads to the same expression for the energy. The boundary condition on the wave function is $\Psi(\vartheta) = \Psi(2\pi + \vartheta)$. This is only true if r is zero or an integer. When r is an integer each state is doubly degenerate (Ψ_r and Ψ_r' have the same energy). Unlike the square-well problem, however, there is a solution with $r = 0$, $\Psi_0 = N$. This is non-degenerate and has zero energy. Since $V = 0$, this means that the kinetic energy is zero, which is only true if the momentum is zero. If the momentum definitely is zero then the position of the particle is infinitely uncertain. What we mean by this is not $\Delta\vartheta = \infty$, but that we have no knowledge about the position of the electron on the ring, and this is clearly the case if Ψ is independent of ϑ. The normalizing constant is found from the condition

$$\int_0^{2\pi} N^2 \sin^2 r\vartheta \, d\vartheta = 1,$$

which gives $N = \pi^{-\frac{1}{2}}$(the cosine function has the same normalizing constant). For $r = 0$ only the cos $r\vartheta$ solution exists and the normalization constant is $(2\pi)^{-\frac{1}{2}}$.

Chapter 3

3.1

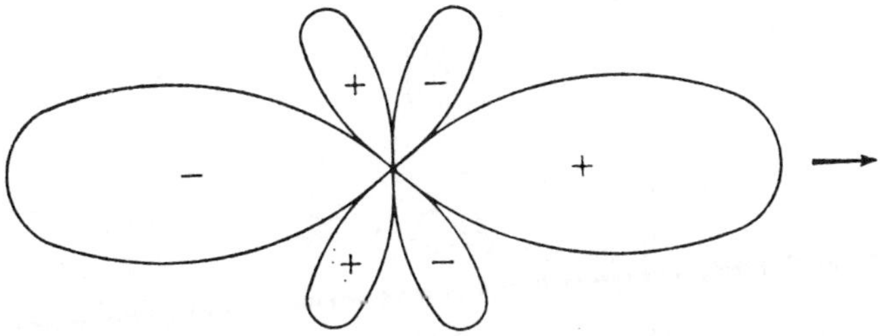

3.2 A p orbital pointing towards the (1, 1, 1) coordinate must have an equal mixture of p_x, p_y and p_z. After normalizing, the wave function is Ψ (1, 1, 1) $= \sqrt{\tfrac{1}{3}}(p_x + p_y + p_z)$. Alternatively: $1/\sqrt{3}$ is the cosine of the angle which the vector (1, 1, 1) makes with each of the coordinate axes.

3.3 From table 3.3 we have

$$\left.\begin{aligned} 1s &= N\mathrm{e}^{-\rho/2} \\ 2s &= N'(2-\rho)\mathrm{e}^{-\rho/2} \end{aligned}\right\} \rho = 2Zr/na_0.$$

Introducing the constant $\sigma = Zr/a_0$, which is independent of n, we have

$$1s2s = NN'(2-\sigma)\mathrm{e}^{-3\sigma/2}.$$

We are required to prove that

$$\int 1s\,2s\,r^2\,\mathrm{d}r\,\sin\vartheta\,\mathrm{d}\vartheta\,\mathrm{d}\varphi = 0.$$

We shall show this by proving that the integration over the radial coordinate is zero. Taking σ as the radial variable throughout we examine

$$\int_0^\infty (2-\sigma)\sigma^2\mathrm{e}^{-3\sigma/2}\,\mathrm{d}\sigma.$$

Using the formula given in the hints, this integral is equal to

$$\frac{2.2!}{(3/2)^3} - \frac{3!}{(3/2)^4} = 0,$$

thus establishing the orthogonality relationship.

3.4 $$\overline{\left(\frac{1}{r}\right)} = \int_0^\infty \int_0^\pi \int_0^{2\pi} (1s)^2 \overline{\left(\frac{1}{r}\right)} r^2 \, dr \sin \vartheta \,.\, d\vartheta \,.\, d\varphi.$$

From table 3.3 we have

$$1s = (Z/a_0)^{3/2} \pi^{-\frac{1}{2}} e^{-Zr/a_0},$$

hence

$$\overline{\left(\frac{1}{r}\right)} = \left(\frac{Z}{a_0}\right)^3 \pi^{-1} \int_0^\infty \int_0^\pi \int_0^{2\pi} r e^{-2zr/a_0} \, dr \sin \vartheta \, d\vartheta \, d\varphi.$$

Now $\int_0^{2\pi} d\varphi = 2\pi$, $\int_0^\pi \sin \vartheta \, d\vartheta = 2$, $\int_0^\infty r e^{-2zr/a_0} dr = \left(\frac{a_0}{2z}\right)^2.$

Hence $\overline{(1/r)} = Z/a_0$. The potential energy is $-Ze^2/r$, hence the mean value of the potential energy is $-Z^2e^2/a_0$. Since the total energy is (from expression 3.21) $-Z^2e^2/2a_0$, this means that the mean kinetic energy is $Z^2e^2/2a_0$, thus establishing the virial theorem $\overline{T} = -E$.

Chapter 5

5.1 $e^2/r = (4{\cdot}8 \times 10^{-10})^2/2{\cdot}5 \times 10^{-8}$ erg $= 5{\cdot}75$ ev.

5.2 At infinite separation Li^+Cl^- has an energy 1·44 ev greater than LiCl. This will be balanced by the coulomb attraction of the ions when

$$\frac{e^2}{r} = 1{\cdot}44 \times 1{\cdot}60 \times 10^{-12} \text{ erg.}$$

This gives $r = 10^{-7}$ cm $= 10$ Å.

5.3 When $r \to 0$ the Morse curve gives $V(0) = D_e (1 - e^{\beta r_0})^2$, which is not infinite as is the correct potential. Differentiating the Morse potential twice gives

$$\frac{d^2V}{dr^2} = D_e(-2\beta^2 e^{-\beta(r-r_0)} + 4\beta^2 e^{-2\beta(r-r_0)})$$

$$\left(\frac{d^2V}{dr^2}\right)_{r_0} = 2\beta^2 D_e = k = 4\pi^2 \nu^2 \mu.$$

Hence $\beta = \nu(2\pi^2\mu/D_e)^{\frac{1}{2}}$.

5.4

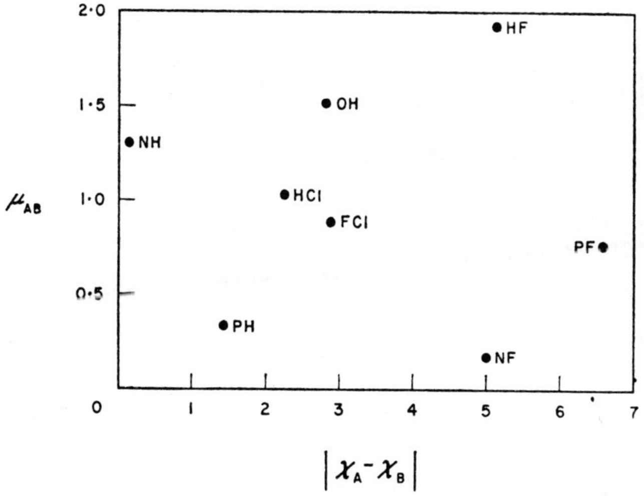

There is no correlation.

5.5

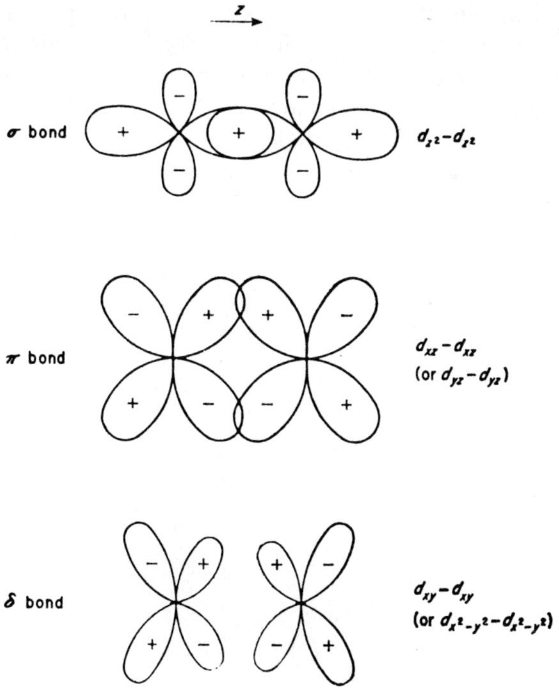

5.6 $\psi = N(1s_a + 1s_b)$.

$$\int \psi^2\, dv = N^2 \int (1s_a{}^2 + 2\,1s_a\,1s_b + 1s_b{}^2)\, dv = N^2(2 + 2S).$$

For this to be unity $N = (2 + 2S)^{-\frac{1}{2}}$.
Let us take the density at a point midway between the two nuclei where $1s_a = 1s_b$. Then ψ^2 at this point has the value

$$4(2 + 2S)^{-1}\, 1s_a{}^2,$$

so that two electrons in this orbital have a total density of $4\,(1 + S)^{-1}\, 1s_a{}^2$. This is greater than the density of two separate hydrogen atoms, $2\,(1s_a{}^2)$, providing $S < 1$, which it is.

5.7 An sp^3 hybrid must have a total of $\frac{1}{4}$ s and $\frac{3}{4}$ p character (see page **50**). If this hybrid is to point along the z axis then by symmetry p_x and p_y cannot contribute to it. It follows that the wave function must be

$$(\tfrac{1}{2})(s + \sqrt{3}p_z).$$

5.8 In butadiene all the carbon atoms are sp^2 hybridized. Each carbon atom has one p orbital not used in forming the sp^2 hybrids and these will form a four-centre delocalized π bond system. In vinyl acetylene the carbon atoms involved in the triple bond are sp hybridized and those involved in the double bond are sp^2 hybridized. There is one delocalized π system spreading over all four carbon atoms and another perpendicular to this

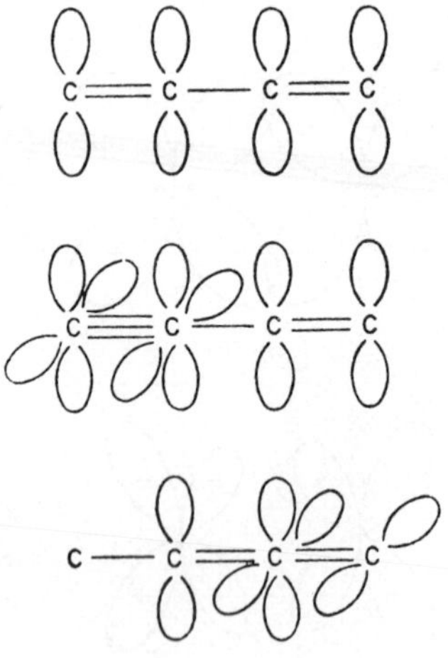

which is localized in the triple bond. In methyl allene the methyl group is sp^3 hybridized, the lone carbon atom sp hybridized and the other two are sp^2 hybridized. It is energetically more favourable to have two localized π bonds formed from orbitals mutually perpendicular to one another as shown in the figure, than to have one delocalized three-centre π system and one non-bonded orbital on the middle carbon atom.

5.9 $\psi_{sp} = \sqrt{\tfrac{1}{2}}(s + p_z), \qquad \int \psi_{sp}\, 1s\, dv = \sqrt{\tfrac{1}{2}}(0{\cdot}57 + 0{\cdot}46) = 0{\cdot}73$

$$\psi_{sp^2} = \sqrt{\tfrac{1}{3}}s + \sqrt{\tfrac{2}{3}}p_z, \qquad \int \psi_{sp^2} 1s\, dv = \sqrt{\tfrac{1}{3}}(0{\cdot}57 + \sqrt{2}\, 0{\cdot}46) = 0{\cdot}70$$

$$\psi_{sp^3} = \tfrac{1}{2}s + \tfrac{\sqrt{3}}{2}p_z, \qquad \int \psi_{sp^3} 1s\, dv = \tfrac{1}{2}(0{\cdot}57 + \sqrt{3}\, 0{\cdot}46) = 0{\cdot}68.$$

These numbers show a correlation with the force constants.

Chapter 6

6.1 If Ψ_j is an eigenfunction of $\mathscr{B}$ then

$$\int \Psi_i^* \mathscr{B} \Psi_j \, dv = \int \Psi_i^* b_j \Psi_j \, dv = b_j \int \Psi_i^* \Psi_j \, dv,$$

which is zero from the orthogonality condition.

6.2 If

$$\mathscr{B}\Psi_j = b_j \Psi_j$$

then

$$\mathscr{B}^2 \Psi_j = \mathscr{B} b_j \Psi_j = b_j \mathscr{B} \Psi_j = b_j^2 \Psi_j.$$

This can be generalized to show that b_j^n is an eigenvalue of $\mathscr{B}^n$.

6.3 The momentum operator is $p_x = (h/2\pi i)(\partial/\partial x)$. This has eigenfunctions e^{kx}

$$\left(\frac{h}{2\pi i}\right) \frac{\partial}{\partial x} (e^{kx}) = \frac{kh}{2\pi i} (e^{kx}),$$

where k is any constant, real or complex. However, if we want an eigenfunction that is finite in the range $x = -\infty$ to ∞, then k must be a pure imaginary quantity, $k = ik'$, say. In this case the eigenfunction $e^{ik'x}$ has a real eigenvalue $k'h/2\pi$.

6.4 For a single electron having a potential energy V

$$\mathscr{H} = -\frac{h^2}{8\pi^2 m}\left\{\frac{\partial^2}{\partial x^2} + \frac{\partial^2}{\partial y^2} + \frac{\partial^2}{\partial z^2}\right\} + V.$$

The only part of $\mathscr{H}$ which does not commute with x is $\partial^2/\partial x^2$, hence

$$(\mathscr{H}x - x\mathscr{H})\Psi = -\frac{h^2}{8\pi^2 m}\left\{\frac{\partial^2}{\partial x^2}.x - x\frac{\partial^2}{\partial x^2}\right\}\Psi = -\frac{h^2}{4\pi^2 m}\frac{\partial \Psi}{\partial x}.$$

Therefore

$$\int \Psi_i^*(\mathscr{H}x - x\mathscr{H})\Psi_j \, dv = -\frac{h^2}{4\pi^2 m}\int \Psi_i^* \frac{\partial}{\partial x}\Psi_j \, dv. \qquad \text{(a)}$$

But $\displaystyle\int \Psi_i^*(x\mathscr{H})\Psi_j \, dv = E_j \int \Psi_i^* x \Psi_j \, dv$

and $\displaystyle\int \Psi_i^*(\mathscr{H}x)\Psi_j \, dv = \int x\Psi_j \mathscr{H}^* \Psi_i^* \, dv = E_i \int x\Psi_j\Psi_i^* \, dv$

$$= E_i \int \Psi_i^* x \Psi_j \, dv.$$

The left-hand side of (a) therefore becomes

$$(E_i - E_j)\int \Psi_i^* x \Psi_j \, dv,$$

6.5

If ϑ_i is a complete set of functions then the function $\mathscr{B}\vartheta_b$ can be expanded as follows

$$\mathscr{B}\,\vartheta_b = \sum_i c_i\,\vartheta_i. \qquad \text{(a)}$$

The coefficients can be found by multiplying from the left by ϑ_j^* (for example) and integrating.

$$\int \vartheta_j^*\,\mathscr{B}\,\vartheta_b \, dz = \sum_i c_i \int \vartheta_j^* \vartheta_i \, dz = c_j. \qquad (b)$$

This follows because ϑ is an orthonormal set. Substituting (b) into (a) gives

$$\mathscr{B}\,\vartheta_b = \sum_i \left\{\int \vartheta_i^*\,\mathscr{B}\,\vartheta_b \, dz\right\}\vartheta_i, \qquad (c)$$

Multiplying this from the left by $\vartheta_a^*\,\mathscr{A}$ and integrating gives the required result.

$$\int \vartheta_a^*\,\mathscr{A}\,\mathscr{B}\,\vartheta_b \, dz = \sum_i \int \vartheta_a^*\,\mathscr{A}\,\vartheta_i \, dz \,.\int \vartheta_i\,\mathscr{B}\,\vartheta_b \, dz. \qquad (c)$$

Chapter 8

8.1 CH_2Cl_2 (($\mathbf{C}_{2v}$); *trans* CHCl=CHCl ($\mathbf{C}_{2h}$); C_2H_4 ($\mathbf{D}_{2h}$); IF_7 ($\mathbf{D}_{5h}$); C_6H_5Cl ($\mathbf{C}_{2v}$); HCN ($\mathbf{C}_{\infty v}$); CO_2 ($\mathbf{D}_{\infty h}$).

8.2

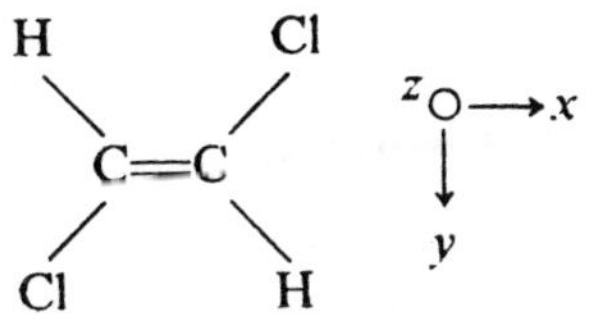

	I	C_2	σ_h	i
I	I	C_2	σ_h	i
C_2	C_2	I	i	σ_h
σ_h	σ_h	i	I	C_2
i	i	σ_h	C_2	I

8.3

$\mathbf{C}_{2h}$	I	C_2	σ_h	i
A_g	1	1	1	1
A_u	1	1	−1	−1
B_g	1	−1	−1	1
B_u	1	−1	1	−1

8.4 (a) $A_1 + A_2 + B_1 + B_2$ (b) $2A_2 + B_1 + 2B_2$.

8.5 To find the representation of the group with the atomic orbitals as basis we ask how many of the orbitals are unchanged on the operations of the group (taking −1 for each one changed in sign).

	I	C_2	σ_h	i	
The 1*s* orbitals of the hydrogen atoms	2	0	2	0	$= A_g + B_u$
The p_x, p_y and p_z orbitals of each chlorine atom	6	0	2	0	$= 2A_g + A_u + 2B_u + B_g$

8.6

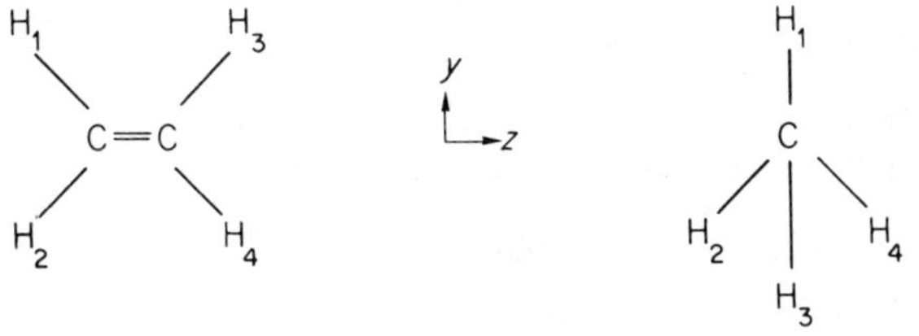

Let $\phi_1 \dots \phi_4$ be the four hydrogen $1s$ orbitals. Following the method described in the previous problem we first find the component *I.R.*'s obtained from the representation based on these orbitals.

Ethylene	I	$C_2(z)$	$C_2(y)$	$C_2(x)$	i	$\sigma(xy)$	$\sigma(xz)$	$\sigma(yz)$
	4	0	0	0	0	0	0	4

$= A_g + B_{3g} + B_{1u} + B_{2u}$

Methane	I	$6S_4$	$3C_2$	$6\sigma_d$	$8C_3$
	4	0	0	2	$1 = A_1 + T_2$

For ethylene there is no complication through degeneracy. We take one of the functions and operate on it with the group operations.

	I	$C_2(z)$	$C_2(y)$	$C_2(x)$	i	$\sigma(xy)$	$\sigma(xz)$	$\sigma(yz)$
ϕ_1 :	ϕ_1	ϕ_2	ϕ_3	ϕ_4	ϕ_4	ϕ_3	ϕ_2	ϕ_1

To find the combinations which transfer like the *I.R.*'s deduced above we multiply this set by the appropriate characters of the *I.R.* and add

$$A_g = \phi_1 + \phi_2 + \phi_3 + \phi_4 + \phi_4 + \phi_3 + \phi_2 + \phi_1 = 2(\phi_1 + \phi_2 + \phi_3 + \phi_4)$$
$$B_{3g} = \phi_1 - \phi_2 - \phi_3 + \phi_4 + \phi_4 - \phi_3 - \phi_2 + \phi_1 = 2(\phi_1 - \phi_2 - \phi_3 + \phi_4)$$
$$B_{1u} = \phi_1 + \phi_2 - \phi_3 - \phi_4 - \phi_4 - \phi_3 + \phi_2 + \phi_1 = 2(\phi_1 + \phi_2 - \phi_3 - \phi_4)$$
$$B_{2u} = \phi_1 - \phi_2 + \phi_3 - \phi_4 - \phi_4 + \phi_3 - \phi_2 + \phi_1 = 2(\phi_1 - \phi_2 + \phi_3 - \phi_4)$$

Each of these combinations is to be multiplied by a normalizing factor of 1/4 (assuming the orbitals do not overlap one another).

If we carry out the operations of the tetrahedron on ϕ_1 we get

	I	$6S_4$	$3C_2$	$6\sigma_d$	$8C_2$
ϕ_1 :	ϕ_1	$2\phi_2 + 2\phi_3 + 2\phi_4$	$\phi_2 + \phi_3 + \phi_4$	$3\phi_1 + \phi_2 + \phi_3 + \phi_4$	$2\phi_1 + 2\phi_2 + 2\phi_3 + 2\phi_4$

The combination which has A_1 symmetry is then obtained by adding all these terms: $6(\phi_1 + \phi_2 + \phi_3 + \phi_4)$.

This is to be multiplied by a normalizing constant 1/12.

A combination which transforms like T_2 is

$$3\phi_1 - (2\phi_2 + 2\phi_3 + 2\phi_4) - (\phi_2 + \phi_3 + \phi_4) + (3\phi_1 + \phi_2 + \phi_3 + \phi_4) + 0(2\phi_1 + 2\phi_2 + 2\phi_3 + 2\phi_4) = 6\phi_1 - 2(\phi_2 + \phi_3 + \phi_4). \qquad \text{(a)}$$

To find two other combinations which transform like T_2 we operate on two other orbitals. For example

operating on ϕ_2 gives $\quad 6\phi_2 - 2(\phi_3 + \phi_4 + \phi_1) \quad$ (b)

operating on ϕ_3 gives $\quad 6\phi_3 - 2(\phi_4 + \phi_1 + \phi_2) \quad$ (c)

The functions (a), (b) and (c) after normalization are

$$\psi_1 = \sqrt{\tfrac{1}{12}}(3\phi_1 - \phi_2 - \phi_3 - \phi_4)$$
$$\psi_2 = \sqrt{\tfrac{1}{12}}(3\phi_2 - \phi_3 - \phi_4 - \phi_1)$$
$$\psi_3 = \sqrt{\tfrac{1}{12}}(3\phi_3 - \phi_4 - \phi_1 - \phi_2).$$

But these are not orthogonal. If we carry out a Schmidt orthogonalization of ψ_2 and ψ_1, we can get a function

$$\psi_2' = \sqrt{\tfrac{1}{6}}(2\phi_2 - \phi_3 - \phi_4),$$

which is orthogonal to ψ_1. Lastly, we can orthogonalize ψ_3 to ψ_1 and ψ_2' and get

$$\psi_3' = \sqrt{\tfrac{1}{2}}(\phi_3 - \phi_4).$$

One can, however, obtain a more symmetrical set of functions. For example,

$$\frac{\sqrt{3}}{2}(\psi_1 + \psi_2) = \tfrac{1}{2}(\phi_1 + \phi_2 - \phi_3 - \phi_4),$$

and since the four functions are equivalent to one another there will be similar combinations of T_2 symmetry obtained by permuting the symbols 1, 2, 3 and 4.

$$\tfrac{1}{2}(\phi_1 - \phi_2 - \phi_3 + \phi_4)$$
$$\tfrac{1}{2}(\phi_1 - \phi_2 + \phi_3 - \phi_4).$$

These are seen to be mutually orthogonal.

8.7 By examination of the character table of the group we find

$A_g . A_g = A_g \qquad A_u . A_u = A_g \qquad B_g . B_u = A_u$

$A_g . A_u = A_u \qquad A_u . B_g = B_u \qquad B_u . B_u = A_g$

$A_g . B_g = B_g \qquad A_u . B_u = B_g$

$A_g . B_u = B_u \qquad B_g . B_g = A_g$

Hence an operator transforming like A_u will mix the following states: $A_g - A_u$, $B_g - B_u$.

Chapter 9

9.1 $$\mathbf{L}_x\mathbf{L}_y = -\left(\frac{h^2}{4\pi^2}\right)\left(y\frac{\partial}{\partial z} - z\frac{\partial}{\partial y}\right)\left(z\frac{\partial}{\partial x} - x\frac{\partial}{\partial z}\right)$$

$$= -\frac{h}{4\pi^2}\left\{y\left(\frac{\partial}{\partial x} + z\frac{\partial^2}{\partial x\partial z} - x\frac{\partial^2}{\partial z^2}\right) - z\left(z\frac{\partial^2}{\partial x\partial y} - x\frac{\partial^2}{\partial y\partial z}\right)\right\}.$$

$$\mathbf{L}_y\mathbf{L}_x = -\left(\frac{h^2}{4\pi^2}\right)\left(z\frac{\partial}{\partial x} - x\frac{\partial}{\partial z}\right)\left(y\frac{\partial}{\partial z} - z\frac{\partial}{\partial y}\right)$$

$$= -\left(\frac{h^2}{4\pi^2}\right)\left\{z\left(y\frac{\partial^2}{\partial x\partial z} - z\frac{\partial^2}{\partial x\partial y}\right) - x\left(y\frac{\partial^2}{\partial z^2} - \frac{\partial}{\partial y} - z\frac{\partial^2}{\partial z\partial y}\right)\right\}.$$

Hence

$$\mathbf{L}_x\mathbf{L}_y - \mathbf{L}_y\mathbf{L}_x = -\left(\frac{h^2}{4\pi^2}\right)\left\{y\frac{\partial}{\partial x} - x\frac{\partial}{\partial y}\right\} = -\frac{ih}{2\pi}\mathbf{L}_z.$$

9.2 $^2D\ (S = \frac{1}{2}, L = 2, \therefore J = \frac{5}{2}, \frac{3}{2})$ gives $^2D_{\frac{5}{2}}$, $^2D_{\frac{3}{2}}$.

$^1G\ (S = 0, L = 4, \therefore J = 4)$ gives $^1G_4$.

$^6S\ (S = \frac{5}{2}, L = 0, \therefore J = \frac{5}{2})$ gives $^6S_{\frac{5}{2}}$

9.3 Two electrons give a singlet and a triplet ($S_1 = 1$ or 0). Combining these with the other electron ($S_2 = \frac{1}{2}$) we have

$$S_1 = 1, S_2 = \tfrac{1}{2}, S = \tfrac{3}{2}, \tfrac{1}{2}.$$
$$S_1 = 0, S_2 = \tfrac{1}{2}, S = \tfrac{1}{2}.$$

We therefore get a quartet and *two* doublets.

9.4

$m =$	1	0	−1	$M_L = \Sigma m$,	$M_S = \Sigma m_s$
1	↑↓	↑↓		2	0
2	↑↓	↑	↑	1	1
3	↑↓	↑	↓	1	0
4	↑↓	↓	↑	1	0
5	↑↓	↓	↓	1	−1
6	↑	↑↓	↑	0	1
7	↑	↑↓	↓	0	0
8	↑↓		↑↓	0	0
9	↓	↑↓	↑	0	0
10	↓	↑↓	↓	0	−1
11	↑	↑	↑↓	−1	1
12	↑	↓	↑↓	−1	0
13	↓	↑	↑↓	−1	0
14	↓	↓	↑↓	−1	−1
15		↑↓	↑↓	−2	0

These arrangements 1–15 have exactly the same M_L and M_S values as those in table 9.3. Hence p^4 gives the same terms as p^2.

9.5

	Li	Be	B	C	N	O	F	Ne
Configuration $1s^2$ +	$2s$	$2s^2$	$2s^22p$	$2s^22p^2$	$2s^22p^3$	$2s^22p^4$	$2s^22p^5$	$2s^22p^6$
Terms	${}^2S$	${}^1S$	${}^2P$	${}^3P$, ${}^1D$, ${}^1S$	${}^4S$, ${}^2D$, ${}^2P$	${}^3P$, ${}^1D$, ${}^1S$	${}^2P$	${}^1S$
Ground term	${}^2S$	${}^1S$	${}^2P$	${}^3P$	${}^4S$	${}^3P$	${}^2P$	${}^1S$

The configuration p^3 can be shown to give the terms ${}^4S$, ${}^2D$ and ${}^2P$ by the method described for the p^2 configuration.

9.6 The energy of the level having quantum numbers J, L, S, relative to the mean energy of the term, is (from 9.20)

$$E_J = \tfrac{1}{2}\{J(J+1) - L(L+1) - S(S+1)\}\frac{h^2}{4\pi^2}.$$

The energy of the $J-1$, L, S level is

$$E_{J-1} = \tfrac{1}{2}\{(J-1)J - L(L+1) - S(S+1)\}\frac{h^2}{4\pi^2}.$$

Hence $\quad E_J - E_{J-1} = J\dfrac{h^2}{4\pi^2},$

which proves the Landé interval rule (page 119).

9.7 A shell which is less than half full gives a normal multiplet (J smallest has lowest energy), but if it is more than half full one gets an inverted multiplet (J greatest has lowest energy). A ${}^2P$ term gives levels ${}^2P_{\frac{1}{2}}$ and ${}^2P_{\frac{3}{2}}$. The lowest level for p^1 is ${}^2P_{\frac{1}{2}}$ and for p^5 is ${}^2P_{\frac{3}{2}}$.

9.8 The pd configuration gives the terms ${}^3F$, ${}^3D$, ${}^3P$, ${}^1F$, ${}^1D$, ${}^1P$. These are split by spin-orbit coupling to give ${}^3F_4$, ${}^3F_3$, ${}^3F_2$; ${}^3D_3$, ${}^3D_2$, ${}^3D_1$; ${}^3P_2$, ${}^3P_1$, ${}^3P_0$; ${}^1F_3$; ${}^1D_2$; ${}^1P_1$. For the configuration $3p3d$, electron-interaction energies will be much larger than spin-orbit coupling energies, hence levels having the same L and the same S will have almost equal energy (the term energy). However, for $3p6d$ the spin-orbit coupling will be larger than the electron interaction and now the energy of the levels depends mainly on the j values of the two electrons. An electron in a p orbital has ($l = 1$, $s = \frac{1}{2}$) $j = \frac{3}{2}$ or $\frac{1}{2}$: an electron in a d orbital has $j = \frac{5}{2}$ or $\frac{3}{2}$. There will now be four groups of levels ($j_1 = \frac{3}{2}, j_2 = \frac{5}{2}; J = 4, 3, 2, 1$), ($j_1 = \frac{3}{2}, j_2 = \frac{3}{2}; J = 3, 2, 1, 0$), ($i_1 = \frac{1}{2}, j_2 = \frac{5}{2}, J = 3, 2$), ($j_1 = \frac{1}{2}, j_2 = \frac{3}{2}, J = 2, 1$). It can be seen that there is a correspondence between the J values deduced from the L–S coupling scheme and those deduced from the j–j coupling scheme.

9.9 Let us give the three p orbitals the symbols ϕ_1, ϕ_0, ϕ_{-1}, the subscript indicating their m value. From table 9.3 it is seen that the arrangement with two electrons in ϕ_1 must be a component of the ^1D term. This has an antisymmetrized wave function which we write $F_{2,2}$ (since $L = 2$, $M_L = 2$)

$$F_{2,2} = |\phi_1\bar{\phi}_1| = \sqrt{\tfrac{1}{2}}\{\phi_1(1)\,\bar{\phi}_1(2) - \phi_1(2)\,\bar{\phi}_1(1)\},$$

where we use the convention of page 83 and indicate the β spin-orbital by a barred function and the α spin-orbital by an unbarred function. Operating on $F_{2,2}$ with $\mathbf{L}_- = \mathbf{L}_x - i\mathbf{L}_y$ will generate the function $F_{2,1}$ which has $L = 2$, $M_L = 1$.
Now $\mathbf{L}_- = \mathbf{L}_-(1) + \mathbf{L}_-(2)$, where (1) and (2) indicate the two electrons and from (9.21) we see that

$$\mathbf{L}_-F_{1,1} = (h/2\pi)\sqrt{2}F_{1,0};\ \mathbf{L}_-F_{1,0} = (h/2\pi)\sqrt{2}F_{1,-1};\ \mathbf{L}_-F_{1,-1} = 0.$$

Hence

$$(\mathbf{L}_-(1) + \mathbf{L}_-(2))\phi_1(1)\bar{\phi}_1(2) = \sqrt{2}\,\frac{h}{2\pi}\,\{\phi_0(1)\bar{\phi}_1(2) + \phi_1(1)\bar{\phi}_0(2)\}$$

and therefore

$$(\mathbf{L}_-(1) + \mathbf{L}_-(2))|\phi_1\bar{\phi}_1| = \sqrt{2}\,\frac{h}{2\pi}\,\{|\phi_0\bar{\phi}_1| + |\phi_1\bar{\phi}_0|\}.$$

But again from (9.21) $\mathbf{L}_-F_{2,2} = 2(h/2\pi)\,F_{2,1}$, and hence

$$F_{2,1} = \sqrt{\tfrac{1}{2}}\{|\phi_0\bar{\phi}_1| + |\phi_1\bar{\phi}_0|\},$$

which is the wave function of the ^1D state with $M_L = 1$.

We now operate on $F_{2,1}$ with $\mathbf{L}_-$. We have

$$(\mathbf{L}_-(1) + \mathbf{L}_-(2))\phi_0(1)\bar{\phi}_1(2) = \sqrt{2}\,\frac{h}{2\pi}\,\{\phi_{-1}(1)\bar{\phi}_1(2) + \phi_0(1)\bar{\phi}_0(2)\},$$

$$(\mathbf{L}_-(1) + \mathbf{L}_-(2))\phi_1(1)\bar{\phi}_0(2) = \sqrt{2}\,\frac{h}{2\pi}\,\{\phi_0(1)\bar{\phi}_0(2) + \phi_1(1)\bar{\phi}_{-1}(2)\}$$

hence

$$(\mathbf{L}_-(1) + \mathbf{L}_-(2))\sqrt{\tfrac{1}{2}}\{|\phi_0\bar{\phi}_1| + |\phi_1\bar{\phi}_0|\} = \frac{h}{2\pi}\{|\phi_{-1}\bar{\phi}_1| + |\phi_1\bar{\phi}_{-1}| + 2|\phi_0\bar{\phi}_0|\}.$$

But $\mathbf{L}_-F_{2,1} = \sqrt{6}\,\dfrac{h}{2\pi}\,F_{2,0}$ hence

$$F_{2,0} = \sqrt{\tfrac{1}{6}}\{2|\phi_0\bar{\phi}_0| + |\phi_{-1}\bar{\phi}_1| + |\phi_1\bar{\phi}_{-1}|\}.$$

Carrying the process a stage further (or noting that $F_{2,-1}$ and $F_{2,-2}$ must have the same form as $F_{2,1}$ and $F_{2,2}$ except for the switching of the functions ϕ_1 and ϕ_{-1}), we have

$$F_{2,-1} = \sqrt{\tfrac{1}{2}}\{|\phi_0\bar{\phi}_{-1}| + |\phi_{-1}\bar{\phi}_0|\}$$

$$F_{2,-2} = |\phi_{-1}\bar{\phi}_{-1}|.$$

9.10 The configuration p^2 gives terms ^3P, ^1D and ^1S; pd gives ^3F, ^1F, ^3D, ^1D ^3P, ^1P. The selection rules for Russell–Saunders coupling are $\Delta S = 0$, $\Delta L = 0, \pm 1$. Hence the allowed transitions are

$$^3P \rightarrow {}^3P,\quad {}^3D,$$
$$^1D \rightarrow {}^1F,\quad {}^1D,\quad {}^1P$$
$$^1S \rightarrow {}^1P.$$

9.11 A ^2P term has levels $^2P_{\frac{1}{2}}$, $^2P_{\frac{3}{2}}$ (see problem 9.7). A small magnetic field H^M will split the $^2P_{\frac{1}{2}}$ into two states with separation $g\beta H^M$ where (from 9.27)

$$g = 1 + \frac{\frac{1}{2}(1+\frac{1}{2}) - 1(1+1) + \frac{1}{2}(1+\frac{1}{2})}{2(\frac{1}{2})(1+\frac{1}{2})} = \tfrac{2}{3}.$$

The $^2P_{\frac{3}{2}}$ will be split into four equally spaced levels with a separation given by $g\beta H^M$ where

$$g = 1 + \frac{\frac{3}{2}(1+\frac{3}{2}) - 1(1+1) + \frac{1}{2}(1+\frac{1}{2})}{2(\frac{3}{2})(1+\frac{3}{2})} = \tfrac{4}{3}.$$

9.12

$$\iint 1s(1)1s(2)\left(\frac{1}{r_{12}}\right)1s(1)1s(2)\,dv_1\,dv_2.$$

We put $(1/r_{12}) = 1/r_>$, according to (9.42), and evaluate

$$\int_0^\infty 1s^2(2)\left\{\frac{1}{r_2}\int_0^{r_2} 1s^2(1)\,dv_1 + \int_{r_2}^\infty 1s^2(1)\frac{1}{r_1}\,dv_1\right\}dv_2. \tag{a}$$

If we now take the $1s$ orbital in the form (3.6 and table 3.2) $1s = 2\zeta^{3/2}e^{-\zeta r}$ $Y_{00}(\vartheta, \varphi)$, where Y_{00} is a normalized spherical harmonic, then the integration over the angular coordinates reduces expression (a) to

$$16\zeta^6\int_0^\infty e^{-2\zeta r_2}\left\{\frac{1}{r_2}\int_0^{r_2} e^{-2\zeta r_1}r_1^{\,2}\,dr_1 + \int_{r_2}^\infty e^{-2\zeta r_1}r_1\,dr_1\right\}r_2^{\,2}\,dr_2.$$

Introducing the new variables $s = 2\zeta r_1$, $t = 2\zeta r_2$ this becomes

$$\frac{\zeta}{2}\int_0^\infty e^{-t}\left\{\frac{1}{t}\int_0^t e^{-s}s^2\,ds + \int_t^\infty e^{-s}s\,ds\right\}t^2\,dt.$$

Integrating by parts gives

$$\int_t^\infty e^{-s}s\,ds = [-e^{-s}(1+s)]_t^\infty = e^{-t}(1+t)$$

$$\int_0^t e^{-s}s^2\,ds = [-e^{-s}(s^2+2s+2)]_0^t = 2 - e^{-t}(t^2+2t+2).$$

The integration over s then reduces the integral to

$$\frac{\zeta}{2}\int_0^\infty [-(t^2+2t)e^{-2t} + 2te^{-t}]\,dt = 5\frac{\zeta}{8}.$$

9.13 The wave function for the ^3S state having spin component $M_S = 1$ is

$$\Psi = |1s\alpha\ 2s\alpha| = \sqrt{\tfrac{1}{2}}\{1s(1)\alpha(1)2s(2)\alpha(2) - 1s(2)\alpha(2)2s(1)\alpha(1)\}.$$

Then

$$\Psi^2 = \tfrac{1}{2}\{1s^2(1)2s^2(2) + 1s^2(2)2s^2(1) - 2.1s(1)2s(1)1s(2)2s(2)\}\alpha^2(1)\alpha^2(2).$$

To find the two-particle probability density we integrate over the spin coordinates, and, since α is a normalized function,

$$P(r_1, r_2) = \tfrac{1}{2}\{1s^2(1)2s^2(2) + 1s^2(2)2s^2(1) - 2.1s(1)2s(1)1s(2)2s(2)\}.$$

Then, since $1s$ and $2s$ are orthogonal,

$$P(r_1) = \int P(r_1, r_2)\,dv_2 = \tfrac{1}{2}(1s^2 + 2s^2)(1),$$

and a similar expression holds for $P(r_2)$. It follows that

$$\begin{aligned}
P(r_1, r_2) - P(r_1)P(r_2) &= \tfrac{1}{2}\{1s^2(1)2s^2(2) + 1s^2(2)2s^2(1) - 2.1s(1)2s(1)1s(2)2s(2)\}\\
&\quad -\tfrac{1}{4}\{1s^2(1)2s^2(2) + 1s^2(2)2s^2(1) + 1s^2(1)1s^2(2) + 2s^2(1)2s^2(2)\}\\
&= -\tfrac{1}{4}(1s^2 - 2s^2)(1)(1s^2 - 2s^2)(2) - 1s(1)2s(1)1s(2)2s(2).
\end{aligned}$$

Chapter 10

10.1 From expression (10.16)

$$E_g = (1+S)^{-1}(\mathscr{H}_{aa} + \mathscr{H}_{ab}),$$

$$\mathscr{H}_{aa} = \int 1s_a\left[-\tfrac{1}{2}\nabla^2 - \frac{1}{r_a} - \frac{1}{r_b} + \frac{1}{R}\right]1s_a\,dv.$$

But $\left(-\tfrac{1}{2}\nabla^2 - \frac{1}{r_a}\right)1s_a = E_H 1s_a.$

Hence

$$\mathscr{H}_{aa} = E_H + \frac{1}{R} + \int 1s_a\left(\frac{-1}{r_b}\right)1s_a\, dv = E_H + \frac{1}{R} + \varepsilon_{aa}.$$

Similarly

$$\mathscr{H}_{ab} = \int 1s_a\left[-\tfrac{1}{2}\nabla^2 - \frac{1}{r_a} - \frac{1}{r_b} + \frac{1}{R}\right]1s_b\, dv = S\left(E_H + \frac{1}{R}\right) + \varepsilon_{ab}.$$

Hence $E_g = E_H + \dfrac{1}{R} + (1 + S)^{-1}(\varepsilon_{aa} + \varepsilon_{ab})$.

In a similar manner one can prove expression (10.19) for the energy of the state ψ_u.

10.2 From expression (10.11), the overlap integral between the two $1s$ orbitals whose centres are 2 atomic units apart is 0·59.

Let the $1s$ orbitals be (10.10)

$$1s_a = \pi^{-\frac{1}{2}}e^{-r_a}, \qquad 1s_b = \pi^{-\frac{1}{2}}e^{-r_b},$$

then $$\psi_g{}^2 = 0{\cdot}314\pi^{-1}(e^{-2r_a} + e^{-2r_b} + 2e^{-(r_a+r_b)})$$

and $$\psi_u{}^2 = 1{\cdot}220\pi^{-1}(e^{-2r_a} + e^{-2r_b} - 2e^{-(r_a+r_b)}).$$

These functions are shown in the following figure.

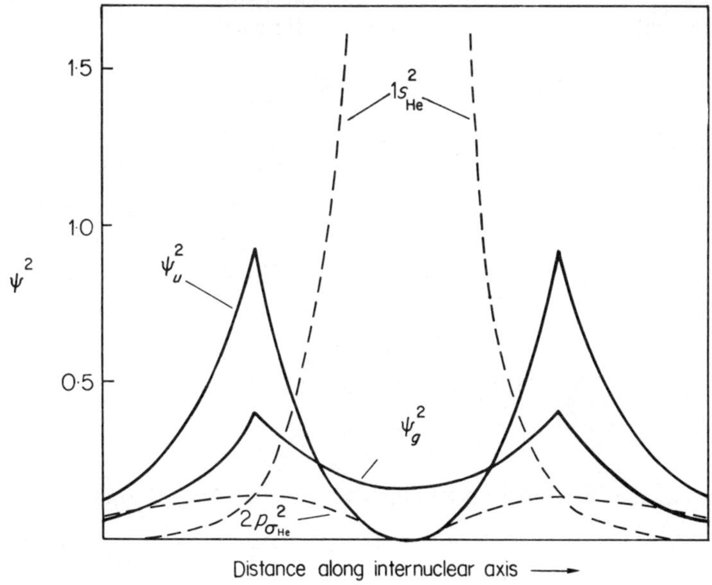

The orbital ψ_g correlates with the $1s$ orbital of He^+, and ψ_u with the $2p\sigma$ orbital of He^+; these have wave functions (from table 3.2)

$$1s = \left(\frac{8}{\pi}\right)^{\frac{1}{2}} e^{-2r}$$

$$2p\sigma = \pi^{-\frac{1}{2}} r e^{-r} \cos \vartheta,$$

and the squares of these orbitals are also plotted in the figure. Notice that the orbital ψ_u is represented fairly well by the united atom orbital.

10.3 $C_2{}^+$ has the electron configuration (from table 10.2)

$$KK(\sigma_g 2s)^2(\sigma_u 2s)^2(\pi_u 2p)^3.$$

There are three electrons in the $\pi_u 2p$ shell, one less than the maximum. This gives rise to the same states as the configuration $(\pi_u 2p)^1$, namely, just the doubly-degenerate state ${}^2\Pi_u$.

10.4 Let us write the wave functions of the two p orbitals in complex form $\pi_g^+ 2p$, $\pi_g^- 2p$ (cf. 11.21). An electron in these orbitals has an angular momentum $+h/2\pi$ or $-h/2\pi$ respectively. The different ways of allocating two electrons to these orbitals are set out on page 149. The ${}^1\Delta$ state has two components whose wave functions are as follows

$${}^1\Delta \begin{cases} |\pi_g^+ 2p \quad \overline{\pi_g^+ 2p}| & \Sigma\lambda = 2 \\ |\pi_g^- 2p \quad \overline{\pi_g^- 2p}| & \Sigma\lambda = -2, \end{cases}$$

where we have used the convention for Slater determinants described on page 83. The ${}^3\Sigma^-$ state has one electron in each orbital, and the three spin components have the form shown on page 123 $(\alpha\alpha, \sqrt{\tfrac{1}{2}}(\alpha\beta + \beta\alpha), \beta\beta)$, hence

$${}^3\Sigma^- \begin{cases} |\pi_g^+ 2p \quad \pi_g^- 2p| \\ \sqrt{\tfrac{1}{2}}\{|\pi_g^+ 2p \quad \overline{\pi_g^- 2p}| + |\overline{\pi_g^+ 2p} \quad \pi_g^- 2p|\}. \\ |\overline{\pi_g^+ 2p} \quad \overline{\pi_g^- 2p}| \end{cases}$$

Likewise, the ${}^1\Sigma^+$ function has the spin function $\sqrt{\tfrac{1}{2}}(\alpha\beta - \beta\alpha)$ hence

$${}^1\Sigma^+ : \sqrt{\tfrac{1}{2}}\{|\pi_g^+ 2p \quad \overline{\pi_g^- 2p}| - |\overline{\pi_g^+ 2p} \quad \pi_g^- 2p|\}.$$

10.5

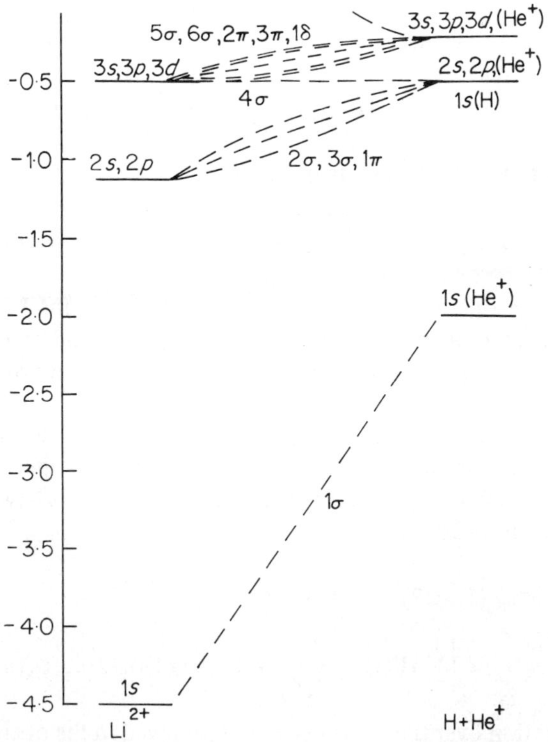

10.6 Labelling the two $1s$ orbitals $1s_a$ and $1s_b$, we need to solve the secular determinant (6.68)

$$\begin{vmatrix} \mathscr{H}_{aa} - ES_{aa} & \mathscr{H}_{ab} - ES_{ab} \\ \mathscr{H}_{ab} - ES_{ab} & \mathscr{H}_{bb} - ES_{bb} \end{vmatrix} = 0.$$

Using normalized orbitals $S_{aa} = S_{bb} = 1$; also $\mathscr{H}_{aa} = \mathscr{H}_{bb}$. On expanding the determinant we have

$$E^2(1 - S_{ab}{}^2) - E(2\mathscr{H}_{aa} - 2\mathscr{H}_{ab}S_{ab}) + (\mathscr{H}_{aa}{}^2 - \mathscr{H}_{ab}{}^2) = 0.$$

This has solutions

$$\begin{aligned} E = \{&2\mathscr{H}_{aa} - 2\mathscr{H}_{ab}S_{ab} \\ &\pm [4(\mathscr{H}_{aa} - \mathscr{H}_{ab}S_{ab})^2 - 4(1 - S_{ab}{}^2)(\mathscr{H}_{aa}{}^2 - \mathscr{H}_{ab}{}^2)]^{\frac{1}{2}}\}/2(1 - S_{ab}{}^2) \\ = \{&\mathscr{H}_{aa} - \mathscr{H}_{ab}S_{ab} \pm [\mathscr{H}_{ab} - \mathscr{H}_{aa}S_{ab}]\}/(1 - S_{ab}{}^2). \end{aligned}$$

That is,

$$E = (1 + S)^{-1}(\mathscr{H}_{aa} + \mathscr{H}_{ab}),$$

or

$$E = (1 - S)^{-1}(\mathscr{H}_{aa} - \mathscr{H}_{ab}),$$

which are expressions (10.16).

10.7 We wish to evaluate

$$E = \int |\psi_a\psi_b \ldots \psi_k| \left\{ \sum_i \mathbf{H}^c(i) + \sum_{i>j} \frac{1}{r_{ij}} + V_{nn} \right\} |\psi_a\psi_b \ldots \psi_k| \, d\tau_1 \ldots d\tau_n.$$

We first write the determinants in the form

$$|\psi_a\psi_b \ldots \psi_k| = \left(\frac{1}{n!}\right)^{\frac{1}{2}} (1 + \mathscr{P})\psi_a(1)\psi_b(2) \ldots \psi_k(n),$$

where $\mathscr{P}$ is an operator which permutes all the electrons 1 ... n amongst the available spin-orbitals and multiples by ± 1 (depending on whether it is an odd or even permutation) so as to preserve the antisymmetry of the function.

Suppose we take just one of the $n!$ terms from the determinant on the right; this must give the same contribution to E as any other term since all electrons are equivalent. Multiplying this one term by $n!$ (which cancels the normalizing constant) gives

$$E = \int (1 + \mathscr{P})\psi_a(1)\psi_b(2) \ldots \psi_k(n) \times \left\{ \sum_i \mathbf{H}^c(i) + \sum_{i>j} \frac{1}{r_{ij}} + V_{nn} \right\} \psi_a(1)\psi_b(2) \ldots \psi_k(n) \, d\tau_1 \ldots d\tau_n.$$

After integration over the coordinates not involved in the operator one has

$$\int \psi_a(1)\psi_b(2) \ldots \psi_k(n) \left\{ \sum_i \mathbf{H}^c(i) + \sum_{i>j} \frac{1}{r_{ij}} + V_{nn} \right\} \psi_a(1)\psi_b(2) \ldots \psi_k(n) \, d\tau_1 \ldots d\tau_n$$

$$= \sum_{r=\hat{a}}^{k} \mathbf{H}_{rr}^c + \sum_{\text{pairs } rs} J_{rs} + V_{nn},$$

which is expression (10.59).

Suppose we now exchange electrons in ψ_a and ψ_b. This gives the term

$$-\int \psi_a(2)\psi_b(1) \ldots \psi_k(n) \left\{ \sum_i \mathbf{H}^c(i) + \sum_{i>j} \frac{1}{r_{ij}} + V_{nn} \right\} \psi_a(1)\psi_b(2) \ldots \psi_k(n) \, d\tau_1 \ldots d\tau_n.$$

Because ψ_a and ψ_b are orthogonal functions the only non-zero part of this integral will arise for the operator $(1/r_{12})$, which gives $-K_{ab}$. There is one such term for all pairs of spin orbitals rs. If we exchange more than two electrons there will be at least three electrons in different spin-orbitals on the left and right of the operation and this can give no contribution to E. The total energy is therefore

$$E = \sum_{r=a}^{k} \mathbf{H}_{rr}^c + \sum_{\text{pairs } r} (J_{rs} - K_{rs}) + V_{nn}.$$

10.8 $$\mathbf{F}_{at} = \int |\psi_a\psi_b \ldots \psi_k| \left\{ \sum_i \mathbf{H}^c(i) + \sum_{i>j} \frac{1}{r_{ij}} + V_{nn} \right\} |\psi_t\psi_b \ldots \psi_k| \, d\tau_1 \ldots d\tau_n.$$

We carry out the same reduction as for problem 10.7 and find

$$\mathbf{F}_{at} = \int (1 + \mathscr{P})(\psi_a(1)\psi_b(2) \ldots \psi_k(n)) \times \left\{ \sum_i \mathbf{H}^c(i) + \sum_{i>j} \frac{1}{r_{ij}} + V_{nn} \right\} \psi_t(1)\psi_b(2) \ldots \psi_k(n) \, d\tau_1 \ldots d\tau_n.$$

Now

$$\int \psi_a(1)\psi_b(2) \ldots \psi_k(n) \left\{ \sum_i \mathbf{H}^c(i) + \sum_{i>j} \frac{1}{r_{ij}} + V_{nn} \right\} \psi_t(1)\psi_b(2) \ldots \psi_k(n) \, d\tau_1 \ldots d\tau_n$$
$$= \mathbf{H}^c_{at} + \sum_r \iint \psi_a(1)\psi_r(i) \frac{1}{r_{1i}} \psi_t(1)\psi_r(i) \, d\tau_1 \, d\tau_i,$$

since only operators involving the coordinates of electron 1 will give a non-zero contribution to the integral. The only exchange terms which will be non-zero are those involving electron 1 and another electron, for example

$$-\int \psi_a(2)\psi_b(1) \ldots \psi_k(n) \left\{ \sum_i \mathbf{H}^c(i) + \sum_{i>j} \frac{1}{r_{ij}} + V_{nn} \right\} \psi_t(1)\psi_b(2) \ldots \psi_k(n) d\tau_1 \ldots d\tau_n$$
$$= -\iint \psi_a(2)\psi_b(1) \frac{1}{r_{12}} \psi_t(1)\psi_b(2) \, d\tau_1 d\tau_2.$$

Summing these terms (for all orbitals such as ψ_b) gives the required expression.

Chapter 11

11.1 $$\int \Psi_+{}^2 \, dv = (2 + 2S^2)^{-1} \int \{1s_a{}^2(1)1s_b{}^2(2) + 1s_a{}^2(2)1s_b{}^2(1) + 2\, 1s_a(1)1s_b(1)1s_a(2)1s_b(2)\} \, dv_1 \, dv_2$$
$$= (2 + 2S^2)^{-1}(1 + 1 + 2S^2) = 1.$$

A similar calculation establishes that Ψ_- is normalized.

11.2

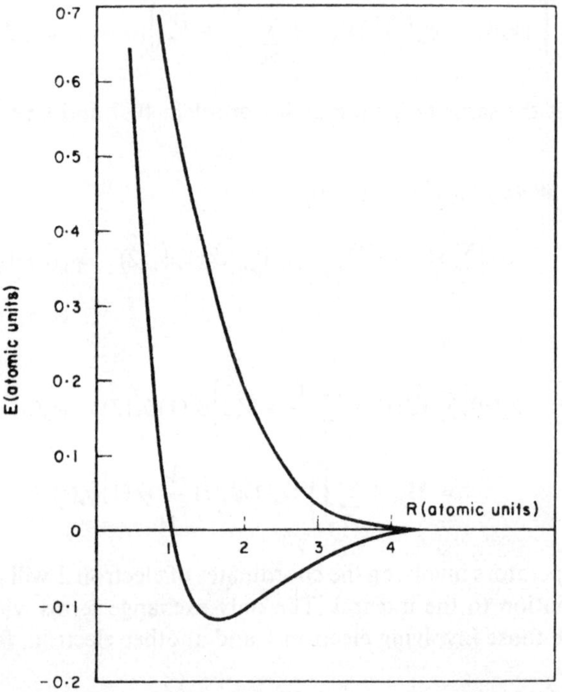

11.3 Carbon has the ground state configuration $1s^22s^22p^2$. We would anticipate that the two $2p$ electrons would give one σ and one π bond, but this would give rise to a a $^1\Pi$ state, and the ground state of C_2 is known to be $^1\Sigma_g{}^+$ (table 10.2). In MO theory the ground state of C_2 has a completely filled $\pi_u 2p$ shell, and the VB analogue of this would be to have both $2p$ electrons giving π bonds. Using the same nomenclature as we adopted for N_2, this would give a wave function

$$\widehat{p_{xa}p_{xb}}\widehat{p_{ya}p_{yb}} = \tfrac{1}{2}\{|p_{xa}\bar{p}_{xb}p_{ya}\bar{p}_{yb}| - |\bar{p}_{xa}p_{xb}p_{ya}\bar{p}_{yb}| - |p_{xa}\bar{p}_{xb}\bar{p}_{ya}p_{yb}| + |\bar{p}_{xa}p_{xb}\bar{p}_{ya}p_{yb}|\}.$$

11.4 With the assumption (a) the normalizing constant of the wave function Ψ is $N = (1 + \lambda^2)^{-\frac{1}{2}}$. The dipole moment is given by a generalization of expression (10.51).

Let μ be the dipole moment operator.

$$\boldsymbol{\mu} = -\sum_i e\mathbf{r}_i + \sum_\mu Z_\mu e\mathbf{r}_\mu .$$

$$\mu = (1 + \lambda^2)^{-1}\left\{\int \Psi(\mathrm{H-X})\boldsymbol{\mu}\Psi(\mathrm{H-X})\,d\tau + \lambda^2\int \Psi(\mathrm{H^+X^-})\boldsymbol{\mu}\Psi(\mathrm{H^+X^-})\,d\tau + 2\lambda\int \Psi(\mathrm{H-X})\boldsymbol{\mu}\Psi(\mathrm{H^+X^-})\,d\tau\right\}.$$

From assumptions (b) and (c) and the orthogonality condition it follows that the first and last terms are zero, hence

$$\mu = \lambda^2(1 + \lambda^2)^{-1}\int \Psi(\mathrm{H^+X^-})\boldsymbol{\mu}\Psi(\mathrm{H^+X^-})\,d\tau,$$

which from assumption (c) reduces to

$$\mu = \lambda^2(1 + \lambda^2)^{-1}eR = P\frac{eR}{100}.$$

From the data given we construct the following table

	HF	HCl	HBr	HI
μ (D)	1·74	1·03	0·78	0·38
eR (D)	4·42	6·10	6·77	7·73
P	39	17	12	5

(a) One can always choose the wave functions to be normalized, but if they are built up in the usual way from non-orthogonal atomic orbitals (cf. 11.30, 11.31) then they will not be orthogonal.
(b) This is not generally true (cf. figure 11.4).
(c) If X has a spherically-symmetrical electron cloud then the dipole moment of H^+X^- is exactly given by eR (in fact the assumption is correct providing the electron cloud of X^- has a centre of symmetry, as can easily be seen if the origin of the electron and nuclear coordinates is taken as the nucleus of X).
(d) This is not true if the wave functions are non-orthogonal and not necessarily true even if they are. The integral is only necessarily zero if the electron densities of H and X^- overlap at no point in space.

11.5 The secular equations are

$$c_1(\mathscr{H}_{11} - ES_{11}) + c_2(\mathscr{H}_{12} - ES_{12}) = 0$$
$$c_1(\mathscr{H}_{12} - ES_{12}) + c_2(\mathscr{H}_{22} - ES_{22}) = 0.$$

Introducing the values given for the integrals the secular determinant is

$$\begin{vmatrix} -9{\cdot}48 - 1{\cdot}19E & -2{\cdot}12 - 0{\cdot}26E \\ -2{\cdot}12 - 0{\cdot}21E & -10{\cdot}19 - 1{\cdot}29E \end{vmatrix} = 0.$$

The solutions of this are $E = -7{\cdot}882, -7{\cdot}863$. Substituting the lower of these energies back into the secular equations one has

$$\frac{c_1}{c_2} = -\frac{\mathscr{H}_{12} - ES_{12}}{\mathscr{H}_{11} - ES_{11}} = 2{\cdot}40.$$

The ground state wave function is then

$$\Psi = N(2{\cdot}40\Psi_1 + 1{\cdot}00\Psi_2).$$

For the function to be normalized we require

$$N^2((2{\cdot}40)^2(1{\cdot}19) + (1{\cdot}00)^2(1{\cdot}29) + (4{\cdot}80)(0{\cdot}26)) = 1,$$

whence $N = 0{\cdot}326$.

$$\Psi = 0{\cdot}782\Psi_1 + 0{\cdot}362\Psi_2.$$

Compare this with the function on page 207.

Chapter 12

12.1 We first examine the transformation properties of the hydrogen $1s$ orbitals and the carbon $2s$ and $2p$ orbitals. We take the z axis to be the C_3 axis. We evaluate the characters of the representations based on these orbitals (see problem 8.5).

	I	σ_h	$2C_3$	$2S_3$	$3C_2'$	$3\sigma_v$	
$1s_1, 1s_2, 1s_3$	3	3	0	0	1	1	$= A_1' + E'$
$2s$	1	1	1	1	1	1	$= A_1'$
$2p_z$	1	-1	1	-1	-1	1	$= A_2''$
$2p_x, 2p_y$	2	2	-1	-1	0	0	$= E'$

Note that on a C_3 rotation $2p_x$ and $2p_y$ go into a mixture of these two orbitals. The appropriate 2×2 matrix which represents the C_3 operation is found on page 99: its character is -1. The $2s$ orbital of the carbon will then form molecular orbitals with the A_1' combination of the hydrogen orbitals; this will give a bonding and an antibonding molecular orbital of A_1' symmetry. Likewise, $2p_x$ and $2p_y$ and the E' combination of the hydrogen orbitals will give a bonding and an antibonding E' molecular orbital. The $2p_z$ orbital will be non-bonding. The methyl radical has seven electrons (apart from the carbon $1s$ electrons), six of these will fill all the bonding molecular orbitals and the seventh will occupy the non-bonding orbital.

The ground state configuration is $(A_1')^2(E')^4(A_2'')$ which has A_2'' symmetry. The antibonding orbitals are of A_1' and E' symmetry, and the excitation of an electron from the A_1' bonding to the A_1' antibonding, or the E' bonding to the E' antibonding will also give a state of A_2'' symmetry. There are therefore two singly-excited configurations having the same symmetry as the ground state.

When the molecule distorts to a $\mathbf{C}_{3v}$ configuration (character table 8.6) then the A_1' and A_2'' orbitals have the same symmetry, A_1, in the lower symmetry group; the E' orbitals are still doubly degenerate, E, in the $\mathbf{C}_{3v}$ group. A distortion will therefore introduce some $2p_z$ character into the C—H bonding molecular orbital of A_1 symmetry and increase its energy, but the non-bonding orbital will take on some $2s$ character and be stabilized.

12.2 The three equivalent orbitals will be orthogonal linear combinations of ψ_1, ψ_2 and ψ_3. The property that these equivalent orbitals are transformed into one another by the operations of the group will be sufficient to define the coefficients in these linear combinations in an exactly similar manner to the way in which symmetry was used to construct sp^2 hybrid orbitals (page 51). One can in fact use the form of the sp^2 hybrids as a short cut to the elucidation of the equivalent orbitals.

The three sp^2 hybrids have the form (5.8–5.10)

$$\sigma_1 = \sqrt{\tfrac{1}{3}}(s + \sqrt{2}p_x)$$
$$\sigma_2 = \sqrt{\tfrac{1}{3}}(s - \sqrt{\tfrac{1}{2}}p_x + \sqrt{\tfrac{3}{2}}p_y)$$
$$\sigma_3 = \sqrt{\tfrac{1}{3}}(s - \sqrt{\tfrac{1}{2}}p_x - \sqrt{\tfrac{3}{2}}p_y).$$

Now ψ_1 has the same transformation properties as an s orbital, ψ_2 the same as p_x and ψ_3 the same as p_y. It follows that the three equivalent orbitals are

$$\vartheta_1 = \sqrt{\tfrac{1}{3}}(\psi_1 + \sqrt{2}\psi_2)$$
$$\vartheta_2 = \sqrt{\tfrac{1}{3}}(\psi_1 - \sqrt{\tfrac{1}{2}}\psi_2 + \sqrt{\tfrac{3}{2}}\psi_3)$$
$$\vartheta_3 = \sqrt{\tfrac{1}{3}}(\psi_1 - \sqrt{\tfrac{1}{2}}\psi_2 - \sqrt{\tfrac{3}{2}}\psi_3),$$

that is,

$$\vartheta_1 = 0{\cdot}44\,2s + 0{\cdot}09\,2p_z + 0{\cdot}51\,2p_x + 0{\cdot}42h_1 - 0{\cdot}07h_2 - 0{\cdot}07h_3$$
$$\vartheta_2 = 0{\cdot}44\,2s + 0{\cdot}09\,2p_z - 0{\cdot}25\,2p_x + 0{\cdot}44\,2p_y + 0{\cdot}42h_2 - 0{\cdot}07h_1 - 0{\cdot}07h_3$$
$$\vartheta_3 = 0{\cdot}44\,2s + 0{\cdot}09\,2p_z - 0{\cdot}25\,2p_x - 0{\cdot}44\,2p_y + 0{\cdot}42h_3 - 0{\cdot}07h_1 - 0{\cdot}07h_2.$$

12.3 The energy of the structure (a) (figure 12.7) in the perfect-pairing scheme is (12.31)

$$E = Q + A_{x,h''} + A_{y,h'} - \tfrac{1}{2}(A_{x,y} + A_{h'h''} + A_{x,h'} + A_{y,h''}),$$

where A are the VB exchange integrals defined as in (11.11), x and y are the oxygen $2p_x$ and $2p_y$ orbitals and h', h'' the two hydrogen $1s$ orbitals.

12.4 We first form the atomic orbitals of Be into *sp* hybrids (5.4)

$$\sigma_1 = \sqrt{\tfrac{1}{2}}(s + p_z) \quad \sigma_2 = \sqrt{\tfrac{1}{2}}(s - p_z).$$

The VB wave function for BeH_2 is then (cf. problem 11.3)

$$\tfrac{1}{2}\{|\sigma_1\bar{h}_1\sigma_2\bar{h}_2| - |\bar{\sigma}_1 h_1\sigma_2\bar{h}_2| - |\sigma_1\bar{h}_1\bar{\sigma}_2 h_2| + |\bar{\sigma}_1 h_1\bar{\sigma}_2 h_2|\}.$$

The valence state of the Be has the wave function

$$\Omega = \tfrac{1}{2}\{|\sigma_1\sigma_2| - |\bar{\sigma}_1\sigma_2| - |\sigma_1\bar{\sigma}_2| + |\bar{\sigma}_1\bar{\sigma}_2|\}.$$

Expanding this in terms of atomic orbitals, and remembering that determinants such as $|ss|$ are zero because both columns are the same, we have

$$\begin{aligned}\Omega = \tfrac{1}{4}\{&-|sp_z| + |p_z s| - |\bar{s}p_z| + |\bar{p}_z s| - |\bar{s}s| + |\bar{p}_z p_z| - |s\bar{p}_z| \\ &+ |p_z\bar{s}| - |s\bar{s}| + |p_z\bar{p}_z| - |\bar{s}\bar{p}_z| + |\bar{p}_z\bar{s}|\} \\ = \tfrac{1}{2}\{&|p_z s| + |\bar{p}_z\bar{s}| + |p_z\bar{s}| + |\bar{p}_z s|\}.\end{aligned}$$

These functions all arise from the configuration of Be $2s2p$, which gives terms ^1P and ^3P. The first two determinants have spins of $+1$ and -1 respectively and must be components of the triplet state. The third is a sum of the singlet and triplet functions having $M_s = 0$

$$\begin{aligned}|p_z\bar{s}| &= \sqrt{\tfrac{1}{2}}\{\sqrt{\tfrac{1}{2}}(|p_z\bar{s}| + |\bar{p}_z s|) + \sqrt{\tfrac{1}{2}}(|p_z\bar{s}| - |\bar{p}_z s|)\} \\ &= \sqrt{\tfrac{1}{2}}\{^3P + {^1P}\}.\end{aligned}$$

Similarly the last term is

$$|p_z\bar{s}| = \sqrt{\tfrac{1}{2}}\{^3P - {^1P}\}.$$

We now sum the squares of the coefficients of the individual atomic state functions to get the total valence state and find that it is $\tfrac{3}{4}\,{^3P} + \tfrac{1}{4}{^1P}$.

Chapter 13

13.1 Fe^{3+} has five $3d$ electrons. In an octahedral complex the ground state will be $^6A_{1g}$ (weak field) or $^2T_{2g}$ (strong field). Since the ion is colourless there are apparently no low energy electronic states which are optically accessible from the ground state. From figure 13.9 it can be seen that if $^2T_{2g}$ were the ground state then there should be other low lying doublet states some of which would be expected to give rise to the absorption of visible light. These would all be *g–g* transitions, but we have seen in section

13.8 that this is not a very strict selection rule. On the other hand, if the ground state is $^6A_{1g}$ then this is the only low lying sextet state and hence all low lying transitions would be spin-forbidden as well as g–g forbidden. From its position in the spectrochemical series F^- is expected to give weak-field complexes.

13.2 The character for the operation of rotation by an angle α is, from (13.4),

$$\sin\frac{7\alpha}{2}\Big/\sin\frac{\alpha}{2},$$

so $\chi(C_2)$. for example, is $\dfrac{\sin 630°}{\sin\ 90°} = -1.$

Similarly, $\chi(C_3) = 1$, $\chi(C_4) = -1$. Because d orbitals are symmetric to inversion $\chi(i) = 7$. The following set of characters is sufficient to show that the component *I.R.*'s are T_{1g}, T_{2g} and A_{2g}.

	I	$6C_4$	$3C_2$	$6C_2'$	$8C_3$	i
$\chi =$	7	-1	-1	-1	1	7

13.3 A d^9 configuration of a free atom gives the term ^2D. In an octahedral field the ground state for this term is ^2E_g, which from table 13.5 has wave functions

$$\sqrt{\tfrac{1}{2}}[(2,2)+(2,-2)] \quad \text{and} \quad (2,0).$$

The configuration of this state is $(t_{2g})^6(e_g)^3$ and the appropriate eigenfunction written as a Slater determinant would be, for example

$$|d_{xy}\ \overline{d_{xy}}\ d_{yz}\ \overline{d_{yz}}\ d_{zx}\ \overline{d_{zx}}\ d_{z^2}\ \overline{d_{z^2}}\ d_{x^2-y^2}|.$$

A d^8 configuration of a free atom gives a ground state term ^3F. In an octahedral field this gives a ground state $^3A_{2g}$, whose wave function is, from table 13.6,

$$\sqrt{\tfrac{1}{2}}[(3,2)-(3,-2)].$$

The strong-field d-orbital configuration is $(t_{2g})^6(e_g)^2$, and one component of the triplet state would have a wave function

$$|d_{xy}\ \overline{d_{xy}}\ d_{yz}\ \overline{d_{yz}}\ d_{zx}\ \overline{d_{zx}}\ d_{z^2}\ d_{x^2-y^2}|.$$

13.4

Frequency (cm^{-1})	Assignment
18,870	$^6A_{1g} \rightarrow {}^4T_{1g}\,(^4G)$
23,120	$^6A_{1g} \rightarrow {}^4T_{2g}\,(^4G)$
24,960	$^6A_{1g} \rightarrow {}^4A_{1g}\,(^4G)$ } or reversed
25,275	$^6A_{1g} \rightarrow {}^4E_g\,(^4G)$ }
27,980	$^6A_{1g} \rightarrow {}^4T_{2g}\,(^4D)$
29,750	$^6A_{1g} \rightarrow {}^4E_g\,(^4D)$

HEIDT *et al.*, *J. Am. Chem. Soc.*, **80**, 6471 (1958).

13.5 The interaction integrals can be obtained from the matrix (13.9). The only non-zero off-diagonal integral is between (2,2) and (2,−2), from which it follows that there are no interaction integrals between different components of the ^3P state. Let us evaluate the crystal-field energy of the first of the eigenfunctions given ($M_L = 1$)

$$\int [\sqrt{\tfrac{2}{5}}|(2,2)(2,-1)| - \sqrt{\tfrac{3}{5}}|(2,1)(2,0)|]^*[V_{\text{oct}}(1) + V_{\text{oct}}(2)]$$
$$[\sqrt{\tfrac{2}{5}}|(2,2)(2,-1)| - \sqrt{\tfrac{3}{5}}|(2,1)(2,0)|]\, dv_1\, dv_2$$

$$= \tfrac{2}{5}\int (2,2)^* V_{\text{oct}}(2,2)\, dv + \tfrac{2}{5}\int (2,-1)^* V_{\text{oct}}(2,-1)\, dv$$
$$+ \tfrac{3}{5}\int (2,0)^* V_{\text{oct}}(2,0)\, dv + \tfrac{3}{5}\int (2,1)^* V_{\text{oct}}(2,1)\, dv$$

$$= \frac{2}{5}\left(\frac{-3\Delta}{10}\right) + \frac{3}{5}\left(\frac{\Delta}{5}\right) = 0.$$

13.6 Label the orbitals as in the figure given in the hints and consider the behaviour of A_z.

	I	$6C_4$	$3C_2$	$6C_2'$	$8C_3$
A_z becomes	A_z	$A_y - A_y$ $+E_x - F_x$ $+D_z + B_z$	$-A_z + C_z$ $-C_z$	$-E_x - D_z$ $+C_y + F_x$ $-C_y - B_z$	—
T_{2g}	3	−1	−1	1	0
Multiply	$3A_z$	$-E_x + F_x$ $-D_z - B_z$	A_z	$-E_x + F_x$ $-B_z - D_z$	
	i	$6S_4$	$3\sigma_h$	$6\sigma_d$	$8S_6$
A_z becomes	$-C_z$	$-C_y + C_y$ $-F_x + E_x$ $-D_z - B_z$	$C_z - A_z$ $+A_z$	$F_x - A_y$ $-E_x + A_y$ $+D_z + B_z$	—
T_{2g}	3	−1	−1	1	0
Multiply	$-3C_z$	$F_x - E_x$ $+D_z + B_z$	$-C_z$	$F_x - E_x$ $+D_z + B_z$	

Add $4(A_z - E_x + F_x - C_z)$.

Normalize $\psi_1(t_{2g}) = \frac{1}{2}(A_z - E_x - C_z + F_x)$.

By symmetry the other components will be

$$\psi_2(t_{2g}) = \tfrac{1}{2}(F_y - D_z - E_y + B_z)$$
$$\psi_3(t_{2g}) = \tfrac{1}{2}(D_x - A_y - B_x + C_y).$$

Chapter 14

14.1 There are two satisfactory structures. The more probable is

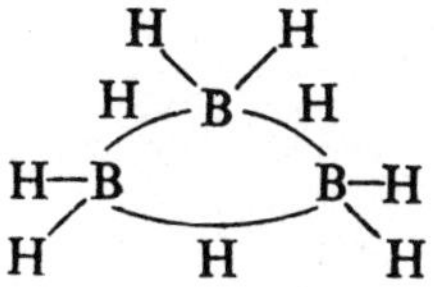

Of the eighteen valence electrons, twelve are involved in the bonding of the terminal hydrogen and six in the three-centred bonds.

The other possible structure is

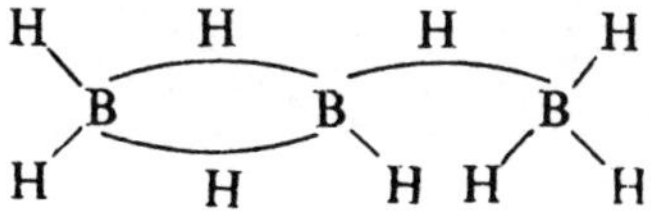

which also has filled bonding orbitals. It is possible, however, that both of these compounds would have more stable disproportionation products. In addition, those reactions which might produce B_3H_9 (pyrolysis of B_2H_6 etc.) are such that further reaction with BH_3 radicals would occur.

14.2 The secular determinant is of the general form

$$\begin{vmatrix} \alpha - E & \beta & \gamma \\ \beta & \alpha - E & \beta \\ \gamma & \beta & \alpha - E \end{vmatrix} = 0,$$

where $H_{aa} = H_{bb} = H_{cc} = \alpha$, $H_{ab} = H_{bc} = \beta$, $H_{ac} = \gamma$.

Expansion of this determinant gives

$$E^3 - 3\alpha E^2 + (3\alpha^2 - 2\beta^2 - \gamma^2)E + \gamma^2\alpha - 2\beta^2\alpha - \alpha^3 + 2\beta^2\alpha = 0,$$

which factorizes as follows

$$(E - \alpha + \gamma)[E^2 - (2\alpha + \gamma)E + (\alpha^2 + \alpha\gamma - 2\beta^2)] = 0.$$

This has roots

$$E_0 = \alpha - \gamma$$

$$E_+ = \alpha + \frac{\gamma}{2} + \left[\left(\frac{\gamma}{2}\right)^2 + 2\beta^2\right]^{\frac{1}{2}}$$

$$E_- = \alpha + \frac{\gamma}{2} - \left[\left(\frac{\gamma}{2}\right)^2 + 2\beta^2\right]^{\frac{1}{2}}.$$

The following figure shows how these energies will vary with the angle $\widehat{ABC}$

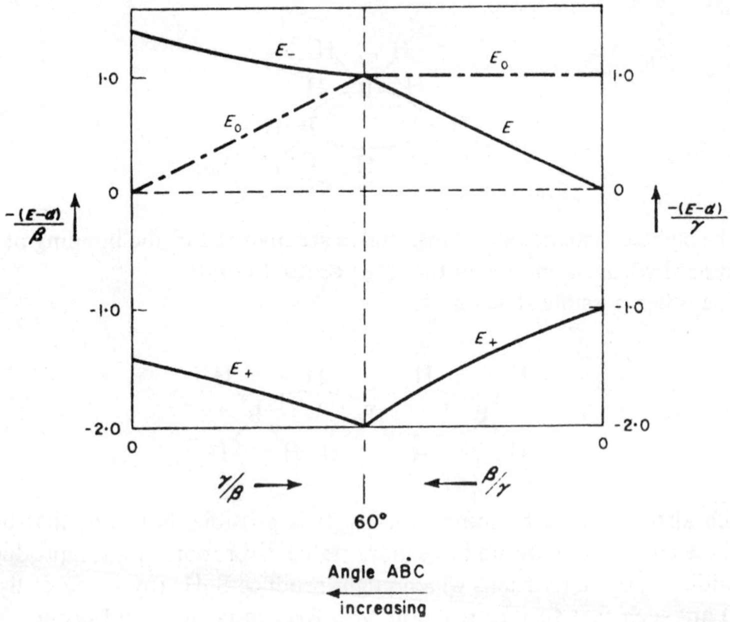

Chapter 15

15.1

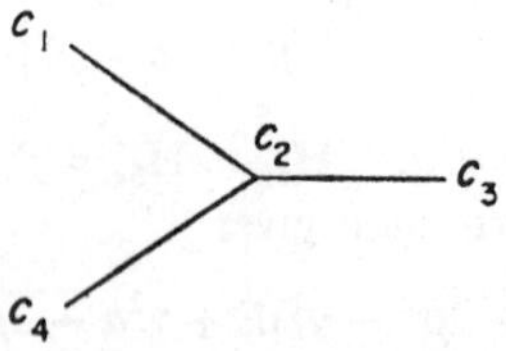

The secular equations are

$$
\begin{aligned}
xc_1 + c_2 \qquad\qquad &= 0 \\
c_1 + xc_2 + c_3 + c_4 &= 0 \\
c_2 + xc_3 \qquad &= 0 \\
c_2 \qquad + xc_4 &= 0.
\end{aligned}
$$

Thus the secular determinant is

$$
\begin{vmatrix} x & 1 & 0 & 0 \\ 1 & x & 1 & 1 \\ 0 & 1 & x & 0 \\ 0 & 1 & 0 & x \end{vmatrix} = 0 \qquad \therefore x^4 - 3x^2 = 0.
$$

The solutions of this are $x = \sqrt{3}, -\sqrt{3}, 0, 0$, where

$$
x = \frac{\alpha - E}{\beta}. \tag{15.4}
$$

The energies of the four Hückel orbitals are

$$
\begin{aligned}
E_1 &= \alpha + \sqrt{3}\beta \\
E_2 &= \alpha \\
E_3 &= \alpha \\
E_4 &= \alpha - \sqrt{3}\beta.
\end{aligned}
$$

Notice that ψ_2 and ψ_3 are degenerate.

To determine the coefficients for ψ_1 we put $x = -\sqrt{3}$ back into the secular equations

$$
\begin{aligned}
-\sqrt{3}c_1 + c_2 &= 0 \qquad & -\sqrt{3}c_3 + c_2 &= 0 \\
c_1 - \sqrt{3}c_2 + c_3 + c_4 &= 0 \qquad & -\sqrt{3}c_4 + c_2 &= 0.
\end{aligned}
$$

Thus $c_1 = c_3 = c_4 = \dfrac{c_2}{\sqrt{3}}$.

The normalization condition is

$$
c_1{}^2 + c_2{}^2 + c_3{}^2 + c_4{}^2 = 1
$$

hence

$$
c_2{}^2 + 3c_1{}^2 = 1.
$$

That is

$$
c_1 = \sqrt{\tfrac{1}{6}}, \quad c_2 = \sqrt{\tfrac{1}{2}}.
$$

To determine the coefficients for ψ_2 we substitute $x = 0$ back into the secular equations, and find $c_2 = 0$, $c_1 + c_3 + c_4 = 0$. Any two orbitals that are orthogonal, and satisfy these conditions will have energies $E = \alpha$. A satisfactory pair are (cf. page 101)

$$\psi_2 = \sqrt{\tfrac{1}{2}}(\phi_1 - \phi_3),$$
$$\psi_3 = \sqrt{\tfrac{1}{6}}(2\phi_4 - \phi_1 - \phi_3).$$

We have four electrons to put into the four Hückel orbitals, i.e. two will go into ψ_1 and applying Hund's rule one each with parallel spins into the degenerate pair ψ_2 and ψ_3 so that the molecule would have a triplet ground state.

The bond order p_{12} is, from (15.9),

$$p_{12} = 1\left(\frac{1}{\sqrt{6}} \times 0\right) + 1\left(\frac{1}{\sqrt{6}} \times 0\right) + 2\left(\frac{\sqrt{3}}{\sqrt{6}} \times \frac{1}{\sqrt{6}}\right)$$
$$= \sqrt{\tfrac{1}{3}}.$$

This value is of general interest since the free-valence index is defined by

$$F_\mu = n_{\max} - n_\mu,$$

(see (15.13)) where n_μ is the sum of the π-electron bond orders for all bonds joining atom μ and $n_{\max}$ is the maximum possible value; which arises from the molecule we have just considered

$$n_{\max} = 3\sqrt{\tfrac{1}{3}} = 1{\cdot}73.$$

15.2 $H_1—H_2—H_3$

The secular determinant is

$$\begin{vmatrix} x & 1 & 0 \\ 1 & x & 1 \\ 0 & 1 & x \end{vmatrix} = 0.$$

This has solution $x^3 - 2x = 0$,

$$\therefore\ x(x^2 - 2) = 0$$
$$\therefore\qquad x = -\sqrt{2}, 0, +\sqrt{2}.$$

The three Hückel orbitals are

$$\psi_1 \qquad E_1 = \alpha + \sqrt{2}\beta$$
$$\psi_2 \qquad E_2 = \alpha$$
$$\psi_3 \qquad E_3 = \alpha - \sqrt{2}\beta.$$

The total Hückel energies are

$$H_3^+ \quad \varepsilon = 2\alpha + 2\sqrt{2}\beta$$
$$H_3^\bullet \quad \varepsilon = 3\alpha + 2\sqrt{2}\beta$$
$$H_3^- \quad \varepsilon = 4\alpha + 2\sqrt{2}\beta.$$

For triangular H_3

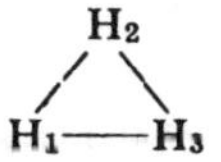

the secular determinant is

$$\begin{vmatrix} x & 1 & 1 \\ 1 & x & 1 \\ 1 & 1 & x \end{vmatrix} = 0 \qquad \therefore \quad x^3 - 3x + 2 = 0, \quad \therefore (x-1)^2(x+2) = 0.$$

This gives three solutions $x = 1$, $x = 1$ and $x = -2$.

As in problem 15.1 we have two degenerate orbitals. The three Hückel energies are

$$E_1 = \alpha + 2\beta$$
$$E_2 = E_3 = \alpha - \beta.$$

The total Hückel energies are

$$H_3^+ \qquad \varepsilon = 2\alpha + 4\beta$$
$$H_3^\bullet \qquad \varepsilon = 3\alpha + 3\beta$$
$$H_3^- \qquad \varepsilon = 4\alpha + 2\beta.$$

Thus, according to this treatment, H_3^+ is more stable in the triangular configuration; for $H_3^\bullet$ the two configurations are of almost equal stability, the triangular being very slightly the more stable; but for H_3^- the linear configuration is substantially the more stable. The result of $H_3^\bullet$ is not in agreement with more exact calculations which predict that the linear molecule is the more stable.

15.3 Fulvene belongs to the $\mathbf{C}_{2v}$ group: the orbitals will either be symmetric or antisymmetric to rotation about the C_2 axis, and as they are π orbitals they will be antisymmetric to reflection in the molecular plane. The possible symmetries are:

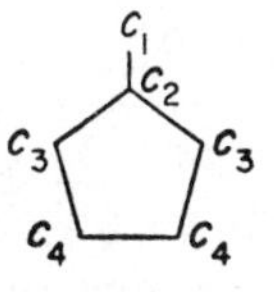

A_1

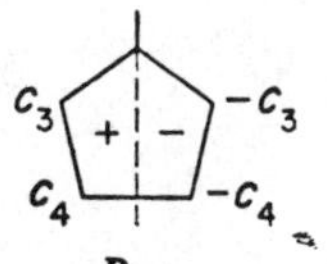

B_2

From these diagrams we can write down the secular determinants directly using the method described on pages 279–282.

$$\begin{array}{c|cccc|}
c_1 & x & 1 & 0 & 0 \\
c_2 & 1 & x & 2 & 0 \\
c_3 & 0 & 1 & x & 1 \\
c_4 & 0 & 0 & 1 & x+1
\end{array} = 0 \qquad \begin{array}{c|cc|}
c_3 & x & 1 \\
c_4 & 1 & x-1
\end{array} = 0$$

$$\therefore x^2 - x - 1 = 0$$

$$\therefore x^4 + x^3 - 4x^2 - 3x + 1 = 0.$$

The solutions of these two equations are as follows

$$x = -2{\cdot}115, \quad -1, \quad -0{\cdot}618, \quad +0{\cdot}254, \quad +1{\cdot}618, \quad +1{\cdot}861.$$

The complete table of coefficients for the bonding orbitals is as follows

	c_1	c_2	c_3	c_4	x
ψ_1	0·245	0·523	0·429	0·385	−2·115
ψ_2	0·500	0·500	0	−0·500	−1
ψ_3	0	0	0·602	0·372	−0·618

The π-electron density $q_v = 2 \sum_{r\,\mathrm{occ}} c_{rv}^2$. (15.8)
This gives

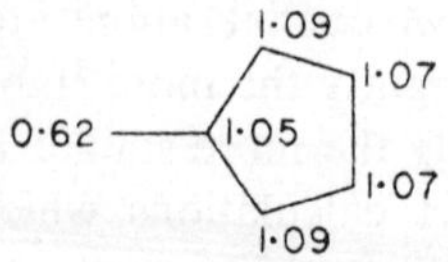

After subtracting a nuclear charge of +1 from each atom, the net charge densities are as follows

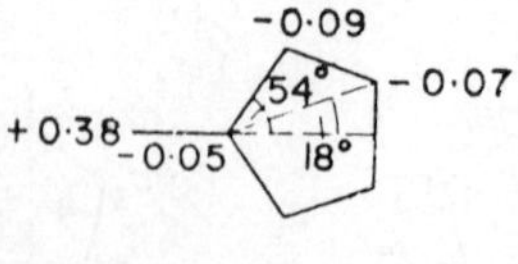

In order to obtain the dipole moment we can take carbon atom 2 as the

origin and, taking all bonds as 1·40 Å (the average length of a single and double bond.

$$\boldsymbol{\mu} = \sum q_v \mathbf{r}_v = 1{\cdot}40 \times 10^{-8}[0{\cdot}38 + 2(0{\cdot}09)\cos 54° + 2(0{\cdot}07)(2\cos 54°)\cos 18°] \times 4{\cdot}8 \times 10^{-10} \text{ esu}$$

$$= 4{\cdot}3 \times 10^{-18} \text{ esu} = 4{\cdot}3 \text{ D}.$$

This is along the direction of the C_2 axis.
Experiment shows that the dipole is only 1·2 D, which emphasizes the limitations of the Hückel method for non-alternants (cf. page 303).

15.4 If $\mathbf{H}'$ is the perturbation to the one-electron Hamiltonian arising from the introduction of the substituent, then the perturbed orbitals are given by

$$\psi_r' = \psi_r + \sum_{s=1}^{n}{}' \frac{\int \psi_r \mathbf{H}' \psi_s \, dv}{E_r - E_s} \psi_s.$$

The prime on the summation indicates that the term with $s = r$ is omitted.

Now,
$$\psi_r = \sum_{\mu=1}^{n} c_{\mu r}\phi_\mu,$$

where the ϕ_μ are atomic orbitals.
Hence

$$\psi_r' = \psi_r + \sum_{s=1}^{n}{}' \frac{\int \sum_{\mu=1}^{n} c_{\mu r}\phi_\mu \mathbf{H}' \sum_{\nu=1}^{n} c_{\nu s}\phi_\nu \, dv}{E_r - E_s} \psi_s.$$

If the perturbation $\mathbf{H}'$ only affects the coulomb integral of atom μ, then we pick out the terms in

$$\int \phi_\mu \mathbf{H}' \phi_\mu \, dv = \delta\alpha_\mu,$$

$$\psi_r' = \psi_r + \left(\sum_{s=1}^{n}{}' \frac{c_{\mu r}c_{\mu s}\delta\alpha_\mu}{E_r - E_s}\right)\psi_s$$

$$= \psi_r + \delta\alpha_\mu \sum_{s=1}^{n}{}' \frac{c_{\mu r}c_{\mu s}}{E_r - E_s} \sum_{\nu=1}^{n} c_{\nu s}\phi_\nu,$$

which we may write as

$$\psi_r' = \psi_r + \sum_{\nu=1}^{n} c_{\nu r}'\phi_\nu,$$

where $$c_{\nu r}' = \delta\alpha_\mu \sum_{s=1}^{n}{}' \frac{c_{\mu r}c_{\mu s}c_{\nu s}}{E_r - E_s}.$$

The change in π-orbital electron density at position ν is

$$\delta q_\nu = 2 \sum_{r\ \mathrm{occ}} [(c_{\nu r} + c_{\nu r}')^2 - c_{\nu r}^2],$$

where the summation is over the occupied orbitals.

$$\delta q_\nu = 4 \sum_{r\ \mathrm{occ}} c_{\nu r}c_{\nu r}' + \text{smaller terms in } (c_{\nu r}')^2.$$

Substituting for $c_{\nu r}'$

$$\delta q_\nu = 4\delta\alpha_\mu \sum_{r\ \mathrm{occ}} \sum_{s=1}^{n}{}' \frac{c_{\mu r}c_{\nu r}c_{\mu s}c_{\nu s}}{E_r - E_s}$$

$$\therefore\ \pi_{\mu,\nu} = \frac{\delta q_\nu}{\delta\alpha_\mu} = 4 \sum_{r\ \mathrm{occ}} \sum_{s=1}^{n}{}' \frac{c_{\mu r}c_{\nu r}c_{\mu s}c_{\nu s}}{E_r - E_s}$$

We can divide up the summation over s into sums over occupied and unoccupied orbitals.

$$\pi_{\mu,\nu} = 4 \sum_{r\ \mathrm{occ}} \left[\sum_{s\ \mathrm{occ}}{}' + \sum_{s\ \mathrm{unocc}}\right] \frac{c_{\mu r}c_{\nu r}c_{\mu s}c_{\nu s}}{E_r - E_s}.$$

There will be no contribution from the double summation over the occupied orbitals, because for every term

$$\frac{c_{\mu a}c_{\nu a}c_{\mu b}c_{\nu b}}{E_a - E_b}$$

there will be one

$$\frac{c_{\mu b}c_{\nu b}c_{\mu a}c_{\nu a}}{E_b - E_a}.$$

It follows that

$$\pi_{\mu,\nu} = 4 \sum_{r\ \mathrm{occ}} \sum_{s\ \mathrm{unocc}} \frac{c_{\mu r}c_{\nu r}c_{\mu s}c_{\nu s}}{E_r - E_s}.$$

15.5 The general case of a regular polygon inscribed in a circle of radius $2|\beta|$ is shown in the diagram.

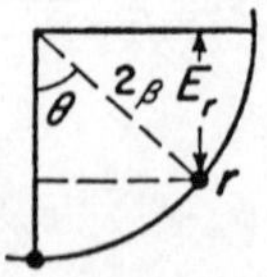

The energy corresponding to apex r shown in the diagram is

$$E_r = 2\beta \cos \theta$$

For a t-sided polygon

$$\theta = \frac{2\pi r}{t},$$

where
$$r = 0 \rightarrow \frac{t}{2} \ (t \text{ even}),$$

or
$$r = 0 \rightarrow \frac{t-1}{2} \ (t \text{ odd})$$

$$\therefore \ E_r = 2\beta \cos \frac{2\pi r}{t}.$$

15.6 If we ignore the interaction between different double bonds then the orbitals are those of three ethylene molecules, three bonding and three antibonding. The molecule has a horizontal symmetry plane; the three bonding orbitals are symmetric to reflection in this plane and the three antibonding orbitals are antisymmetric. It follows that if we introduce an interaction term β' between orbitals of different double bonds then the bonding orbitals can interact together and the antibonding orbitals can interact together but the bonding and antibonding orbitals will not interact because they have different symmetries. Although the interaction between the bonding orbitals will change the energy of the bonding orbitals, it will not change the total energy of the three bonding orbitals, and hence this interaction will give no added stability to the system.

15.7 The atom polarizabilities for butadiene could be computed directly from (15.40), but in practice it is easier to simplify this expression by making use of the properties of the even alternant character of butadiene.

$$\pi_{\mu,\nu} = \frac{\partial q_\nu}{\partial \alpha_\mu} = 4 \sum_{r \text{ occ}} \sum_{s \text{ unocc}} \frac{c_{\mu r} c_{\nu r} c_{\mu s} c_{\nu s}}{E_r - E_s}. \tag{15.40}$$

We have seen that when the number of carbon atoms n is even, the molecular orbitals occur in pairs with equal and opposite energies. If we label the Hückel orbital energies in order from E_1 to E_n so that

$$E_1 < E_2 \cdots < E_{\frac{1}{2}n} < \alpha < E_{\frac{1}{2}n+1} \cdots < E_n,$$

it follows that

$$E_{n-r+1} = -E_r.$$

The n π electrons will occupy the lowest $n/2$ molecular orbitals. Thus (15.36) simplifies to

$$\pi_{\mu,\nu} = \pm 4 \sum_{r\,\mathrm{occ}} \sum_{s\,\mathrm{occ}} \frac{c_{\mu r}c_{\nu r}c_{\mu s}c_{\nu s}}{E_r + E_s}.$$

The positive sign is to be taken when μ and ν are both starred or both unstarred, and the negative sign when one is starred and the other unstarred.

Thus for carbon atoms 1 and 2 in butadiene we have

$$\pi_{1,2} = -4\left(\frac{c_{11}c_{21}c_{11}c_{21}}{E_1 + E_1} + \frac{c_{11}c_{21}c_{12}c_{22}}{E_1 + E_2} + \frac{c_{12}c_{22}c_{11}c_{21}}{E_2 + E_1} + \frac{c_{12}c_{22}c_{12}c_{22}}{E_2 + E_2}\right).$$

Taking the values for the energies and the coefficients from page 273 we have

$$\pi_{1,2} = -4\left(\frac{0{\cdot}37 \times 0{\cdot}60 \times 0{\cdot}37 \times 0{\cdot}60}{1{\cdot}62 + 1{\cdot}62} + \frac{0{\cdot}37 \times 0{\cdot}60 \times 0{\cdot}60 \times 0{\cdot}37}{1{\cdot}62 + 0{\cdot}62}\right.$$
$$\left. + \frac{0{\cdot}60 \times 0{\cdot}37 \times 0{\cdot}37 \times 0{\cdot}60}{0{\cdot}62 + 1{\cdot}62} + \frac{0{\cdot}60 \times 0{\cdot}37 \times 0{\cdot}60 \times 0{\cdot}37}{0{\cdot}62 + 0{\cdot}62}\right)\left(\frac{1}{\beta}\right)$$
$$= -0{\cdot}402\,\beta^{-1}.$$

Completing the calculation for the other substituents we have

$\pi_{\mu,\nu}$	$\nu = 1$	$\nu = 2$	$\nu = 3$	$\nu = 4$
$\mu = 1$	0·626	−0·402	0·045	$-0{\cdot}268\,\beta^{-1}$

The electron densities q_ν are given by

$$q_\mu = 1 + \pi_{\mu,x}(\alpha_{\mathrm{X}} - \alpha_{\mathrm{C}}). \tag{15.41}$$

For acrolein we replace carbon 1 by oxygen and $\alpha_{\mathrm{X}} = \alpha_{\mathrm{C}} + h_{\mathrm{X}}\beta$, and let $h_{\mathrm{O}} = 1$ we have $\alpha_{\mathrm{O}} - \alpha_{\mathrm{C}} = \beta$, thus q_μ for acrolein

```
1·63    0·59    1·05    0·73
 O———————C———————C———————C
```

15.8 The molecular orbitals of ethylene are determined by symmetry to be

$$\psi_1 = \sqrt{\tfrac{1}{2}}(\phi_a + \phi_b), \qquad \psi_2 = \sqrt{\tfrac{1}{2}}(\phi_a - \phi_b).$$

The ground state has two electrons in the bonding orbital with opposite spins

$$|\psi_1\bar{\psi}_1|.$$

The energy of this function is, from (10.79),

$$E = 2\mathbf{H}^c_{11} + J_{11} + V_{nn}.$$

Then expanding the molecular orbital in terms of atomic orbitals this gives

$$\mathbf{H}^c_{11} = \tfrac{1}{2}\int (\phi_a + \phi_b)\mathbf{H}^c(\phi_a + \phi_b)\,dv = \mathbf{H}^c_{aa} + \mathbf{H}^c_{ab},$$

$$J_{11} = \tfrac{1}{2}\iint (\phi_a + \phi_b)(1)(\phi_a + \phi_b)(2)\frac{1}{r_{12}}(\phi_a + \phi_b)(1)(\phi_a + \phi_b)(2)\,dv_1\,dv_2$$
$$= \tfrac{1}{2}(\gamma_{aa} + \gamma_{ab}),$$

where the γ are defined by (15.48).

We write $\mathbf{H}^c_{ab} = \beta$, and from (15.56)

$$\mathbf{H}^c_{aa} = I - \gamma_{ab}.$$

Hence we have

$$E = 2I - 2\gamma_{ab} + 2\beta + \tfrac{1}{2}(\gamma_{aa} + \gamma_{ab}) + V_{nn}$$
$$= 2I + 2\beta + \tfrac{1}{2}\gamma_{aa} - \tfrac{3}{2}\gamma_{ab} + V_{nn}.$$

15.9 From (15.63) the energy of the orbital r is given by

$$E_r = \frac{h^2 r^2}{8md^2(2n+1)^2}.$$

A polyene with $2n$ carbon atoms has $2n$ electrons and these fill the lowest n molecular orbitals. The highest occupied orbital has an energy

$$E_n = \frac{h^2 n^2}{8md^2(2n+1)^2},$$

and the lowest vacant orbital has an energy

$$E_{n+1} = \frac{h^2(n+1)^2}{8md^2(2n+1)^2}.$$

The difference between these energies is

$$E_{n+1} - E_n = \frac{h^2}{8md^2(2n+1)^2}[(n+1)^2 - n^2] = \frac{h^2}{8md^2(2n+1)}.$$

This can be equated with the energy required to excite the molecule to the first excited state, and this will be associated with an absorption band of wavelength

$$\lambda = \frac{hc}{E_{n+1} - E_n} = \frac{8md^2c(2n+1)}{h},$$

which establishes the relationship $\lambda \propto n$.

Chapter 17

17.1

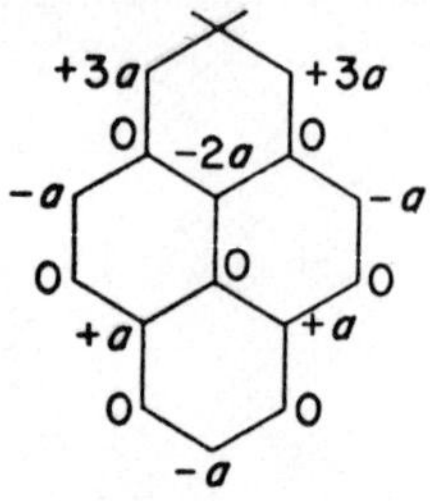

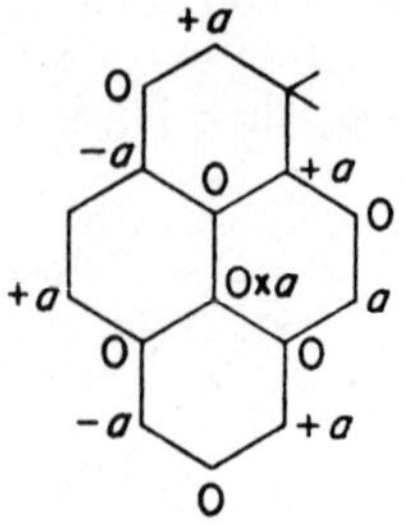

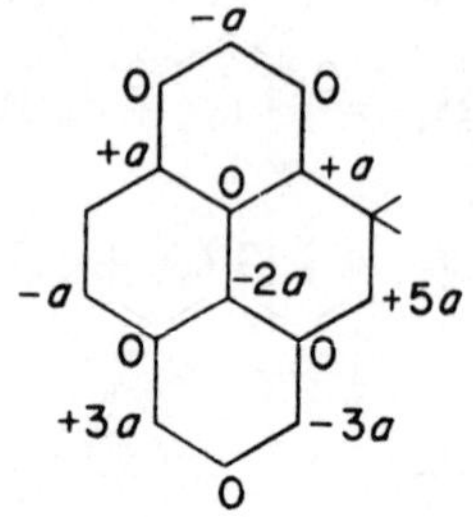

$$a = \frac{1}{3\sqrt{3}} \quad N_4 = 2{\cdot}31 \qquad a = \frac{1}{\sqrt{7}} \quad N_3 = 1{\cdot}51 \qquad a = \frac{1}{\sqrt{51}} \quad N_1 = 1{\cdot}68$$

The lower the Dewar number, the lower the localization energy and the easier the attack. The predicted rates of attack on pyrene are thus $3 > 1 \gg 4$.

17.2 We replace the negatively charged oxygen atom in the naphtholate anions by a methylene group and we get the following charge densities from the coefficients of the non-bonding orbitals.

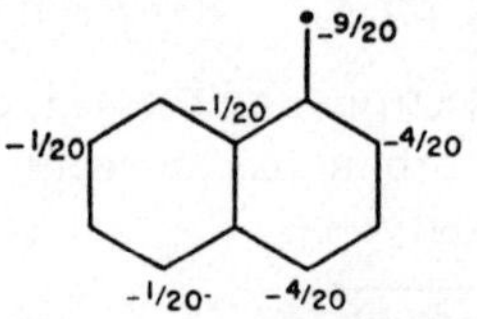

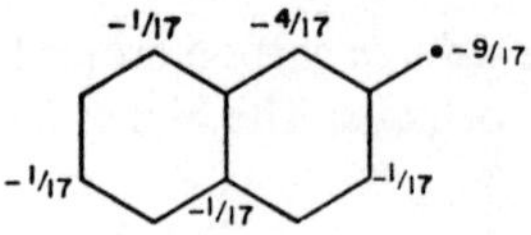

Thus we predict that α-naphthol will couple with diazonium salts with equal facility in the 2- and 4-positions while β-naphthol will couple exclusively in the 1-position (these predictions are in good accord with experiment). The isolated-molecule approach works well here probably because in the naphtholate anions we are concerned with a charged species and hence there are large differences in π-electron density between one carbon atom and another.

17.3 If we have two bases B_1 and B_2 with basic dissociation constants K_1 and K_2, and we wish to compare these bases in a given solvent at a given temperature, we can write

$$K' = \frac{K_1}{K_2} = \frac{[B_1H^+]}{[B_1]} \times \frac{[B_2]}{[B_2H^+]} \times \frac{f_{B_1H^+} f_{B_2}}{f_{B_2H^+} f_{B_1}},$$

where f is an activity coefficient.

We can thus relate the equilibrium constant K to the standard free-energy change for the transfer of a proton from B_1 to B_2

$$\Delta G^\circ = -RT \log K'.$$

Hammett has shown that to a good approximation the ratios of the activity coefficients for two similar bases in the same medium are equal, i.e.

$$\frac{f_{B_1H^+}}{f_{B_1}} = \frac{f_{B_2H^+}}{f_{B_2}}.$$

Making this assumption we can equate ΔG° with the potential energy change in transferring a proton from B_1 to B_2. We will assume that this change in potential energy is principally due to the difference in π-electron energy of B_1H^+ and B_2H^+, i.e.

$$\begin{aligned} \delta E_2 - \delta E_1 &\approx \Delta G^\circ \\ &= -RT \log K' \\ &= 2{\cdot}303RT(pK_1 - pK_2). \end{aligned}$$

The first question we must ask is, on which nitrogen in an aminoquinoline will the proton be preferentially attached? If it is attached to the ring nitrogen atom, to a first approximation we only alter the value of α for this atom; if, however, it is attached to the exocyclic NH_2 group we are taking two electrons out of the π-electron system, hence protonation of the ring nitrogen involves less loss of π-electron energy than protonation of the exocyclic nitrogen atom.

We have seen that if the coulomb integral of an atom μ is altered by a small amount of $\delta\alpha_\mu$ then the π-electron energy change is given by

$$\delta E = q_\mu \delta\alpha_\mu = 1{\cdot}2\beta q_\mu.$$

In an alternant hydrocarbon the electron density is unity at each carbon atom. We can compute the electron density at carbon atom 1 in the substituted aminoquinolines by treating the amino groups as CH_2^- groups and remembering that we can calculate the coefficients of the non-bonding molecular orbital very simply.
The values for c_o we obtain from the following diagrams.

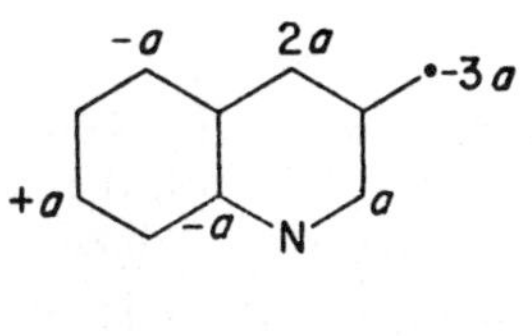

Model for 3-aminoquinoline

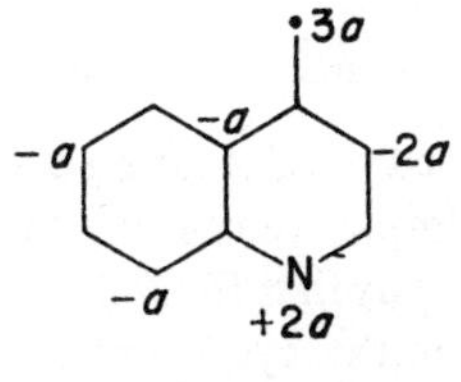

4-aminoquinoline

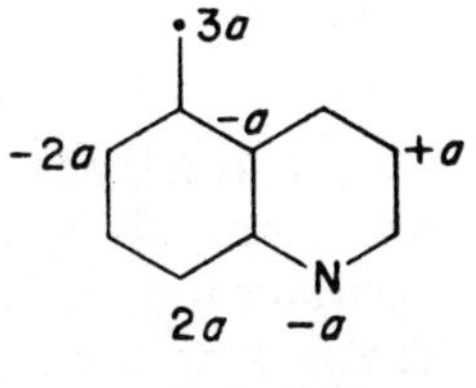

5-aminoquinoline

Thus we have for 3-aminoquinoline $\delta E_2 - \delta E_1 = 0$

$$pK_{3\text{-amino}} \approx pK_{\text{quinoline}}.$$

For 4-aminoquinoline ($R = 1{\cdot}98$ cal degree^{-1} mole^{-1})

$$\delta E_2 - \delta E_1 = -1{\cdot}2 \times 20{,}000 \times \tfrac{4}{20} = -2{\cdot}303 \times RT(pK - 5)$$
$$\therefore pK = 3{\cdot}6 + 5 = 8{\cdot}6.$$

For 5-aminoquinoline

$$\delta E_2 - \delta E_1 = -1{\cdot}2 \times 20{,}000 \times \tfrac{1}{20} = -2{\cdot}303 \times RT(pK - 5)$$
$$\therefore pK = 0{\cdot}9 + 5 = 5{\cdot}9.$$

The experimental values are 3-aminoquinoline 5; 4-aminoquinoline 9·2; and 5-aminoquinoline 5·5.

17.4 In the previous problem the addition of a proton to the quinoline bases involved no change in the total number of π electrons and we calculated the change in π-electron energy in terms of a change in α. In the present examples going from toluene to the benzyl anion or from cyclopentadiene to the cyclopentadienyl anion we are concerned with an increase in the number of π electrons. If we suppose the total π-electron energy of a hydrocarbon AH is $2n\alpha + a\beta$ and that of the anion (which has one more conjugate centre and two more π electrons) is $(2n + 2)\alpha + b\beta$, we can express the change in π-electron energy $\Delta\varepsilon$ as

$$\Delta\varepsilon = (2n + 2)\alpha + b\beta - 2n\alpha - a\beta - E_\sigma$$
$$= 2\alpha + (b - a)\beta - E_\sigma,$$

where E_σ is the C—H σ-bond energy broken in the ionization process and which we may consider to be approximately constant for different C—H σ bonds. If we now consider another hydrocarbon A'H, the difference in ionization energy for the two hydrocarbons is given simply by

$$\Delta\varepsilon - \Delta\varepsilon' = [(b - a) - (b' - a')]\beta.$$

Using the same arguments as in the previous question we have the following expression

$$\Delta\varepsilon - \Delta\varepsilon' \approx \Delta G^\circ = 2{\cdot}303RT(pK' - pK).$$

The problem is now reduced to calculating a and b, and a' and b'. For cyclopentadiene this is very simple, since, according to Hückel theory, the π-electron energy is the same as that for butadiene (see page 274) i.e. 4·48 β. The π-electron energy for the cyclopentadienyl anion is obtained from expression (15.26). We have two electrons in ψ_1 ($E_1 = \alpha + 2{\cdot}00\beta$) and four electrons in the two degenerate orbitals ψ_2 and ψ_3 ($E_2 = E_3 = \alpha + 0{\cdot}62\,\beta$). Thus

$$b - a = 6{\cdot}48 - 4{\cdot}48 = 2{\cdot}00.$$

The π-electron energy of toluene, according to the approximations we are using, is the same as that of benzene and from (15.24) $\varepsilon = 6\alpha + 8\beta$. There is no simple way of determining the orbital energies of the benzyl anion. We can slightly simplify the calculation by the use of symmetry. Benzyl belongs to the $\mathbf{C}_{2v}$ group and we can thus write the following diagrams

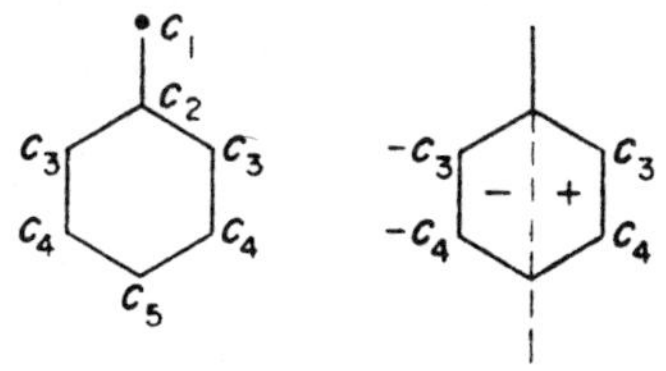

A_1 B_1

Writing down the secular determinants as usual and solving for x; (we are only interested in values of x which are negative or zero since we are only concerned with the filled orbitals) we obtain the following values

$$x = -2{\cdot}10; \quad -1{\cdot}26; \quad -1{\cdot}00; \quad 0.$$

We could have deduced that the highest filled orbital would be a non-bonding orbital, from the fact that benzyl is an odd alternant. The total π-electron energy of the benzyl anion is $8\alpha + 8{\cdot}72\beta$. Combining these results with those for benzene we have

$$b' - a' = 8{\cdot}72 - 8{\cdot}00 = 0{\cdot}72.$$

Our final expression is therefore

$$\mathrm{p}K' - \mathrm{p}K \approx \frac{1{\cdot}28 \times 20{,}000}{2{\cdot}303 \times 1{\cdot}987 \times 298}$$
$$\approx 18{\cdot}5.$$

The experimental value is 21.

Index